U.S. Environmental Protection Agency
Region III Information Resource
Center (3PM52)
841 Chestnut Street
Philadelphia, PA 19107

AF763775

TD
193.5
.P74
1992

Advances in
Modern Environmental Toxicology

Series Editor, M.A. Mehlman, Ph.D.

VOLUME XX

Predicting Ecosystem Risk

edited by:

John Cairns, Jr., B.R. Niederlehner, and David R. Orvos

Princeton Scientific Publishing Co., Inc.
Princeton, New Jersey

Copyright © 1992 by Princeton Scientific Publishing Co., Inc. All rights reserved. No part of this publication may be reproduced, stored in a retrieval system, or transmitted, in any form or by any means, electronic, mechanical, photocopying, recording, or otherwise, without permission in writing from the publisher.

Printed and bound in the United States of America.

PRINCETON SCIENTIFIC PUBLISHING CO., INC.
P.O. Box 2155
Princeton, New Jersey 08543
Tel: 609/683-4750
Fax: 609/683-0838

LIBRARY OF CONGRESS CATALOG NUMBER: 92-080326
ISBN: 0-911131-27-2

CONTENTS

FOREWORD

Some edited books attempt to achieve a certain degree of uniformity of style. In a very real sense, this suppresses the individuality of each author. In this volume, authors have been encouraged to use the style most comfortable to them with the expectation that many readers will find the contrasts stimulating and engrossing. Similarly, while guidelines were given for length, they were not adhered to rigidly. In some cases, the author was able to communicate with less than the allotted pages; in others, more were necessary. Backgrounds and disciplines vary, which is appropriate in a volume in which various disciplines are contributing. Naturally, contributors accustomed to a wide variety of professional journals should not be hampered by prescriptive requirements in delivering their message. Obviously, the volume could be an order of magnitude larger and still not adequately illustrate all the various aspects of estimating ecosystem risk. Regrettably, however, even the original range chosen could not be fulfilled. I have long followed the practice of not delaying a book because some authors could not meet a deadline set well over a year in advance. Delays mean that the efforts of all other authors are becoming outdated while the delinquent authors attempt to catch up. In two previous instances when I did delay publication time for a delinquent author, neither of the contributions showed up. In other cases, authors drop out without any prior warning when it is too late to seek a replacement. Under such circumstances, it still seems advisable to proceed without them if a critical number of contributions are in hand. Once I had a chapter in an edited volume held up for over four years while one editor scrambled for replacements and attempted to get contributions from delinquent authors. I decided at that time that never would I permit such a thing to happen with a book in which I was involved as an editor. All of the contributors to this volume have extremely busy schedules, and I am indebted to them for meeting the deadline and allowing this volume a timely publication.

John Cairns, Jr.

PREFACE

Why develop a robust predictive capability for ecosystem risk? Simply and most directly, because our lives depend on it! We may not realize the extent of our dependence because ecosystems do not go on strike when they are mistreated. Members of human societies are becoming increasingly isolated from natural ecosystems, yet they are simultaneously causing great destruction to these systems. Our isolation stems from the fact that, in this century, for the first time in history, nearly half the earth's population lives in cities or highly urbanized areas. Furthermore, even most rural populations live in markedly altered ecosystems. The vast grain-raising plains of the United States are sparsely populated, yet even that ecosystem has been drastically altered. A Kansas farmer may grow crops, but this controlled activity is not a reflection of the ecosystem's natural form as it existed before European settlement in North America.

The few societies who do have symbiotic relationships with comparatively pristine natural systems are themselves endangered. Natives of the threatened Brazilian rainforest appeal to their government and to other societies to save their threatened ecosystems. The Kankadu in Australia are, at best, holding their own. These societies have treated their natural ecosystems in a way that enables a sustained, non-exploitative relationship with their environment over millennia. Members of such societies, however, are presented to the inhabitants of developed countries as oddities; their way of life and relationship to their ecosystems can at best arouse a dim appreciation in outside onlookers who live in isolation from their natural environment. Inhabitants of developed countries must realize that societal well-being is as much influenced by ecological as by economic conditions. The goal of all societies should be to develop a relationship with natural ecosystems that permits them to flourish indefinitely.

We are clearly far from reaching that goal. An overwhelming amount of evidence indicates that we are living on our "ecological capital" (Ehrlich and Ehrlich 1990). We are losing topsoil at a frightening rate; groundwater aquifers that were maintained in a relatively stable condition through millennia are being depleted four feet or more annually; old growth forests, once a dominant feature of our land masses, are disappearing; and, impoverishment of the world's biotic diversity is occurring at a rate that is a cause for serious concern, if not controlled panic (Wilson 1988). The cumulative effect of such global changes could cause even greater alterations in the world's ecosystems than what is now occurring.

In order to prevent or restrict further damage, we need to develop an ability to predict ecosystem risk and to validate models for a system that is being altered at a phenomenal rate on a planetary basis. One thing is clear—we lack a comprehensive methodology for the precise estimation of ecosystem risk. We must investigate every currently available means that appears to have merit.

This book calls attention to the urgency of our need to develop a system for estimating ecosystem risk. Diverse chapters illustrate many facets of the problem; the book is not intended to be encyclopedic in its coverage. Specific topics include the economic consequences of our poor predictive capabilities and the drastically differing problems that face developing versus developed countries. Various views of ecosystem risk analysis and the investigation of promising methods of detecting ecosystem risk are discussed.

REFERENCES

EHRLICH, P.R. and EHRLICH, A.H. (1990). *The Population Explosion*. Simon & Schuster, New York. 320 pp.

WILSON, E.O. (ed.) (1988). *Biodiversity*. National Academy Press, Washington, DC. 521 pp.

DEDICATION

This book is dedicated to Dr. Ruth Patrick, Francis Boyer Chair of Limnology and Senior Curator, Academy of Natural Sciences, Philadelphia, Pennsylvania, who was a pioneer in predicting ecosystem risk well before these particular words were actually used. Ruth had worked on freshwater diatoms for a number of years before she began her pioneering work in stream pollution in 1948. At that time, most pollution assessment was carried out by what were then called sanitary engineers (waste treatment specialists) and chemists. The accepted procedure was that if certain limited chemical/physical conditions were met, particularly with regard to dissolved oxygen concentration, BOD, and pH, there was little or no need to examine the biota. In fact, Ruth spent much time talking to water pollution professionals in engineering and chemistry, trying to persuade them that biological information was useful in determining ecosystem condition. I realize that younger readers may find this statement so incredible as to be unbelievable, but most biologists were not especially interested in environmental pollution in those days, and the assessment and prediction of risk went by default to those involved with water treatment and waste treatment systems (predominantly sanitary engineers and chemists). In those days, pollution biology was strongly influenced by the "indicator species" saprobian system developed in Europe primarily for sewage. One of Ruth's major contributions was to examine assemblages of organisms in the now famous Conestoga survey. I was privileged to serve as the junior protozoologist (the senior one was Dr. Mary Gojdics) on one of the two field teams studying the Conestoga Basin and its tributaries with the home base at Lehigh University in Lancaster, Pennsylvania. Persuasive evidence soon became available that aquatic community structure was dramatically changed not only by organic pollution (sewage) but also by a variety of types of industrial pollution. Moreover, these patterns were demonstrated in different streams at different times even though the species composition was different. In short, there was a response at a higher level of biological organization than the single species that transcended the particular species present. These responses were subsequently tested in a variety of temperate zone freshwater rivers and streams in North America and were eventually tested in the upper Amazon.

A few years after the Conestoga survey, Ruth pioneered the use of artificial substrates (then called the Catherwood Diatometer) to show that diatom communities fit a log-normal distribution in non-polluted, freshwater, temperate zone situations and that the log-normal distribution was significantly and predictably altered by various degrees of pollutional stress. That is, the correspondence between the degree of pollution and the shape of the distribution curve was high.

Ruth Patrick has pioneered in a number of other areas such as highly acclaimed books on such diverse topics as diatom taxonomy and groundwater contamination. In addition, she is the first woman to serve on the board of directors, E. I. du Pont de Nemours and Company. Ruth was a scientist when women scientists were exceedingly uncommon; a leader in the development of the scientific team approach to problem solving when "lone wolf" scientific specialists were dominant; a pioneer in the systems approach of looking at entire drainage basins; and, perhaps most importantly, living proof that theoretical and applied science cannot only coexist, but are commonly synergistic. In fact, one might go further and say that predicting ecosystem risk is the ultimate test of the robustness of ecological theory. Additionally, if we truly understand how ecosystems work, we should not only be able to construct them on paper, but to construct them in the "real world." By bringing theoretical and applied science together, Ruth Patrick forced the validation of ecological theory.

Recognition for all of these unique efforts was a long time coming! But when it did finally start, the recognition was appropriately as diverse and prestigious as Ruth's pioneering efforts.

Honors and Awards

Recipient, Founders Award, Society of Environmental Toxicology and Chemistry, 1982
Fellow, American Academy of Arts and Sciences, 1976
Recipient, Icko Iben Award, American Water Resources Association, 1976
Recipient, Second Annual International John and Alice Tyler Ecology Award (regarded by many as the highest award in ecology), 1975
Elected to the American Philosophical Society, 1974
Recipient, Eminent Ecologist Award, Ecological Society of America, 1972
Recipient, Merit Award, Botanical Society of America, 1971
Elected to the National Academy of Sciences, 1970

Honorary Degrees

D.Sc., Clemson University, 1989
D.Sc., University of South Carolina, 1989
D.Sc., Wake Forest University, 1986
D.Sc., Emery University, 1986
D.Sc., Temple University, 1985
D.Sc., University of Pennsylvania, 1984
D.Sc., Lehigh University, 1983
D.Sc., Princeton University, 1980
D.Sc., University of Massachusetts, 1980
D.Sc., S. Lawrence University, 1978
D.Sc., Rensselaer Polytechnic Institute, 1976
D.Sc., Bucknell University, 1976
D.Sc., University of New Haven, 1975
D.Sc., Swarthmore College, 1975
D.Sc., Drexel University, 1975
D.Sc., Medical College of Pennsylvania, 1975
D.Sc., Hood College, 1975
L.H.D., Chestnut Hill College, 1974
D.Sc., Wilkes College, 1974
D.Sc., Cedar Crest College, 1974
D.Sc., Philadelphia College of Pharmacy and Science, 1973
L.LD., Coker College, 1971
D.Sc., PMC Colleges, 1971 (now Widener College)
D.Sc., Beaver College, 1970

Ruth is not one to rest on her laurels—abundant and prestigious as they may be. She now has several books and a variety of research activities underway and is a consultant to high ranking government officials, state organizations, private foundations, and the like. Although she claims to have cut down on her travel, just a few weeks ago I had to be fairly nimble to catch her in the one-day interval between a flight from England and a trip to the U.S. Midwest for a meeting of a board of trustees on which she serves. I was also privileged to know Ruth's husband Charles, a professor of entomology at Temple University and her son Charles, a physician.

One must know Ruth personally to fully appreciate her zest for scientific problem solving, which has been the keystone of her success. I know that the awards—particularly election to the National Academy of Sciences—meant much to her, but could not have been acquired without the continued dedication to science that she has demonstrated throughout her lifetime. It is a great pleasure to make this small tribute to a remarkable woman whose professional career continues to flourish.

John Cairns, Jr.

ACKNOWLEDGMENTS

We are grateful to various colleagues and contributors who have given their time and energy to produce this volume. Special thanks are due to Darla Donald, Editorial Assistant in the University Center for Environmental and Hazardous Materials Studies, who has seen this book through all the stages of its publication. Teresa Moody transcribed dictation on all correspondence and for the book chapters produced by the editors. We appreciate the authors who met the deadlines and made the production of this book a reality.

CONTRIBUTORS

Robert B. Atkinson
University Center for Environmental and Hazardous Materials Studies
Virginia Polytechnic Institute and State University
Blacksburg, VA 24061-0415

Roderick Bally
Department of Zoology
University of Fort Hare
Alice 5700
South Africa 27-404-32011

Scott E. Belanger
Ivorydale Technical Center
Environmental Safety Department
The Procter & Gamble Company
5299 Spring Grove Avenue
Cincinnati, OH 45217

Gordon Brady
Environmental Studies
Sweet Briar College
Sweet Briar, VA 24595

P. B. Bush
Cooperative Extension
University of Georgia
Athens, GA 30602

John Cairns, Jr.
University Center for Environmental and Hazardous Materials Studies
1020 Derring Hall
Virginia Polytechnic Institute and State University
Blacksburg, VA 24061-0415

J. S. Clark
Department of Botany
University of Georgia
Athens, GA 30602

W. David Conn
University Center for Environmental and Hazardous Materials Studies
201-A Architecture Annex
Virginia Polytechnic Institute and State University
Blacksburg, VA 24061-0113

Anne and Paul Ehrlich
Department of Biological Sciences
Stanford University
Stanford, CA 94305

Charles H. Hocutt
Center for Environmental and Estuarine Studies and The University of Maryland Eastern Shore
Horn Point Laboratories
P. O. Box 775
Cambridge, MD 21613

Bernd Kahn
Environmental Research Center
Office of Interdisciplinary Programs
Georgia Institute of Technology
Atlanta, GA 30332

Larry Morris
School of Forest Resources
University of Georgia
Athens, GA 30602

Donald I. Mount
AScI Corporation
112 East Second Street
Duluth, MN 55805

B. R. Niederlehner
University Center for Environmental and Hazardous Materials Studies
Virginia Polytechnic Institute and State University
Blacksburg, VA 24061-0415

David R. Orvos
SmithKline Beecham Pharmaceuticals
Environmental Research Laboratory, L-25
709 Swedeland Road, PO Box 1539
King of Prussia, PA 19406

Ruth Patrick
Francis Boyer Chair of Limnology
Patrick Center for Environmental Research
Academy of Natural Sciences
19th and the Parkway
Logan Square
Philadelphia, PA 19103

James R. Pratt
School of Forest Resources
213 Ferguson Building
Pennsylvania State University
University Park, PA 16802

Richard C. Rich
University Center for Environmental and Hazardous Materials Studies
606 McBryde Hall
Virginia Polytechnic Institute and State University
Blacksburg, VA 24061-0130

Richard E. Sparks
Illinois Natural History Survey
River Research Laboratory of the Forbes Biological Station
P. O. Box 599
Havana, IL 62644

Jay R. Stauffer, Jr.
School of Forest Resources
Pennsylvania State University
University Park, PA 16802

Gordon Tullock
College of Business and Public Administration
Department of Economics
Building #23
University of Arizona
Tucson, AZ 85721

ABOUT THE EDITORS

Dr. John Cairns, Jr. is University Distinguished Professor of Environmental Biology and Director of the University Center for Environmental and Hazardous Materials Studies at Virginia Polytechnic Institute and State University, Blacksburg. He received his PhD and his MS from the University of Pennsylvania and an AB from Swarthmore College, and completed a postdoctoral course in isotope methodology at Hahnemann Medical College, Philadelphia. He was Curator of Limnology at the Academy of Natural Sciences of Philadelphia for 18 years, and has taught at various universities and field stations. Professional certifications include Qualified Fishery Administrator with the American Fisheries Society and Certified Environmental Professional with the National Association of Environmental Professionals.

Among his awards are Fellow, American Association for the Advancement of Science in 1969; the Presidential Commendation in 1971; the Charles B. Dudley Award in 1978 for excellence in publications from the American Society for Testing and Materials; Superior Achievement Award, U. S. Environmental Protection Agency in 1980; the Founder's Award of the Society for Environmental Toxicology and Chemistry in 1981; the Icko Iben Award from the American Water Resources Association in 1984; the B. Y. Morrison Medal in 1984; Fellow, American Academy of Arts and Sciences in 1988; the United Nations Environmental Programme Medal in 1988; the American Fisheries Society Award of Excellence in 1989; the 1991 Life Achievement Award in Science by the Commonwealth of Virginia and the Science Museum of Virginia; and Member, National Academy of Sciences, 1991. Dr. Cairns has been consultant and researcher for the government and private industries, and has served on numerous scientific committees. His most recent publications are *Rehabilitating Damaged Ecosystems* (CRC Press, 1988), *Functional Testing of Aquatic Biota for Estimating Hazards of Chemicals* (ASTM, 1989), *On Global Warming* (1989), and *Integrated Environmental Management* (Lewis Publishers, 1991). His slightly over 1000 publications include 33 books with two currently in press. Over one-fourth of all publications are sole authored. Cairns presently serves on nine editorial boards and is chair of the National Research Council Committee on Restoration of Damaged Ecosystems.

B. R. Niederlehner is Laboratory Specialist and Staff Scientist in the University Center for Environmental and Hazardous Materials Studies at Virginia Polytechnic Institute and State University. She has an MS degree in Biology (aquatic toxicology) and 19 years experience in assessing the effects of stress on populations, communities, and ecosystems. Niederlehner is the author of approximately 30 articles within the areas of hazard assessment and microbial community ecology.

David R. Orvos is presently an Associate Senior Investigator in the Environmental Research Laboratory of SmithKline Beecham Corporation,

King of Prussia, PA. His interests include the assessment of environmental risk, fate and effects of hazardous substances, emergency response to hazardous spills, and the effects of genetically engineered organisms. Previously, Orvos was Assistant to the Director at the University Center for Environmental and Hazardous Materials Studies at Virginia Polytechnic Institute and State University and was also Manager of Roy F. Weston's Environmental Fate and Effect Laboratory in West Chester, PA.

Orvos received his BS in Biology from Eastern Mennonite College, Harrisonburg, VA, a MS in Biology from James Madison University, Harrisonburg, VA., and the PhD in Biology from Virginia Polytechnic Institute and State University, Blacksburg, VA. During the PhD degree, Orvos was awarded the 1987 Society for Environmental Toxicology and Chemistry Pre-Doctoral Fellowship sponsored by Procter & Gamble Company and the 1988 Cunningham Research Fellowship from Virginia Polytechnic Institute and State University. He has also received awards from Sigma Xi and the American Chemical Society. Orvos is the author or co-author of over 30 articles, abstracts, and book chapters.

CHAPTER 1

COMMUNICATING ABOUT ECOSYSTEM RISKS

W. David Conn and Richard C. Rich
University Center for Environmental and Hazardous Materials Studies
Virginia Polytechnic Institute and State University
Blacksburg, Virginia

INTRODUCTION

Most other chapters in this book are concerned with the prediction, by best available scientific techniques, of the magnitude of ecosystem risks—that is, the likelihood of deleterious changes in ecosystems resulting from human activities. Assuming some degree of success in making these predictions, this chapter explores issues associated with the communication of the resulting information to the public.

Prior to the early 1980s, not much attention was given to the communication of risk information as a subject worthy of study in its own right. The arrival (for the second time) of William Ruckelshaus in 1983 as Administrator of the United States Environmental Protection Agency (USEPA) helped change this situation, for he brought to USEPA a newly stated emphasis on risk-based decision making and on public involvement in the making of these decisions (Andrews 1990). Since that time, interest in the topic of risk communication has grown considerably. In 1989, a committee of the National Research Council (NRC) published a comprehensive report that brought together many of the findings to date, interpreted them for a general audience, and made recommendations for improving the risk communication process. Since then, the field has continued to advance, with the issuance of many scholarly publications as well as professional manuals and reports. This chapter does not attempt to review this literature in detail; however, it does summarize some of the findings and consider their significance in relation to the communication of ecosystem risks.

SOME DEFINITIONS AND DISTINCTIONS

To avoid confusion in the use of certain terms, some definitions and distinctions are given here. Rather than inventing new ones, those used in the NRC report (1989, Appendix E) are restated.

A *hazard* is "an act or phenomenon posing potential harm to some person(s) or thing(s); the magnitude of the hazard is the amount of harm that

---

1. Corresponding Author: W. David Conn, University Center for Environmental and Hazardous Materials Studies, Virginia Polytechnic Institute and State University, Blacksburg, VA 24061-0113.
2. Key Words: risk, communication, ecosystem.

might result, including the seriousness and the number of people exposed." In contrast, *risk* "adds to the hazard and its magnitude the probability that the potential harm or undesirable consequence will be realized."

Risk assessment is "the characterization of potential adverse effects of exposures to hazards; includes estimates of risk and of uncertainties in measurements, analytical techniques, and interpretive models; quantitative risk assessment characterizes the risk in numerical representations." *Risk control assessment* is the "characterization of alternative interventions to reduce or eliminate the hazard and/or unwanted consequences." *Risk management*, on the other hand, is "the evaluation of alternative risk control actions, selections among them (including doing nothing), and their implementation."

Risk communication is "an interactive process of exchange of information and opinion among individuals, groups, and institutions; often involves multiple messages about the nature of risk or expressing concerns, opinions, or reactions to risk messages or to legal and institutional arrangements for risk management."

Some comments about these definitions and distinctions are in order. Distinguishing risk assessment from risk management has given rise to the issue of the separation of "fact" from "value." Some have argued that risk assessment is (or can be) scientific, objective, and value free, while risk management is invariably unscientific, subjective, and value laden. Others have responded that risk assessment may indeed be scientific, but that science itself can never be devoid of values. They point out that the choices involved in selecting areas for investigation, hypotheses, methodologies, includable data, and so on reflect values and biases. Scientists often disagree over supposed facts, in some cases (it appears) because of differences of opinion regarding "how we know what we know" and what constitutes acceptable evidence (e.g., Mazur 1973; Rushefsky 1982). Furthermore, scientists rarely have access to all the facts they need, especially in matters dealing with risk. Frequently, investigators must substitute their judgment which, regardless of good intentions, inevitably (once again) reflects values and biases.

The definition of risk communication given by the NRC is significant in that—for reasons that should become apparent below—it departs from the traditional notion, still held by some, that risk communication is simply the development (by technical experts) and delivery (to non-experts) of a one-way message describing or characterizing hazards, risks, or risk-reducing actions. The definition emphasizes, instead, that risk communication is an interactive process that is inclusive of all messages and interactions bearing on risk decisions. As the NRC committee points out, risk communication is a component of risk *management* (emphasis added).

USES OF RISK COMMUNICATION

In the simplest terms, risk communication is typically used for either or both of two purposes: to assist in the *making* or the *selling* of decisions (considered here in reverse order).

The Selling of Decisions

In matters that involve a significant amount of technical information, such as environmental protection, public officials have traditionally tended to use a "decide, announce, defend" approach to handling decisions. Once the decisions have been made, officials have attempted to sell these decisions to the public; that is, they have tried to persuade the public that the decisions should be accepted, or at least not opposed.

In this situation, efforts by these officials to communicate have typically been one way and, when viewed against the NRC definition, incomplete. Repeatedly, these efforts have also proved unsuccessful. Public response has often been quite vehement, causing the decisions in a growing number of cases to be reconsidered, modified, or reversed. For example, decisions made behind closed doors to locate waste treatment facilities in particular communities have frequently been overturned due to subsequent public opposition. As a result, time has been lost, resources have been wasted, relationships have been damaged, and the future credibility of public officials has been undermined.

When this has happened to decisions that were firmly grounded in supposedly "objective" risk assessments, the officials and their technical advisers have often expressed dismay at the apparent ignorance, inability to comprehend, and/or irrationality of the public who (in the eyes of the officials) has acted either contrary to its own self-interests or, at the least, in a manner inconsistent with its other actions. Why, for example, would the public so strongly oppose the siting of a municipal waste incinerator when at least some "experts" have assured the people that the risks associated with the construction and operation of such a facility are far less than those associated with many activities (such as driving a car) in which each engages routinely?

The flaws in this reasoning have been exposed by a considerable and growing body of research, some of whose conclusions will be summarized only briefly here. Lay people, it seems, typically understand risk differently than experts (e.g., Covello et al. 1983). The problem was originally characterized as a divergence between the "perceptions" of laypeople and the "actual" risks measured by experts. However, it is not clear that the word "actual" is meaningful in this context. Instead, differing views of the *significance* of risk are held by most laypeople and experts. While experts tend to focus on the quantitative attributes of risk, notably magnitude and likelihood, citizens add (and act in response to) more qualitative dimensions such as familiarity, fairness, and controllability (e.g., Otway and Von Winterfeldt 1982; Raynor and Cantor 1987; Sandman 1989). A waste facility may be opposed not simply on the basis of the estimated likelihood of ill effects, but also because (for example) the risks will not be assumed voluntarily, the risks are man-made rather than natural, those who face the risks will have little or no control over their generation, and the benefits of the facility will largely accrue to others.

Another important factor is the cultural context in which the risk-related decision is being made (Douglas and Wildavsky 1982). Officials and their technical advisers may see the issues as being solely technical in nature; they can detach themselves from the cultural context in a way that the public is unlikely to do. Plough and Krimsky (1987; also Krimsky and Plough 1988) distinguish between "technical rationality" and "cultural rationality." In the former, "logical consistency is an imperative ... two events that have an identical risk profile are treated the same—they are interchangeable." However, an understanding of the latter requires that "anthropological and phenomenological issues" be addressed as well as behavioral ones. The cultural approach holds that the public's perception of the risk will be affected by "the circumstances under which the risk is identified and publicized, the place of the individual in his or her community, the values of the community at large, and many other factors" (Hadden 1991).

The public may, it seems, be concerned as much with the *process* of decision making as with the nature of the decision itself. One of the reasons why efforts at selling decisions have often proved unsuccessful is that people have reacted negatively to the underlying premise of the communication. They may not be persuaded by the technical analysis behind the decision to build an incinerator, for example—no matter how reassuring it is on technical grounds—if they object to the way in which the decision was made (specifically, their exclusion from that process).

Making Decisions

This brings us to the second use of risk communication, that of assisting in making decisions. If, instead of being sold a decision after it has been made by others, the public is to play a meaningful role in decision making, then arguably the technical experts must ensure that the public has access to—and can comprehend—the best available technical knowledge, presented in as unbiased a manner as possible. Armed with this information, the public may still make or call for decisions that *to the experts* seem ignorant or irrational; however, a democracy gives the public the prerogative to do so. Science may suggest, for example, that hauling untreated waste to a landfill not only will be more expensive than constructing an incinerator but also will pose greater risks. However, if the public knowingly chooses hauling over constructing the incinerator, then officials must respect that decision.

The key, of course, is to ensure that the public is fully informed, or at least has all the knowledge that it wishes to have. The challenge is to provide access to the best available science in a comprehensible format to those with little or no scientific training (and, in some cases, with little or no education of any kind). This may be very difficult to do successfully. The nature of the hazard and its effects may be technically complex, and it may be virtually impossible to develop an explanation that is both complete (in conveying what is necessary for the decision making) and yet simple enough for someone lacking a specialized education to follow the scientific reasoning.

The probability of a hazard occurring may be outside the range of probabilities with which most people have some experience. It is not easy, for example, for anyone to grasp the significance of a probability of 1 in 10 million, or to differentiate such a probability from (perhaps) 1 in 100 million. Furthermore, the task of communication is complicated by the observation that people tend to employ "cognitive heuristics" or mental rules of thumb in thinking about probabilities, which may be responsible for consistent errors in this thinking (Tversky and Kahneman 1974). Factors, such as how recently and how frequently an event has been experienced and how much emotion has been associated with its occurrence, seem to be significant in this respect (e.g., Fischhoff et al. 1981; Slovic et al. 1980).

Uncertainty adds yet another complication. The best available scientific data may be (and usually are) quite incomplete, and even the most highly qualified experts may disagree on interpretation of data. The conflicting messages that result can be especially confusing for the layperson, who lacks the capability of evaluating competing claims (Hattis 1987).

Simply capturing people's attention, so that the knowledge may be imparted, can be a major task. Many environmental risks seem distant in space and time to the public-at-large, who has its hands full in dealing with what it perceives as more immediate problems. Moreover, modern communication techniques flood people with messages that claim to be important; consequently, drawing attention to any one message is difficult.

ECOSYSTEM RISKS

Specific mention of ecosystem risks is infrequent in the existing literature on risk communication. In instances where particular risks have been featured (for example, in case studies of risk-based decision making or in controlled experiments testing different modes of communication), they typically have been those affecting human health and safety. This situation is consistent with the primary emphasis placed on human health in most political debate about the environment and in most of the environmental legislation that has ensued (Burger 1990). According to the Science Advisory Board of USEPA (1990), "the Agency has considered the protection of public health to be its primary mission, and it has been less concerned about risks posed to ecosystems." Of course, since humans are components of ecosystems, risks to human health are really a subset of ecosystem risks; however, "other" environmental risks (not *directly* impacting on human health), though consistently rated in public surveys as important (e.g., Gillroy and Shapiro 1986), nevertheless seem to be less pressing as priorities in political and legislative action.

Efforts to communicate information on ecosystem risks generally should address all the problems that have so far been discussed. The need to do so may be more acute than in efforts to communicate risks to human health alone. Science is frequently even more complex and often requires tracing multiple tiers of cause and effect. Data are even more scarce, and the need to exercise technical judgment is even greater. Uncertainty abounds, and with it the opportunities for "legitimate" disagreements between experts.

The public is generally less aware of (and knowledgeable about) the nature and significance of ecosystem risks, other than direct human health risks. Additonal education is required to specify terms and produce understanding. While people typically have a sense of what it means to become ill or die, for example, they are less likely to understand the implications of the disappearance of a species from an ecosystem or the loss of functioning of a wetland. This may be a difficult problem to overcome. A videotape, for example, can easily capture the image of a chemical explosion (which directly and immediately threatens human lives) but is at a loss to show a slowly degenerating ecosystem (having considerably less obvious short- or even medium-term symptoms and effects).

Ecosystem risks are likely to appear remote to many people and less deserving of their attention than other matters of (seemingly) greater urgency. Individuals may feel that they have greater control, in general, over human health risks than over other kinds of ecosystem risks and that, because their personal ability to make a difference with regard to ecosystem risks is apparently so small, the latter are not worth bothering with. People are more likely to ignore risk messages, it seems, if they perceive that they are not in a position personally to do much about them.

CONCLUSIONS

Faced with all these problems, what is the risk communicator to do? So far, it must be emphasized, no one has come up with a panacea. As the NRC committee has pointed out: "risk messages necessarily compress technical information, which can lead to misunderstanding, confusion, and distrust." Nevertheless, a number of recommendations have emerged from research to date, many of which have been incorporated into various risk communication "manuals" published by USEPA and other organizations (e.g., Covello et al. 1988; Hadden and Bales 1989; Hance et al. 1988; McNeil et al. 1989). These recommendations, in our opinion, apply as much—if not more—to the communication of ecosystem risks as they do to other kinds of risks.

Routine advice is to begin early (that is, long before any decisions are made) and give as much attention to the process as to the message. The process should be interactive rather than one-way. The intended audience should be carefully identified and studied, so that the message can be designed especially for that audience. People having credibility (or who are able to build credibility) with the intended audience should be chosen as communicators, and multiple channels of communication should be used. Artificial barriers to participation in the process should be eliminated (e.g., meetings should be held at convenient times in accessible locations). Messages should be complete and honest and should address concerns of the recipients. This may require developing a variety of different messages and delivering them through different channels to reach all of the various segments of the audience. Scientific uncertainty, where it exists, should be openly acknowledged.

Expectations should not be misplaced. While a successful risk communication effort might produce some convergence in people's knowledge and understanding of the nature of a hazard and the probability of its occurrence, it will not necessarily achieve any convergence in their feelings about this risk *nor should it be expected to do so*. Attempts to *sell* a decision may find no buyers, even among knowledgeable and rational people. Using risk communication to help in *making* a decision may also be problematical, if convergence does not occur, but this is an inevitable problem of decision making in a pluralistic society.

ACKNOWLEDGMENTS

The authors thank Holly Larson Lesko for assisting in the preparation of this chapter. Any opinions expressed are those of the authors alone and do not reflect an official view of Virginia Polytechnic Institute and State University.

REFERENCES

ANDREWS, R.N.L. (1990). Risk assessment: Regulation and beyond. In: *Environmental Policy in the 1990s* (N.J. Vig and M.E. Kraft, eds), pp. 167-186. CQ Press, Washington, DC.

BURGER, E.J., Jr. (1990). Health as a surrogate for the environment. Daedalus 119:133-153.

COVELLO, V.T., LAMM, G., RODERICKS, J., and TARDIFF, R., eds. (1983). Analysis of Actual vs. Perceived Risks. Plenum Press, NY.

COVELLO, V.T., SANDMAN, P., and SLOVIC, P. (1988). Risk Communication, Risk Statistics, and Risk Comparisons: A Manual for Plant Managers. Chemical Manufacturers Association, Washington, DC.

DOUGLAS, M. and WILDAVSKY, A. (1982). Risk and Culture: An Essay on Selection of Technical and Environmental Dangers. University of Calfornia Press, Berkeley, CA.
FISCHHOFF, B., LICHTENSTEIN, S., SLOVIC, P., DERBY, S. L., and KEENEY, R. L. (1981). Acceptable Risk. Cambridge University Press, Cambridge, England.
GILLROY, J.M. and SHAPIRO, R.Y. (1986). The polls: Environmental protection. Public Opinion Quarterly 50:270-279.
HADDEN, S.G. (1991). Public perception of hazardous waste. Risk Anal. 11:47-57.
HADDEN, S.G. and BALES, B.V. (1989). Risk Communication About Chemicals in Your Community: A Manual for Local Officials, USEPA 230/09-89-066. U.S. Environmental Protection Agency, Washington, DC.
HANCE, B., CHESS, C., and SANDMAN, P. (1988). Improving Dialogue with Communities: A Risk Communication Manual for Government. New Jersey Department of Environmental Protection, Trenton, NJ.
HATTIS, D. (1987). What's wrong with quantitative risk assessment? In: *Effective Risk Communication* (V.T. Covello, D. B. McCallum, and M.T. Pavlova, eds), pp. 45-49. Plenum Press, NY.
KRIMSKY, S. and PLOUGH, A. (1988). Environmental Hazards: Communicating Risks as a Social Process. Auburn House, Dover, MA.
MAZUR, A. (1973). Disputes between experts. Minerva 11:243-262.
MCNEIL, C., ARKIN, E.B., and MCCALLUM, D. (1989). Toxic and Hazardous Substances, Title III and Communities: An Outreach Manual for Community Groups, EPA 56-1-89-002. U.S. Environmental Protection Agency, Washington, DC.
NATIONAL RESEARCH COUNCIL (1989). Improving Risk Communication. National Academy Press, Washington, DC.
OTWAY, H.J. and VON WINTERFELDT, D. (1982). Beyond acceptable risk: On the social acceptability of technologies. Policy Sci. 14:247-256.
PLOUGH A. and KRIMSKY, S. (1987). The emergence of risk communication studies: Social and political context. Sci. Tech. Human Values 12:4-10.
RAYNOR, S. and CANTOR, R. (1987). How fair is safe enough? The cultural approach to societal technology choice. Risk Anal. 7:3-9.
RUSHEFSKY, M. (1982). Technical disputes: Why experts disagree. Policy Stud. Rev. 1:676-685.
SANDMAN, P. M. (1989). Hazard vs. outrage in the public perception of risk. In: *Effective Risk Communication* (V.T. Covello, D.B. McCallum, and M.T. Pavlova, eds), pp. 45-49. Plenum Press, NY.
SLOVIC, P., FISCHHOFF, B., and LICHTENSTEIN, S. (1980). Facts and fears: Understanding perceived risk. In: *Societal Risk Assessment: How Safe is Safe Enough?* (R.C. Schwing and W.A. Albers, Jr., eds), pp. 181-214. Plenum Press, NY.
TVERSKY, A. and KAHNEMAN, D. (1974). Judgement under uncertainty: Heuristics and biases. Science 185:1124-1131.
U.S. ENVIRONMENTAL PROTECTION AGENCY, SCIENCE ADVISORY BOARD (1990). Reducing Risk: Setting Priorities and Strategies for Environmental Protection, SAB-EC-90-021. Washington, DC.

CHAPTER 2

ECOSYSTEM RISKS ASSOCIATED WITH THE POPULATION EXPLOSION

Anne H. Ehrlich and Paul R. Ehrlich
Department of Biological Sciences
Stanford University
Stanford, California

Virtually all major threats to ecosystems are tied in one way or another to the scale of human activities, and the size of the human population in turn is a major factor in that scale. This chapter briefly outlines the connections between population size and ecosystem deterioration and makes some comments on the assessment of ecosystem risks.

INTRODUCTION

Since their emergence a few million years ago, human beings have had impacts on their environments, although for millennia the effects were negligible, largely because populations were very small.[a] All human activities have unavoidable effects on natural ecosystems. In order to live, people take oxygen from the environment and release carbon dioxide into it. Human beings, like other animals, must acquire energy directly or indirectly from the chemical bonds that plants manufacture in the process of photosynthesis, and they must return some of that energy to the environment in the form of waste heat. Every move a person makes changes the environment; the changes may be imperceptible, but they occur. Consequently, the question is not whether human beings should affect their environment—the questions all center on the *scale and consequences* of that influence (Ehrlich and Ehrlich 1991). The scale of human activities has become immense as the 21st century approaches. This situation is in no small part due to the population explosion. As a consequence, ecosystems are being exposed to rapidly escalating risks worldwide.

THE HISTORY OF ECOSYSTEM IMPACTS

From the earliest days of *Homo sapiens*, environmental changes, deliberate or not, usually had the effect of increasing Earth's long-term carrying capacity for humanity, albeit at some cost to other organisms with which

1. Corresponding Author: Anne H. Ehrlich, Department of Biological Sciences, Stanford University, Stanford, CA 94305.
2. Key Words: risk, population, ecosystem, impacts.

[a]Some of the material in this chapter is adapted from chapter 8 of the Ehrlich's book *Healing the Planet* (1991, Addison-Wesley, NY).

people competed for living space, food, or other resources. As the human population grew and spread across the land surface, anthropogenic risks to ecosystems also inevitably increased. Improved human hunting prowess, for instance, was almost certainly a major cause of the extinctions of numerous large mammals in the late Pleistocene (Diamond 1991; Martin and Klein 1984; Martin and Wright 1967).

Ironically, the decline of large mammals as a food source may have been a factor leading to the invention of agriculture. When people first began to cultivate most of their food rather than hunt and gather it some 10,000 years ago, many times more people could be supported on a given area of productive land than previously. In the next 8000 years, the human population expanded from a few million to several hundred million. After the development of agriculture, human impacts on ecosystems made a quantum leap, as huge tracts of grassland and forest were transformed into farmland.

By the 20th century, further improvements in agricultural efficiency and the ability to mobilize resources enabled billions (rather than millions) of people to be supported on Earth. Humanity became a truly planetary ecological force, its influence going far beyond the transformation of landscapes to include altering the composition of the atmosphere, modifying climates globally, interfering with planetary nutrient cycles, and exterminating other life-forms on a grand scale.

Such actions create enormous risks for the future of humanity because they pose serious threats to the natural ecosystems that support society. Through damage to those ecosystems, the risks associated with human-caused environmental impacts feed back onto society itself. The penalties of making environmental mistakes surely plagued early human beings from the time the first fire, set to drive game toward hunters, got out of control and killed someone. However, until very recently, such penalties were borne only by a few or, at worst, a few million people, locally or regionally. Today they are borne globally by billions of people.

In historical perspective, the switch from a modest to an overwhelming scale of impact, and from a positive to a negative impact on carrying capacity, was almost instantaneous and monumental, occurring in little more than a century. The switch was largely driven by an unprecedented surge in human population growth. From 1850 to 1992, the human population expanded more than fivefold, from about 1 billion to 5.5 billion (Ehrlich and Ehrlich 1990). To support its burgeoning population and increasing affluence, humanity began shifting away from its almost complete dependence on the income provided by energy from the sun and made available by photosynthesizing organisms—the energy source for virtually all other organisms. Increasingly, that income was supplemented by using "capital"—a heritage of resources that were not renewed continuously by sunlight, such as deposits of metals and the geophysically stored energy in fossil fuels. This capital also included "renewable" resources, such as fertile soils, groundwater, and biodiversity (defined as "the variety of life-forms, the ecological roles they perform, and the genetic diversity they contain" [Wilcox 1984]), that are increasingly used at rates that make them essentially non-renewable.

As a result of the momentous shift from living on income to living on capital and the quintupling of the population since 1850, global energy use (a useful rough measure of environmental impact) has multiplied some 13-fold (Holdren 1990). The environmental impact (I) of a society, and thus the degree of risk it imposes on ecosystems, can generally be thought of as the product of its population times its per capita consumption of resources (or affluence) times the technology used for each unit of consumption—summarized in the equation P (population) X A (affluence) X T (technology). The I = PAT equation highlights the roles of both population growth and resource consumption in generating pressure on Earth's life support systems (Ehrlich and Ehrlich 1991).

THE SITUATION TODAY

By the 1990s, the combination of explosive population growth and degradation of renewable resources had led to the present predicament. Over 90 million additional people each year need to be supported by food grown with approximately 24 billion tons less topsoil, trillions of gallons less groundwater, and an unmeasured deterioration of ecosystem services due to losses of biodiversity (Ehrlich and Ehrlich 1991).

The human takeover of land and appropriation of its biological productivity (net primary production—NPP—all the energy fixed in photosynthesis, less that used by the photosynthesizers for their own life processes) indeed have resulted in an estimated reduction in terrestrial NPP just since the middle of the century of about 13%, as well as direct consumption or diversion into human controlled systems of nearly a third of it (Vitousek et al. 1986). Taken together, human impacts on terrestrial NPP amount to approximately 40% of the total. This is hardly a reassuring circumstance for a population apparently destined to double its size at least once more before its growth can be halted and reversed humanely—that is, by limiting the number of births (United Nations 1991).

Human activities now appear to be lowering Earth's long-term carrying capacity for human life and incurring risks on a scale unimaginable less than a lifetime ago. Those risks arise primarily from an escalating degradation of ecosystems that can be traced in substantial part to the growth in human numbers.

WHY SHOULD WE BE CONCERNED?

The most serious consequence of humanity's heedless placing of ecosystems at risk will be the loss of ecosystem services. Natural ecosystems—and the myriad organisms that comprise them—provide a series of indispensable free services to humanity (Ehrlich and Ehrlich 1981). Most of these services are not replaceable with human technology; in most cases, we do not know how. These services include maintaining the quality of the atmosphere; moderating weather and climate; running the hydrological cycle; recycling nutrients in soils and aquatic systems; replenishing and preserving soils; disposing of wastes; pollinating plants (including many crops); controlling the vast majority of pests and disease vectors; supplying foods (such as fish and game), forest products, and innumerable other useful materials; and providing a vast genetic library of biotic resources from which the very basis of civilization has been drawn. Earth's biodiversity is part and parcel of humanity's life support system.

The services provided by natural ecosystems have often been overlooked by people interested in exploiting ecosystem resources or modifying them for their own purposes. Biologically productive areas are attractive places for human beings to settle, clear away native vegetation, and start farming or engaging in other activities—and they have done so with a vengeance.

POPULATION GROWTH AND THE RISK TO ECOSYSTEMS

The most direct risk to natural ecosystems posed by population growth is the outright destruction of ecosystems in the process of making room for additional people. Natural systems are paved over to construct homes, offices, factories, schools, roads, and other facilities to serve more and more people. Ecosystems are also cleared away and plowed under to increase food production to nourish more and more human beings (although this trend has slackened considerably in the last few decades, since most of the world's suitable land is already being cultivated [Brown et al. 1991]). Few areas remain today in the temperate or subtropical regions that have not been substantially modified by human action, often beyond recognition. In many regions, only tiny fragments of the original ecosystem may remain. In the last few decades, people have increasingly moved into the moist tropical forests, home to the greatest diversity of life-forms, and attempted to turn them to direct use as well.

Approximately two-thirds of Earth's land surface—generally the most productive areas—now is utilized to produce food and materials for civilization. While only about 2% is covered with buildings or pavement, another 11% is used to grow crops for people and livestock; 25% more is used as pasture; and much of the 30 to 40% that is forest or woodland is heavily exploited or even "farmed" (Food and Agricultural Organization [FAO] 1990). The fraction that remains as virgin forest is swiftly being cleared, especially in the moist tropics (Myers 1989).

This wholesale conversion of natural ecosystems to human directed ones accounts for the marked reduction in terrestrial NPP that has occurred in recent decades, since cropland and pastures in most cases are less biologically productive than the systems they replaced. Many formerly productive lands also have been degraded through overgrazing or overcultivation and have thereby lost productivity. Moreover, the worldwide replacement of natural systems with simpler, human directed ones also goes far to explain the accelerating global loss of biodiversity (Wilson 1988).

INDIRECT ASSAULTS

Outright displacement and degradation of habitats are only the most obvious human assaults on natural ecosystems and the organisms that comprise them. Others include exposure to human produced toxic substances, such as air pollutants, synthetic fertilizers and pesticides; the acidification of entire systems, such as lakes and forests; direct exploitation through collection or hunting of desired species; or deliberate extermination of certain species perceived as threats to people, livestock, or crops. Many of the above actions apply as well to oceanic systems, especially pollution and overexploitation of fisheries. Habitat alteration is a serious problem in coastal and estuarine areas, which often function as "nurseries" for valuable seafood species or their food resources and are among the most biologically productive ecosystems.

Unnoticed losses of species and genetically distinct populations, mostly of obscure organisms, are occurring daily. Estimates vary, but it is plausible to biologists that 10% of the world's original complement of species will have vanished by 2025, and 25% or more might be lost by 2050 if serious efforts are not undertaken to prevent it (Ehrlich and Wilson 1991). Perhaps more important, billions of genetically distinct populations may disappear in the same period, greatly adding to the degradation of ecosystem services. Current and projected losses have been compared in severity to the extinction event at the boundary between the Cretaceous and Tertiary periods some 65 million years ago in which more than half the estimated animal species then extant are estimated to have disappeared (Raup 1988). Recovery from that event, of course, entailed a dramatic change in terrestrial fauna and several million years were required to restore the previous level of diversity.

Among the most serious connections between increasing human numbers and threats to ecosystems is the alteration of the vital trace-gas content of the atmospheric portions of ecosystems (Houghton et al. 1990). Each additional human being adds to the carbon dioxide content of the atmosphere, not just through his/her life processes but to a much greater degree in most cases by participating in energy economies based on the burning of fossil fuels. The cutting and burning of tropical forests is also a major contributor to the carbon dioxide buildup, as is the unsustainable use of traditional fuels—both of which also constitute direct attacks on ecosystems.

Population growth also fuels the buildup of methane, another important greenhouse gas, which is increasing twice as fast as carbon dioxide. Methane reaches the atmosphere by numerous routes, some of them resulting from increased agricultural activities (particularly rice cultivation and cattle production) to feed growing numbers of people (Ehrlich 1990; Ehrlich and Ehrlich 1991). Numbers of people are also implicated in the buildup of nitrous oxide in the atmosphere, although the sources of the buildup are not well understood. Nitrogen fertilizers and land clearing are among the possible causes.

Even the release of chlorofluorocarbons (CFCs) to the atmosphere, which not only threaten the stratospheric ozone shield but also are potent greenhouse gases, is related to human population growth (Jones and Wigley 1989). CFCs could be *relatively* easily controlled by technological means, and they probably will be, but not before significant damage has occurred from both effects. Of course, if there were only a billion people today, all else being equal, the threat to the ozone layer would be minimal, and significant greenhouse warming would be much less imminent.

By contributing to both global warming and ozone depletion, population growth threatens the integrity of virtually all terrestrial ecosystems. The scientific consensus is that global warming will lead to worldwide changes in climates that are unpredictable in detail, but that will occur many times more rapidly than any climate changes in recent geological history (Houghton et al. 1990). The rate of climate change expected with human caused global warming is an order of magnitude greater than even the fastest rate inferred during the past 100,000 years; a temperature rise of 4°C—near the high end of the range of current estimates for around 2050—would produce the warmest climates in several million years.

Such rapid changes in climate will be imposed on terrestrial ecosystems that are already under various kinds of assault from human activities and have been fragmented and disrupted by land conversions to human uses (Wyman 1991). Even if plant and animal species were otherwise able to shift their distributions quickly enough to accommodate changing climates, they would face barriers created by huge numbers of people in most cases—vast tracts of farmland, pastures, managed forests, cities, and highways. Rising sea levels would disrupt or obliterate many coastal systems as well. Rapid climatic change and sea level rise thus would lead to massive extinctions that could greatly impair the ability of those ecosystems to function and deliver services to humanity (Wyman 1991).

Population growth increases the risks of ecosystem degradation in other ways—the more people, the more toxic effluents produced by society (Colburn et al. 1990). Mining, manufacturing, and the transport of dangerous materials will all be increased to meet the needs and desires of additional people. Air and water pollutants not only threaten the health of human beings directly, but they also damage plant and animal life. Effects of air and water pollutants on microflora and -fauna (which often play crucial roles in maintaining ecosystems, especially soil organisms) are little known but might well be substantial. Air pollutants that form acid precipitation are particularly damaging to both aquatic and forest ecosystems. They kill susceptible organisms through an insidious process of gradual acidification and often interact with other human induced environmental changes (World Resources Institute 1986). Lakes, streams, and forests in many parts of Europe and North America have been seriously affected, and many other regions are at risk of similar destruction, especially in developed nations.

More people necessitate intensified agriculture, leading to the application of more pesticides and inorganic fertilizers, which inevitably spread from farms into natural ecosystems and cause damage (Colburn et al. 1990; Ehrlich et al. 1977). Expanding human groups in arid and semiarid regions may cause overgrazing or overcultivating of soils, leading to desertification—a progressive degradation of the land's productive potential (World Resources Institute 1988). Not only does such degradation lead to lower crop or livestock yields, it also reduces NPP and impoverishes the ecosystem.

Human beings also directly exploit many species, not infrequently leading to severe reductions in population sizes or, in extreme cases, their extinction. Many oceanic fisheries and whaling are obvious cases in point, as well as the cases of elephants and rhinos. Attempts to exterminate species to eliminate competitors or predators is the other side of that coin: consider wolves, tigers, prairie dogs, and various raptors (Ehrlich and Ehrlich 1981).

A larger population also means increased recreational use of fragile habitats such as deserts, coral reefs, and alpine forests. Even a seemingly innocuous activity (one that brings indirect benefits by increasing the public's appreciation of nature) can lead to surprisingly severe damage as the numbers of human beings seeking to escape temporarily from urban areas increase (Ehrlich and Ehrlich 1981).

ASSESSING RISKS, COSTS, AND BENEFITS

In the next several decades, humanity will be faced with many decisions about the risks involved in present and future alterations of Earth's life support systems. These risks cannot readily be expressed in monetary terms; for instance, risks of the extinction of a million species or of accelerating climate change (either or both) could possibly result in the loss of tens or hundreds of millions of human lives. Too often, society and decision makers have ignored large environmental and sociopolitical risks because they are difficult to monetize, but leaving them out may be a big mistake. The trap of ignoring what cannot easily be expressed in dollars has been described by economist Herman Daly as concentrating on things that are countable rather than things that count (Holdren 1983).

The qualitative, ethical, and moral contexts in which risks, costs, and benefits are assessed are nevertheless frequently at least as important as economic considerations, albeit harder to measure—and environmental issues often include such qualitative aspects (Daly 1977; Daly and Cobb 1990). Quantitative analyses can also lead to what energy analyst John Holdren calls "the tyranny of illusory precision"—using quantitative risk estimates, divorced from the underlying assumptions and uncertainties associated with the estimates, to announce firm conclusions that are unjustified (Holdren 1983).

A "zero-infinity" problem is one in which the probability of a consequence may be extremely small, almost zero. However, should the unlikely event occur, it would be considered infinitely bad. The likelihood of a full scale nuclear war is an example; the risk has dwindled recently, but the deaths of hundreds of millions of people and the severe disruption of natural and agricultural ecosystems are nonetheless likely if the event occurs (Ehrlich and Ehrlich 1989).

Some other environmental risks, however, seem to be moving (as the number of *Homo sapiens* increases) toward what could be called a "likely-infinity" situation—one in which the probabilities might be 20 to 50% or more, with hundreds of millions or even billions of lives hanging in the balance. Rapid climate change due to global warming is in this category (Houghton et al. 1990), as ozone depletion will be if the Montreal protocol to end the use of CFCs and halons is not complied with, or if some still-unknown factor in atmospheric chemistry, such as the unanticipated ice-cloud chemistry that created the Antarctic ozone hole, renders the protocol inadequate (Jones and Wigley 1989).

Scientific disputes related to the environment are often misinterpreted by the public, who may conclude that the answer lies "in the middle" or that uncertainties about risks justify inaction. Global warming is a case in point. How serious the risks are will be settled by the gigantic experiment humanity is running, and scientists who judged the threat to be severe (rather than nonexistent or minor) may well be proven correct. Much the same can be said for debates over such questions as the potential damage from acid precipitation (will forests covering huge areas be killed?) and ocean pollution (can the seas be pushed over a threshold that, among other things, will cause even more widespread fisheries collapses?).

To date, the record indicates that scientists who warned of hazards in environmental issues have been correct more often than those who took a more complacent view. The growing prominence of environmental problems, the continuing rise in the absolute numbers of hungry people worldwide, and the failure of technological miracles to materialize (with the arguable exception of the green revolution) all generally support this conclusion.

The media often emphasize novel views and controversy that convey an impression of far more uncertainty and disagreement about many issues than actually exists among the scientists studying them, as has happened over the issue of global warming (Schneider 1990). The doubts of the handful of scientists who are skeptical about global warming projections made by climatologists have been exaggerated and continuously quoted by people with a vested interest in the status quo (Gribbin 1990). The resulting, confused reporting places an even greater burden on citizens and decision makers who are attempting to evaluate an issue already made difficult by incomplete knowledge.

Our personal view is that global risks should be handled *conservatively*. We are exceedingly risk-averse with regard to the health of Earth's ecosystems. Great caution should be exercised when potentially irreversible changes are being made in the only home of humanity, or when chances are taken with hundreds of millions (or even billions) of lives, including those of future generations.

Unfortunately, a different attitude is widely shared by people in business and government and even in some segments of the scientific community (Dyson 1975; Ehrlich 1975). These attitudes normally appear in the form of statements such as: "if you can't *prove* that doing X will do serious harm to the environment, then we should go ahead and reap the vast benefits of doing X." Like the accused in the criminal justice system, technologies are innocent until proven guilty.

The risk of global warming was anticipated but long ignored. Damage caused by acid precipitation was recognized late, and too little action was taken against it even then. The atmospheric threats posed by CFCs, though fairly easily resolved, were addressed only after a delay of about 15 years. Society's performance in dealing with recognized risks could certainly be improved, but what about risks that remain unknown long after the technologies generating them have been deployed? The time for costs to appear after decisions are made can be impressive, as the histories of acid rain and CFCs clearly illustrate. Even when risks are anticipated or discovered by a few people, such discoveries may remain unappreciated by decision makers for a dangerously long time.

Risk assessments and decisions are usually made for societies by a few individuals, and most people (including many leaders) are unaware that the assessments and decisions are being made. Concern for intergenerational equity often enters into individual risk assessments (most pregnant women will avoid consuming substances they think may harm their children), but it should play a greater role than it does in social risk taking. Social systems have adapted to the consequences of inevitable, individual deaths. However, should this generation's "decision" to eliminate much of Earth's biodiversity be carried out, the consequences could plague our descendants for 100,000 generations or more—roughly 2.5 million years. *At least* that length of time would be necessary for an impoverished biota to be significantly re-enriched by the process of speciation. Recovery from the extinction event at the end of the Cretaceous period took tens of millions of years.

Therefore, the consequences of mistakes in social risk taking can be much more widespread and long lasting than those of individual risk taking. As has often been said about the current extinction crisis, humanity is not just causing the deaths of many other organisms, it is causing an end to their births. To most people, the latter is a much grimmer consequence, whether for rhinos or for humanity. Since the probability of the extinction of myriad organisms is very high and the consequences to the functioning of ecosystems and indirectly to civilization are deemed very serious, the risk being run by not dealing effectively with the extinction crisis is gigantic.

Social risk-benefit decisions differ from personal ones in another important way. Risks faced daily tend to be familiar and easily perceived, as are the benefits. However, few people recognize the benefits of sharing Earth with millions of kinds of other organisms or think much about the importance of a reasonably stable climate. The problem becomes more complex because the most critical environmental risks are difficult to perceive. With the possible exception of someone observing the death of the last survivor in captivity, no one has ever watched one species become extinct, let alone thousands of species. Nor has anyone ever observed the results of heating the planet by several degrees over 50 years. Costs associated with individual risks are usually paid in minutes or months; those associated with environmental risks may be paid over millennia or more. Costs may be virtually infinite, but they are amorphous and distant in time.

Risks to ecosystems generally appear in forms that are not easily perceived. A swerving car or a snarling dog are the sorts of stimuli that millions of years of biological and cultural evolution have prepared people to recognize as threats. But the gradually climbing, zig-zag line on the chart of atmospheric CO_2 concentrations recorded at Mauna Loa or statistics on habitat destruction are not. For most of human history, people could do nothing about environmental changes occurring over time spans of decades. People did not cause the changes, nor could they stop them. Consequently, human perceptual systems evolved that focused poorly on gradual changes in the environmental backdrop of their lives (Ornstein and Ehrlich 1989).

Despite these difficulties of perception, civilization is now faced with weighing the risks and benefits of actions with delayed and uncertain consequences. Suppose very rapid climate change in the next century could result in hundreds of millions of premature human deaths (Daily and Ehrlich 1991). How much effort should be expended to slow that change if its probability is 1%? 5%? 25%? 50%?

These are not easy questions to answer; there is no *scientific* basis for answering them; and, of course, slowing global warming would not be cost free. However, many of the measures required, from greatly increasing energy efficiency to land reform in less developed nations, would carry enormous additional benefits. Moreover, since the probability of unprecedented climate change appears to be approximately 50%, a very large social effort, one incurring substantial costs, seems not merely justified but virtually mandatory. Yet, the discussion of risks, costs, and benefits associated with possible responses to global warming has hardly begun. Progress clearly will be fraught with disputes, given the unevenness with which benefits and costs will be distributed and the unequal ability of different societies to take effective remedial measures.

Substantial costs will inevitably be incurred in dealing with the entire suite of environmental dilemmas now facing civilization, regardless of whether they are paid to prevent damage or to cure it later. The monetary costs and benefits of either strategy will be difficult to calculate and will doubtless always be somewhat in dispute, although rough orders of magnitude can be discussed.

CONCLUSIONS

Regardless of whether we can predict risks, costs, and benefits with precision, rough estimates indicate that *we can afford to insure ourselves against the most severe risks to ecosystems; that is, to insure against environmental catastrophe*. Not to do so would be bizarre for societies that routinely insure lives, limbs, homes, automobiles, and businesses against risks whose probabilities of occurrence are much smaller than those of the destruction of the remaining ecosystems in the world caused by the human takeover of the planet's productive areas, abetted by assaults from pollution and unprecedented climate change due to global warming.

Unfortunately, the role of continuing population growth in both generating the problems and increasing the risks remains largely unrecognized. However, one essential way to insure against risks to ecosystems is to take whatever humane steps are required to bring the human population explosion to a halt and to start a slow decline toward the number of people who can be sustained over the long term. There is no reason to believe that ecosystem risks can be greatly reduced by operating on the A and T factors of the I = PAT equation, if P is allowed to continue growing. Since controlling the size of the human population will take longer than adjusting consumption patterns or adopting more environmentally benign technologies, it seems fair to say that the greatest single risk to Earth's ecosystems is the population explosion.

REFERENCES

BROWN, L.R., DURNING, A., FLAVIN, C., FRENCH, H., JACOBSON, J., LENSSEN, N., LOWE, M., POSTEL, S., RENNER, M., RYAN, J., STARKE, L., and YOUNG, J. (1990). State of the World 1991. W.W. Norton, NY.

COLBURN, T.E., DAVIDSON, A., GREEN, S.N., HODGE, R.A., JACKSON, C.V., and LIROFF, R.A. (1990). Great Lakes, Great Legacy? Conservation Foundation, Washington, DC.

DAILY, G. and EHRLICH, P.R. (1990). An exploratory model of the impact of rapid climate change on the world food situation. Proc. R. Soc. (B) 241:232-244.

DALY, H.E. (1977). Steady-state Economics. Freeman and Co., San Francisco, CA.

DALY, H.E. and COBB, J.B. (1990). For the Common Good. Beacon Press, Boston, MA.

DIAMOND, J. (1991). The Rise and Fall of the Third Chimpanzee. Radius, London.

DYSON, F. (1975). The hidden cost of saying no! Bull. Atomic Sci. 31:23-27.

EHRLICH, A.H. (1990). Agricultural contributions to global warming. In: *Global Warming: The Greenpeace Report* (J. Leggett, ed), pp. 400-420. Oxford University Press, NY.

EHRLICH, P.R. (1975). The benefits of saying yes! Bull. Atomic Sci. 31:49-51.

EHRLICH, P.R. and EHRLICH, A.H. (1981). Extinction: The Causes and Consequences of the Disappearance of Species. Random House, NY.

EHRLICH, P.R. and EHRLICH, A.H. (1989). Intelligent planning for safety. Society 27:15-16.

EHRLICH, P.R. and EHRLICH, A.H. (1990). The Population Explosion. Simon and Schuster, NY.

EHRLICH, P.R. and EHRLICH, A.H. (1991). Healing the Planet. Addision-Wesley, NY.

EHRLICH, P.R. and WILSON, E.O. (1991). Biodiversity studies: Science and policy. Science 253:758-762.

EHRLICH, P.R., EHRLICH, A.H., and HOLDREN, J.P. (1977). Ecoscience: Population, Resources, Environment. Freeman & Co., San Francisco, CA.

Food and Agriculture Organization (FAO). (1990). 1990 Production Yearbook. FAO, Rome.

GRIBBIN, J. (1990). An assault on the climate consensus. New Sci. 15December:26-31.

HOLDREN, J. (1983). The risk assessors (book review). Bull. Atomic Sci. June/July:33-37. (See also the exchange of correspondence in August/September:59-60.)

HOLDREN, J. (1990). Energy in transition. Sci. Am. September:157-163.

HOUGHTON, J., JENKINS, G. and EPHRAUMS, J., eds. (1990). Climate Change: The IPCC Scientific Assessment. Report of the Intergovernmental Panel on Climate Change (IPCC), World Meteorological Organization (WMO) and United Nations Environmental Programme (UNEP), Geneva. Cambridge University Press, Cambridge, UK.

JONES R. and WIGLEY, T. (1989). Ozone Depletion: Health and Environmental Consequences. John Wiley & Sons, NY.

MARTIN, P.S. and WRIGHT, H.E., JR. (1967). Pleistocene Extinctions: The Search for a Cause. Yale University Press, New Haven, CT.

MARTIN, P.S. and KLEIN, R.G., eds. (1984). Quaternary Extinctions: A Prehistoric Revolution. Arizona University Press, Tucson, AZ.

MYERS, N. (1989). Deforestation Rates in Tropical Forests and Their Climatic Implications. Friends of the Earth, London.

ORNSTEIN, R. and EHRLICH, P.R. (1989). New World New Mind. Doubleday, NY.

RAUP, D.M. (1988). Diversity crises in the geologic past. In: Biodiversity (E. O. Wilson, ed), pp. 51-57. National Academy Press, Washington, DC.

SCHNEIDER, S.H. (1990). Cooling it. World Monitor July:30-38.

UNITED NATIONS FUND FOR POPULATION ACTIVITIES. (1991). State of the World's Population, 1991. United Nations, NY.

VITOUSEK, P.M., EHRLICH, P.R., EHRLICH, A.H., and MATSON, P.A. (1986). Human appropriation of the products of photosynthesis. BioScience 36(6):368-373.

WILCOX, B.A. (1984). In situ conservation of genetic resources: Determinants of minimum area requirements. In: *National Parks, Conservation, and Development: The Role of Protected Areas in Sustaining Society* (J.A. McNeeley and K.R. Miller, eds). Proceedings of the World Congress on National Parks, Bali, Indonesia, 11-12 October 1982.

WILSON, E.O., ed. (1988). Biodiversity. National Academy Press, Washington, DC.

WORLD RESOURCES INSTITUTE. (1986). World Resources 1986. Basic Books, NY.

WORLD RESOURCES INSTITUTE. (1988). World Resources 1988-89. Basic Books, NY.

WYMAN, R., ed. (1991). Global Climate Change and Life on Earth. Chapman and Hall, NY.

CHAPTER 3

ENVIRONMENTAL RISKS—FACTORS TO CONSIDER IN THEIR CONTROL

Ruth Patrick
Academy of Natural Sciences
Philadelphia, Pennsylvania

Environmental risks are produced by the activities of humans and by the plants and animals of the natural world as well as the physical forces making up the natural world. Examples of the natural sources of environmental risk are volcanoes; hurricanes; tornadoes; droughts; floods; chemical products produced by animals, such as venoms; and plants, such as alkaloids. Risks produced by humans are extremely varied and are steadily growing. These risks may be chemical and physical in nature, or they may be structures such as dams, high buildings without proper fire protection, etc.

The evaluation of the importance of risk by humans, that is, how large or how small it is, has a logical, technical base and a perceived base. Unfortunately, what the public perceives as environmental risks is costing a great deal more money in most cases to alleviate than are the real risks produced by the structures or chemical substances. This disparity indicates that environmental scientists need to educate and communicate with the public.

Environmental education should start in the primary schools and continue through the college years. However, this regime may not be enough, and adult education and communication of various types through television, lectures, articles, and even night school should be available. Without such education and communication, proper judgments of risk will not exist. For example, radon is ranked by USEPA as third highest of the 11 top concerns about environmental risks (Roberts 1990). However, the public ranked radon risk as 28, and it was only mentioned by 17% of the people asked (Kunreuther and Patrick 1991). On the other hand, active and abandoned hazardous waste sites were ranked first and second by the public, but were not on the USEPA list of the 11 top concerns.

People tend to be concerned about risk only as it affects human health. Risk to the universe of plants and animals is scarcely considered. However, what happens to the environment may be far more important for future

---

1. Corresponding Author: Ruth Patrick, Francis Boyer Chair of Limnology, Academy of Natural Sciences, 19th and the Parkway, Philadelphia, PA 19103.

2. Key Words: risk, human perception, human health, hazardous materials, radioactive chemicals, biodegradation, groundwater, education, communication, incentives, public perception.

generations (and even for present life over a span of 70-80 years) than what happens to a relatively few human beings.

Concern about human health usually is centered on a relatively few individuals, whereas environmental pollution may affect whole towns and regions. If the ecosystems that cleanse air and water and destroy wastes and provide nutrients for the myriad plants and animals as well as humans are destroyed, civilization would not be maintained and ultimately would be destroyed.

Considering the effect of a hazardous or toxic material requires examination of its fate as well as its concentration at the moment. Also important to consider, for example, is how it interacts with other chemicals and physical conditions in the air, land, and water since these three media all interact. For example, SO^x emitted from a smokestack in the Midwest will interact with water molecules forming small aerosols in the atmosphere and produce acid that falls out on the landscape as acid rain. These not only affect the land but also lakes and streams. Thus, the entire pathway and fate of a toxic material must be considered in determining its hazard to the environment and humans (Patrick 1991).

How organisms affect toxic materials must also be considered. For example, organisms may reduce the toxicity by oxidation and thus change the chemical composition of the toxicant and make it less toxic. Or, in contrast, organisms may make the toxicant more toxic. For example, certain bacteria, and certain protozoans, can take in metallic mercury, methylate it in their metabolism, and produce the much more toxic substance methylated mercury. These types of interactions must be understood and used in a holistic approach to evaluate environmental risks.

Other technical factors enter into the evaluation of a toxic or hazardous material. For example, the kind of radiation present must be known for a radioactive chemical. Gamma radiation is easier to detect than alpha or beta radiation, which have severe effects on the human body. The half-lives of radionuclides are very important in considering their risk. Iodine-131, which has a half-life of a few days will not have nearly the impact on human life, plants, or animals as will Iodine-129, which has a half-life of a great many years.

How a toxicant or hazardous material moves in the environment must also be considered. Is the substance of such a chemical nature that it is quickly sorbed onto the surface of colloids and, therefore, becomes particles and, hence, may fall out of solution? Is it dissolved and moving freely with water? Will it interact with other chemicals to form precipitants and be deposited in the sediment? These types of interactions often require some knowledge of the half-life of the substance. For example Cesium-137 in sediments has a very long half-life; a serious problem may exist with it now and in the future. This is also true for many metals. However, some hazardous materials, such as certain organic chemicals, are easily broken down by organisms and do not form a hazard for long periods.

One great concern is to know the toxicity of a hazardous material in the ground, such as in a waste site. Consequently, its movement and interaction with other chemicals is most important. For example, few people consider the fact that zinc and copper together are far more toxic than zinc or copper by themselves. If zinc and copper were added to a waste site, not knowing that the other element was present, a substance might be produced that was much more toxic than if these atoms were not mixed.

Environmental hazards should also be evaluated according to the areas they cover. Does the hazard occur all over a country or all over a continent? How frequently is the substance producing the hazard used by the public or industry? For example, small amounts of a substance, not used very often, will not present the same hazard as a substance commonly used over a widespread area.

Specifically, point sources of hazardous materials can be handled much more effectively than nonpoint sources. Methods of reducing such materials are by eliminating their use in products; by treating waste containing these substances before they are discharged into holding basins where they may be biologically destroyed; or by mitigating their toxicity through chemical or physical interactions, such as precipitation or coagulation, so that they are more easily handled. Nonpoint sources are harder to control because the sources are often difficult to identify. Several methods have been tried, such as controlling by catchment basins the runoff from factory areas, town parking lots, etc. Other methods to control nonpoint sources have been planting vegetation along edges of lakes and streams that may reduce the entrance of nonpoint source discharges into surface waters.

Point and nonpoint sources of hazardous materials may affect groundwater. Nonpoint sources are particularly difficult to control since hazardous materials are often distributed onto the ground in widely dispersed areas, and the materials subsequently enter groundwater. These toxic materials are utilized to control pests on golf courses and various agricultural crops and are often sprayed by airplanes. Thus, the method of application and the area that is affected are widespread, and the runoff is difficult to control from a pollution standpoint.

When a toxicant or hazardous material enters groundwater, treatment is often very difficult, because the material may be hard to locate, and may behave in various ways in groundwater. If the material dissolves, it moves with the groundwater. If it is light or its specific gravity is much less than that of water, it will float and gradually dissolve. Consequently, two forms of the contaminant must be dealt with: the floating part of the material and the dissolved part. If the specific gravity is greater than that of water, the material will sink to the bed of the aquifer. If the bed of the aquifer is on a slant, the material may flow in the opposite direction from the water if the water is under pressure, as is often the case in confined aquifers (Cherry 1984).

Aquifers bound by limestone often have small cavities known as lacunae. If the toxicant deposits there, complete flushing from these areas may take years and low amounts of toxicity continue to contaminate groundwater.

Surface waters are usually more easily cleansed of a toxicant or hazardous material than is groundwater. This is particularly true for point source discharges that can be treated before they are discharged by precipitation, coagulation, or biodegradation. Nonpoint discharges are more difficult to treat, but some success has been realized by channeling runoff from parking lots, golf courses, and airport strips into lined retention basins where the runoff is then treated. As mentioned previously, vegetated corridors along the edges of streams and lakes have also proved effective for controlling some hazardous materials. Some small headwater streams have most of their flow during drought conditions composed of sewage or effluent from a factory. In such cases, the whole flow should be treated before allowing it to proceed downstream.

Groundwater contamination from waste sites is much more difficult to treat than surface contamination because of the complexity of groundwater movement. However, various means of treatment have been devised that provide for reductions in the concentrations of chemicals originating in hazardous or toxic wastes sites. A way to treat such chemicals is *in situ*. If the chemicals are volatile, a horizontal pipe with air under pressure bubbled through it may be placed in the aquifer or unsaturated zone. At the surface, the volatile chemicals can be collected under plastic sheets and destroyed. Often, vacuum is added to hasten the loss of the volatile materials from the unsaturated soil above the aquifer. This method is particularly useful when chemicals such as benzene and toluene are of concern (Corey and Kauer 1990). The oldest method of treating is to pump the toxic water from an aquifer, treat it at the surface, and then, depending upon its chemical composition, return it to the aquifer or further treat it, and collect any unwanted residue.

More recently, an old method of confining hazardous leachate by placing a wall above and around a dump has been modified. Often in the past, the wall material has been concrete. However, fungi, worms, and roots of plants permeated these walls over time and leachate escaped. Some new experiments using long lasting pesticides within the walls are being considered so that the wall will not deteriorate.

Perhaps the most promising new discovery, particularly for extremely hazardous substances such as PCBs and TCEs, is that these substances are biodegradable, especially under aerobic conditions. Bouwer (1988) points out that many trace organics can be largely removed under aerobic conditions by biodegradation. Examples of these are chlorobenzene, 1,2,4-trichlorobenzene, 1,4-dichlorobenzene, naphthalene, and styrene. Many other toxic materials can be removed anaerobically in a methanogenic environment. They include such chemicals as 1,1,1-trichloroethane, carbon tetrachloride, dibromo chloropropane, dibromochloromethane, and ethylene dibromide. Other toxic materials may degrade under anaerobic conditions containing sulfate. They are chemicals such as bromochloroethane, carbon tetrachloride, and bromoform (Suflita 1988). PCBs may degrade under aerobic conditions. However, the proper mix of nutrients must be present in order that the organisms causing the degradation can proliferate and thus destroy considerable amounts of the hazardous or toxic material.

Thousands of undiscovered organisms must exist that have unusual metabolic pathways and that may be able to bring about degradation of substances formally regarded as impervious to any kind of treatment. The fact that many thousands of such organisms exist in the soil has recently been validated by studies at the Savannah River Site—more than 2000 physiologically different bacteria were present at depths of 700 feet and greater (Corey and Kauer 1990).

One question about hazardous materials in the environment that has not been thoroughly addressed concerns the effects of hazardous materials on bacteria, fungi, and various kinds of organisms, such as worms and protozoa, that live in the soil and are so important in transforming debris that enters the soil into soluble forms that can be utilized by plants and animals. Many of the most important plants that humans use have roots that need particular soil faunas and floras to enhance their production. Any hazardous or toxic material reducing the activities of these organisms that normally function to increase the nutrients in the soil for plant life will alter the productivity of the plants. This is but another incident of the importance of understanding interactions in the whole ecosystem; that is, taking a holistic approach toward solving the problem.

One approach to solving such problems in some parts of the world has been through laws and regulations. To date, most laws and regulations dealing with hazardous or toxic materials have been generated to control point sources. Controlling nonpoint sources has not been adequately addressed. Incentives, as well as laws and regulations, must be developed for controlling wastes from nonpoint sources, as well as controlling hazardous materials from scattered point sources, particularly small industries and municipalities. One suggestion is that permits be given and that dischargers pay specified amounts for the pollution they emit. For example, a discharger would pay more for a small amount of a very toxic substance emitted than for a larger amount of a less toxic substance that could be assimilated by the natural environment. Hazardous and toxic materials particularly would be the object of such permits. Perhaps allowing a tax deduction for the reduction of toxic materials produced in a waste stream would be wiser than granting permits.

In rural and residential areas, considerable amounts of toxic materials enter surface waters as well as groundwater from nonpoint sources. Incentives must be found, as well as new technologies developed, to control these sources. Of course, taxing could always be used—samples of runoff could be taken and the farmer or golf club could be charged for the amount of runoff each releases. Nonpoint sources from private homes or farms would be harder to control; however, perhaps people residing in a given area could be taxed according to the amount of toxic or hazardous materials entering storm sewers from the streets of such areas. Perhaps the most important approach to the nonpoint source problem is education and communication of the facts concerning the maintenance of lawns and golf courses and the production of crops by methods other than using hazardous, toxic chemicals. Considerable research and methods of communications are being developed to alleviate these problems.

Some of the major problems in developing the most effective methods of control and rank ordering of substances to be controlled are public perceptions. Often, as mentioned previously, public perceptions of risk are different from the technological characteristics that a substance might indicate.

Resolving the public perception of risk is much more difficult than solving the problems concerned with the technical aspects of risk. One important consideration in correcting the public perception of a risk is to have the public involved. One example would be to conduct open discussions at the conception of a potential risk, such as the siting of a waste dump. The public must be included; facts must be communicated in a laymen's language; and risks must be clearly set forth and discussed. The public wants to be involved in measures that should be taken to increase the security of the site and in monitoring any effluent, air or water, from the site. Therefore, classes should be set up to teach the public how to monitor, when to monitor, and the significance of the findings. Instructors for such classes should be well informed people who have sympathy for the public and its inadequate knowledge (Hadden 1991).

In general, two kinds of risk exist. One involves a person making a decision to take the risk, such as taking an airplane flight or riding on a motorcycle or in an automobile. The other involves suddenly encountering the risk because of the actions of someone else or a situation over which the person has no control.

Dealing with the problem of hazardous and toxic wastes necessitates giving the public all possible information about the pros and cons of the waste site; teaching them how to monitor the effluents from the waste site; and how to form inspection teams and oversight boards. This will help alleviate the feeling of risk. For example, the public has accepted incinerators as a result of designating a group of citizens and local officials to be present during any burns in order to ensure that optimal operations temperatures are achieved and maintained (Hadden 1991). These citizen groups were trained to monitor a risky facility, and the feeling of risk was diffused, even though the groups did not actually perform the monitoring (Hadden 1991; Kunreuther and Patrick 1991).

People have recently lost respect for authority. Truths espoused by officials are not automatically accepted by the public. However, if the citizenry understands the problem, can realize costs/benefits of the problem, and can help in monitoring and controlling risks that might arise, then the public is more willing to accept the risk. In this age of increased instrumentation and impersonalization of operation of hazardous facilities, the public must thoroughly understand what is being done, and what will be done to mitigate risk.

How the environment functions and mankind's dependency upon it for a healthy life need to be emphasized in basic education from the primary grades through college. Only through understanding how human actions affect other people and the whole environment can the public appreciate the difference between environmental insults that are quickly mitigated by natural processes and those that accumulate. This knowledge is all important in being able to judge the significance of any human action. People must be taught not only how to prevent the production of products and conditions that are hazardous to humans and the environment, but also how to mitigate them so that they can be absorbed by the natural functioning of the environment.

For these reasons, remediation and solution of these immense problems should be a product of not only technical remediation research and new ways to prevent waste generation, but also of efforts to better educate all ages and segments of the population. These problems and their solutions should be discussed and debated, and solutions should be formulated by interactions in the home, in the classroom, and by the community as well as governmental agencies.

REFERENCES

BOUWER, E. (1988). Biotransformation of contaminants in groundwater. In: *Proceedings of the Third National Water Conference*, pp. 133-153. Academy of Natural Sciences, Philadelphia, PA.

CHERRY, J.A. (1984). Contaminant migration in groundwater: Processes and problems. In: *Proceedings of the Second National Water Conference*, pp. 65-94. Academy of Natural Sciences, Philadelphia, PA.

COREY, J.C. and KAUER, E.U. (1990). Bioremediation, a fertile field for cooperative research. In: *Proceedings of the Fourth National Water Conference*, pp. 167-179. Academy of Natural Sciences, Philadelphia, PA.

HADDEN, S.G. (1991). Public perception of hazardous waste. Risk Anal. 11(1):47-57.

KUNREUTHER, H. and PATRICK, R. (1991). Managing the risks of hazardous wastes. Environment 33(3):12-15.

PATRICK, R. (1991). The importance of the technical characteristics of hazardous materials in considering environmental risk. Risk Anal. 11(1):27-29.

ROBERTS, L. (1990). Counting on science at EPA. Science 249:616.

SUFLITA, J. (1988). The biodegradation of groundwater pollutants when oxygen is unavailable. In: *Proceedings of the Third National Water Conference*, pp. 271-295. Academy of Natural Sciences, Philadelphia, PA.

CHAPTER 4

WHAT EVIDENCE OF ECOSYSTEM RISK IS NECESSARY TO INFLUENCE REGULATORY AND INDUSTRIAL DECISIONS?

Donald I. Mount
AScI Corporation/AScI-Duluth Environmental Testing Division
Duluth, Minnesota

INTRODUCTION

Environmental regulatory decision making as a process has been evolving at a rapid rate since the late 1960s. Earth Day 1970 and the time immediately before and after is the obvious pivotal period when a marked shift in public priorities and concern occurred. Before this period, water pollution (as distinct from air pollution, for example) concerns were overwhelming in what would have been called *environmental protection* if such a term had been coined at that time. As history is viewed backward from 1970 though, even the water pollution concerns rapidly narrow to only those of human health. In fact, before 1950 essentially the only noteworthy issue was whether people got sick or not from water supplies! Remember, chemical carcinogenicity was not really broadly recognized until after 1950. Therefore, the human health concerns really consisted only of waterborne diseases—not chemical toxicity.

During the 1940s, my father took me fishing most weeks during the summer months. Our home was between Cincinnati and Dayton, OH, and I recall deteriorating fishing conditions in one stream after another. That the cause was waste discharges never even entered our minds. Such matters were then the responsibility of state health departments and they, as their name implies, were concerned about human health and not the many other concerns now commonplace.

Perhaps you wonder whether such "ancient history" has anything to do with the amount of data needed for regulatory environmental decision making today. It does, because the experience and thinking from the early days of water pollution control was the "mother hen" that molded present-day environmental regulatory philosophy. Therefore, any analyses of current decision making must be done with this understanding in mind.

Returning briefly, once more, to history, the discovery and use of the chlorinated hydrocarbon insecticides, beginning with DDT and then fol-

1. Corresponding Author: Donald I. Mount, AScI Corporation/AScI-Duluth Environmental Testing Division, 112 East Second Street, Duluth, MN 55805-1701.
2. Key Words: decision making, scientific data, environmental regulation.

lowed by the organo-phosphorus insecticides, were extremely significant in focusing attention on "toxic chemicals" and, more importantly, on pollution matters other than human health. The period of increasing use of chemical insecticides coincided with the recognition of chemically caused carcinogenicity. The two were powerful forces that ushered in the concern over "toxics," influencing legislation as well as regulations and the concern of the public.

Modern environmental decision making, therefore, is grounded in regulating chemicals. Only recently has a radically different concern over traditionally "nontoxic chemicals" emerged, i.e., carbon dioxide and other atmospheric contaminants. The stakes, the considerations, the evidence, and the magnitude connected with these emerging problems are extremely different from anything thus far encountered, and the unfolding regulatory picture will be fascinating to watch.

SCIENCE AND DECISION MAKING

Probably nowhere is there a stronger feeling that every decision must be scientifically based than in the environmental area. Two primary reasons perhaps account for this feeling. One reason is related to the water pollution based history. Since human sickness was the motivation for those early actions, we still look for the technical basis on which to justify regulatory action. The second reason is more subtle but probably more important. Since much environmental decision making today is proactive in nature, i.e., trying to prevent impact before it occurs, opinion (some prefer to call it judgment) is the key ingredient of action. Anyone who has held a job requiring constant decision making based on subjective reasoning knows how demanding and emotionally draining it is. When officials of public agencies must make subjective decisions on monumental issues, such as lead in the environment, fear of the consequences can drive the outcome. To diffuse some of that pressure, decision makers desperately search for some objective support. And what better objectivity can be found than a "scientific" basis?

Thus it is that a very often used statement has emerged in environmental decision making—"We must have enough data to make a scientifically defensible decision." If we really think about this statement, the meaninglessness of it is obvious. For any decision, there is no fixed, specified, or commonly accepted amount of data that is deemed the quantity of data needed for a scientific judgment. Rather, what makes any decision scientifically based is the degree of confidence and statement of uncertainties that accompanies the interpretation of the data. Some might argue that, when a decision can be made using statistically based 95% confidence levels, the decision becomes scientifically defensible. However, such a view simply shifts the subjective judgment to a *choice* of the appropriate probability level. One can make equally valid scientifically based judgments whether there is one data point or a million—only by properly stating the confidence level that can be attached.

So we should discard all hope of quantifying the amount of data needed to make a scientific decision. Science can only help after someone has made the decision as to the proper confidence that must be achieved in the decision to be made. Then, and only then, can the amount of data needed be quantitated.

Perhaps some comment about the choice of confidence levels should be made. The 95% confidence level is very widely used for statistical evaluation. The 95% level is often used as though, if there is 95% or better confidence, the truth is established, and, if less, then there is no confidence. Without judging the wisdom of choosing the 95% level, there is clearly no black and white, nor is 95% always the best choice. A statistical difference at 94% confidence is not measurably different from 95%, but the former level will almost always be interpreted as not different.

ACCEPTABLE RISK

If the decision is really one of confidence choice, then obviously the acceptable risk factors into the decision in a heavy way. Many books and articles have been written about risk perception, and this discussion does not dwell on it in depth. For purposes here, I note several types of risk. Probability risk measures the likelihood that there was a chance association and is rather factual; that is, the probability that, given some exposure, adverse effects will occur. Such data are probably well understood by most decision makers, and numerical values can usually be given for such risk. In addition to risk of sickness, injury, or some adverse biological effect, the factors of the economic, political, and moral essence of risk must be considered.

Each of these risks will be judged differently by various individuals or groups. For example, a commercial viewpoint will give economic risk more weight than an environmental activist group who is little affected by the profitability that is at stake for the company directly affected. An elected official or a political appointee is likely to consider the political impact of a given decision much more important than a secure federal bureaucrat well protected by the civil service system. Some environmental activists have made a religion of environmental protection. To these folks, protection is a matter of right and wrong, and economics and such should not even enter into the decision.

One axiom can be stated with complete confidence—no decision is made without the influence of bias. To illustrate, there is a solid reason why people who work for regulatory agencies generally support regulatory actions and people who work for dischargers generally support the decision of the dischargers. The support has nothing to do with science nor the ability of the individual or group to evaluate scientific data. The reason is clearly a matter of bias, which makes each view the various risks differently. If you are making decisions and are not affected by economic costs, why would you let such costs be a big factor in deciding acceptable risk? Many meetings in which I have been a participant resulted in complete agreement as to the technical issues and total disagreement in the decision; the reason—bias.

The word *bias* carries an undesirable connotation, but this should not be true in this discussion. After all, a democracy is built on bias—bias of the majority. Consequently, in a democracy such as the United States, bias that drives decisions in the environmental area is not wrong. The danger arises when the bias may ignore long-term implications of decisions. Some types of bias are likely to favor the short-term issues. This is especially true of economic bias. For example, the hue and cry of many industrial and marketing groups is that energy conservation will force large segments of the economy into non-profitable positions. However, every bit of data available shows that the economic health of a country is directly proportional to its efficiency of energy use! The short-term economic adjustment obliterates the long-term effect. Sometimes scientific bias strengthens to such a level that it overpowers all other biases. The smoking-lung cancer issue is just such an example. Even in this issue, however, the scientific bias is still inadequate for some. This indicates that the scientific evidence for not smoking is still poorly understood by a segment of the general public.

Environmental decisions would be made easier if the debate focused more on the effect that various factors—economic, moral, political, and scientific—have on acceptable risk instead of trying to attribute the decision entirely to the scientific defensibility of environmental impact data. We are all taught from elementary school on, and especially in training for scientific disciplines, to be objective and non-biased. Therefore, it is heresy for a decision maker to admit openly that economic benefit exceeds the adverse effect of cost in life or injury. Yet isn't use of the automobile a vivid example of most people placing lower the risk of injury and even death than the ability to get somewhere faster?

So, the crux of good environmental decision making lies in proper weight being given to various biases. More or better scientific data will not always allow more correct weight to be given to bias—even scientific bias. For example, being even more certain that detrimental impact will occur for some distance below a discharge and being able to state accurately exactly which species will disappear does not ease the difficulty of judging what weight should be given to the economic damage of remedial action.

If environmental decisions could be made by recognizing and getting biases into perspective, at least the issue could be addressed instead of a charade game where we are lead to believe it is all "scientific." There is, underlying the thinking of the public and environmental regulatory groups, a subconscious belief that the human race can live, even the life style of the United States, without changing the environment. Few people even connect a long hot shower in the morning with the greenhouse effect and climate change. I am convinced that most believe that scientifically based decision making can allow us to live and not impact the environment. If I am right, then there is little wonder we have such a strong conviction that scientific data should be the basis for decision making.

Lest I be misunderstood, I do not short change the value of scientific data in decision making. There is no substitute for a thorough understanding of the problem from the scientific point of view. My point is that other factors as outlined earlier are equally valid considerations. The decision then has to be a result of a subjective weighting of all these considerations. When the weighting process is subjective, we then deal with bias.

I cannot conclude this discussion of bias without commenting on an attitude held by so many regulatory people. Many, many such folks truly believe that, when it comes to environmental regulatory decisions, they are unbiased because they are constraining pollution of the environment. This attitude seems to grow and spread like cancer as people work in regulatory affairs. Certainly the regulated community has its biases too, but this bias is well recognized—just ask the regulators. Often the bias of the regulated is economic in nature and the first response is a threat of plant closure. Usually that threat does not materialize and then the regulators feel even more unbiased! So, the decision about acceptable risk is made based on weightings of biases, and scientific data are only one such bias.

ACCOUNTABILITY

As my time in USEPA grew to 30 years of service, I became more and more impressed with the role of relatively low level staff in shaping decisions of widespread geographic scope and even policy. Because bias does play such a decisive role in decision making, motivated individuals can wield untold power in shaping regulatory action. The list of decisions that were so shaped would fill volumes.

There is almost no need to observe that the motivated individual is the key to the power that environmental activist groups hold. Such people, whether in public agencies or in pressure groups, rely heavily on the saying "You cannot unring a bell." We are fortunate to have had the 1970 Earth Day era. Without it, cleanup in the United States might not have happened or might have been slowed to a snail's pace. The power of "the motivated individual" is clearly present in regulatory agencies as well. These people probably chart the agency course and set policy to a greater extent than the policy makers themselves.

More directly, with regard to accountability, a careful study of who really makes most of the environmental decisions would reveal a characteristic about the decision maker not generally appreciated. Most of the day-to-day decision making is performed by mid-level to entry-level staff in regulatory agencies who will feel no direct effect of the decisions made. They are almost totally insulated by management from the consequences. The end result of this situation is that the individual making the decision feels no true responsibility for the consequences of the decision. Many times I have discussed with such folks their actions. Their response showed how detached they themselves felt from the impact of the decision. To them they had a mandate, given or self-imposed, to be as hardnosed on dischargers as possible and "By God that is what they were doing." When that happens, management and nominal decision makers feel frustrated and out of control. They then spend all their energy on the defensive, protecting the policy they get locked into and never really decide if the decision is correct.

The accountability issues involved here are not unlike a politician making decisions—but, a politician who never has to face re-election! I hope, as environmental decision making matures and more experienced staff emerge, that this lack of accountability can be corrected. If it can be, the United States will move to a much more lofty plateau of sophistication—a benefit to the environment for certain.

EXERCISING JUDGMENT

There is an aspect of working in a public agency that I'm sure is not seen by many outside. This aspect may seem trivial but, in fact, explains a great deal about the usual negotiation between regulator and regulated. In the private sector, there are rewards of salary, promotion, and praise when one uses good judgment and the outcome is good for the company. On the contrary, in a regulatory agency, such rewards are nonexistent. Even worse, if judgment is exercised and the outcome is not good, severe repercussions are likely to follow. There is a cadre of citizens who watch for regulatory decisions that deviate from regulations, regardless of their wisdom, and sue or take other legal or disciplinary action. If you work in a public regulatory agency and want a clean record, you will be safe only if you follow the letter of the law and never use common sense judgment. It's the only way to not foul up your track record. Many people have asked me about what they felt were ridiculous regulatory decisions that seemed to make no common sense. Usually, the fear of losing that protection, from following rules exactly, was the basis. Until this threat is removed, use of common sense and good judgment will not occur.

NON-SCIENTIFIC BASIS FOR DECISION MAKING

This subject might be considered a type of bias, but maybe is different enough to be viewed separately. In the more wealthy nations of the world, more money and other resources are spent on non-necessities than on necessities. We, in the United States, spend billions of dollars on lawn care, useless trinkets, and whims. People do many things just because they like it, such as mowing the medians of freeways as though they were lawns. Yet, an environmental decision maker would find it impossible to make a decision on aesthetics with some exceptions, such as haze in the Grand Canyon. Even then, there would probably be an attempt to show loss of tourist dollars or some impact on the economy to justify action. In the battle with the Reserve Mining Company over the discharge of tailings to Lake Superior, there was essentially no threat discovered until much later when asbestos fibers were found in the tailings. But the public simply didn't want the discharge there, even before asbestos was discovered. As one drove up the north shore of Lake Superior through forest and around cliff, the mining operation stood out as a scar, a structure totally out of place in the vacation land of so many citizens. However, no amount of urging would convince USEPA attorneys to pursue the court case on aesthetic grounds. They felt certain they had to have "scientifically defensible" evidence of impact in the lake to proceed. Doesn't the public have the right to decide that certain decisions should be made because—well, just because?

SUMMARY

After reading this chapter, one may wonder if the answer to the question "How much data are needed for regulatory decision" is answered. In short, the answer is that there is no identifiable amount needed. Other factors often cause the decision to be made, and whatever data are available are used to support the decision. In the many instances where scientific evidence uncovers further need for data before regulation, other pressures, i.e. economic, moral, political, will time the decision, and the data available will be used as best it can be to support that decision.

Bias is present in every decision, and it manifests itself largely in judging the acceptable risk, of whatever type. If only scientific data were used for decision making, then technically it would not be a democratic decision. It might not even be a better environmental decision. Since one such bias-free decision has never been made, this statement is hard to prove, true or false!

CHAPTER 5

AN ENVIRONMENTAL ASSESSMENT PRIMER FOR LESS DEVELOPED COUNTRIES, WITH AN EMPHASIS ON AFRICA

Charles H. Hocutt[1], Roderick Bally[2], and Jay R. Stauffer, Jr.[3]

A primer on environmental assessment protocol is provided for less developed countries (LDCs), with an emphasis on the African continent. Topics include the evolution of a conservation ethic, which is the basis for environmental assessment (EA) activities; the development of a world conservation strategy; the legal mandate for the EA process in the United States, which serves as a model for many LDCs; and definitions commonly used in environmental analyses. Examples from Africa provide an overview of cause/effect relationships of specific program impacts, many of which were ill-conceived from a western world perspective.

INTRODUCTION

> Today we are witnessing the extension of the industrial juggernaut into every corner of the globe. Soils are washing into the sea; toxic chemicals are polluting surface and ground waters; chainsaws (sic) and bulldozers are decimating tropical forests and with them exterminating a significant portion of the Earth's complement of species; acid rain is withering the forests and sterilizing the lakes in temperate regions of the northern hemisphere; chlorofluorocarbons are eroding the planet's protective ozone shield; and fossil fuel consumption is loading the Earth's atmosphere with carbon dioxide.
>
> Callicott (1991)

Callicott's (1991) cataclysmic statement is insightful of the changing biosphere, but the vision is incomplete. The functioning of the earth is more

1. Corresponding Author: Charles H. Hocutt, Center for Environmental and Estuarine Studies and University of Maryland Eastern Shore, Cambridge, Maryland 21613.

2. Department of Zoology, University of Fort Hare, Alice 5700, South Africa 27-404-32011.

3. School of Forest Resources, Pennsylvania State University, University Park, Pennsylvania 16802.

4. Key Words: Africa, case histories, conservation ethic, environmental assessment, legal mandate, less developed countries (LDCs).

5. Abbreviations: AMCEN, CERES, EA, EIA, EIS, GEF, GIS, IEM, IUCN, LDC, NEPA, NGO, OAU, UNDP, UNECA, UNEP, USAID, WWF.

intimately linked to man's historical conservation ethic and stewardship of renewable resources than can be imagined. The British chemist James Lovelock advanced the "Gaia hypothesis" in 1972: the earth as a whole is alive, its functioning being a sum of all the living biota at one time, and it is being impacted by a loss of structural or trophic redundancy through man's activities. The hypothesis remains controversial (Mann 1991); however, a number of scientists have called for global climate/ecosystem monitoring (e.g., Cairns 1990) with the primary objectives of (1) maintaining balanced biological communities, (2) protecting the structural and functional integrity of ecosystems, and (3) protecting biodiversity.

Developed countries spawned the industrial revolution, but even today controlling the respective environmental impacts remains unfeasible or impractical in many instances. In the process of Neo-colonialism, third world countries have often acquired technological development, but they have also inherited the associated problems, the effects of which can be magnified due to lack of expertise or financial resources to cope with a dilemma. Successful development by its very nature is structured, thus a sense of order is present in the the corresponding impacts, albeit the effects can be synergistic or cumulative in their fashion. However, unpredicted or unmanageable catastrophic events can occur that are either natural or anthropogenic in origin (e.g., El Nino and the Iraqi/Kuwait War), the influence of which can be global and especially traumatic for less developed countries (LDCs). In the latter cases, LDCs can do little more than respond to a domino effect. However, it is becoming increasingly clear that development within a LDC must not be at the sake of sustaining natural heritage.

CONSERVATION ETHIC

Callicott (1991), emphasizing North America, rationalized that a substantial shift in the philosophy of the conservation ethic has occurred in recent years with the acceptance of mankind as a component *of* the ecosystem, rather than being apart from it. He summarized this philosophical revolution in three major transitional stages.

(1) The *romantic-transcendental ethic* exemplified by Ralph Waldo Emerson and Henry David Thoreau, who recognized "the sanctity of Nature";
(2) The *resource conservation ethic*, considered to be the school of Gifford Pinchot, who viewed the use of resources by man as "the greatest good for the greatest number for the longest time";
(3) The *evolutionary-ecological land ethic* of Aldo Leopold, who was concerned with the "preservation of the integrity, stability and beauty of the ecosystem."

Conservation biology is a synthetic discipline that emerged with the publication of Soule and Wilcox's (1980) *Conservation Biology: An Evolutionary-Ecological Perspective*. Conservation biology is focused on the preservation of biodiversity, and, as such, virtually all the sub-disciplines of biology can contribute to its theory and practice. Brussard (1991) believed that Soule and Wilcox's book had at least three major impacts: (1) it stimulated basic biologists to apply their expertise to conservation problems; (2) it provided a broader perspective to scientists already active in the field; and (3) it defined a new interdisciplinary area.

WORLD CONSERVATION STRATEGY

The joint publication in 1980 of the *World Conservation Strategy* by the International Union for the Conservation of Nature and Natural Resources (IUCN), United Nations Environment Programme (UNEP),and World Wildlife Fund (WWF) stressed the critical need to balance the protection of the natural environment with the equally vital need to raise the standard of living of the world's poor. The aim of the *World Conservation Strategy* was to help advance the achievement of sustainable development through the conservation of living resources via

(1) maintenance of essential ecological processes and life support systems (such as soil regeneration and protection, recycling of nutrients, and cleansing of waters) on which survival and development depend;
(2) preservation of genetic diversity (on which depends the functioning of many of the above processes and life support systems; breeding programs necessary for the protection and improvement of cultivated plants, animals, and microbial organisms; and the security of many industries dependent on renewable resources); and
(3) insured sustainable use of species and ecosystems (notably fish, wildlife and forests that support millions of rural communities and industries worldwide).

More important, the *World Conservation Strategy* recognized that conservation of resources and development are not incompatible, but indeed they are mutually dependent. For instance, the dependence of rural communities on renewable resources is direct and immediate, providing fuel, food, and shelter. However, by increased stripping of trees and shrubs and by burning dung and crop wastes, soils become less valuable for cultivation (due to nutrient loss) and more susceptible to erosion. Concurrently, throughout the developing world, the life times of hydroelectric power stations and water supply systems are being shortened as reservoirs become filled with silt, siltation being a process of deforestation, overgrazing, and other unwise land uses. Thus, the *World Conservation Strategy* noted that unless development is guided by ecological, as well as by social, cultural, and ethical considerations, it will continually fail to meet or sustain its desired economic objective.

Particularly for LDCs, environmental policies that attempt to anticipate or predict significant economic, social, and ecological impacts, rather than simply react to them, are becoming increasingly necessary for the achievement of certain important goals: the satisfaction of basic needs, such as food, clothing, sanitation, and shelter; the development of a high quality environment; the optimum use of available resources; and the control of pollution and other forms of environmental degradation. Such anticipatory environmental policies involve actions to ensure that conservation and other environmental considerations are taken fully into account at the earliest possible stage of any decision that is likely to affect the environment.

It is enlightening to see that non-government organizations (NGOs) such as the World Wildlife Fund are no longer alone in their attempts to preserve ecosystems and endangered species. In recognition of the need to assist developing countries in relieving pressures on global ecosystems, the World Bank, the United Nations Development Programme (UNDP), and the United Nations Environment Programme have cooperated to establish the Global Environment Facility (GEF) program. The GEF is funded for over US$1 billion for 1991-94. The purpose of the program is to support international environmental management and to transfer environmental technologies in order to sustain fragile ecosystems. Specific goals include (1) reducing greenhouse gases, (2) preserving biodiversity and maintenance of natural habitats, (3) eliminating the pollution of international waters, and (4) protecting the ozone layer from further depletion. Equally as positive is the role that development organizations (e.g., the U.S. Agency for International Development [USAID]) are having where bilateral development agreements now have concern for environmental impact mitigation.

On another front, the Coalition for Environmentally Responsible Economies (CERES) is an alliance of investors and environmentalists that was formed within months of the Exxon Valdez spill into Prince William Sound, Alaska. Members of the alliance hold millions of shares in stocks and control $150 billion in pension and mutual funds. Through their financial clout, CERES is attempting to force corporations into accepting responsibility for their environmental conduct (Hinden 1990). CERES has prepared a 10-point code of conduct, the Valdez Principles, that asks corporations to disclose their environmental problems and to act judiciously in resolving them to protect biotic diversity and sustain the use of renewable resources.

AFRICAN MINISTERIAL CONFERENCE ON THE ENVIRONMENT

The first African Ministerial Conference on the Environment (AMCEN) was held in December 1985 in Cairo, Egypt, sponsored by UNEP in cooperation with the United Nations Economic Commission for Africa (UNECA) and the Organization for African Unity (OAU). The focus was the continuous depletion of the continent's natural resources with increased environmental degradation. The Conference adopted the Cairo Programme for African Co-operation to strengthen cooperation between African governments in economic, technical, and scientific activities to seek African approaches and African solutions to the problems and challenges of the region. Specific goals were (1) to halt environmental degradation, (2) to enhance the food producing capability of the continent, (3) to achieve energy self sufficiency, and (4) to rectify the imbalance between population and resources. These objectives were to be achieved through regional networks specializing in environmental monitoring (with the coordinating office in Nairobi, Kenya), climatology (Bujumbura, Burundi), soils and fertilizers (Kumasi, Ghana), water resources (Cairo), energy (Dakar, Senegal), and education and training (Senegal); in addition, genetic resources and science and technology centers are to be established.

THE SOUTH AFRICAN EXPERIENCE

Independent of AMCEN, largely because of historical political policies and consequent isolation from the rest of the world, are parallel South African programs. The Republic's scientists may have much to offer the rest of Africa once South Africa again becomes an accepted member of the African community. Unlike much of the advice given by the over 80,000 expatriate consultants present at any time in Africa (Timberlake 1985), the South African body of knowledge stems from several decades of direct experience.

South Africa presents an interesting combination of first and third world societies and conditions that co-exist uneasily in a very ecologically diverse country. South Africa is in the process of introducing Integrated Environmental Management (IEM). This is a procedure that is broader and more comprehensive in nature than the EIA/EIS system, covering all aspects and stages of the allocation of environmental resources. It includes the early stages of conceptualization, planning, assessment of possible effects, making and implementation of decisions, and a subsequent monitoring of results. There are four formal stages to IEM: proposal generation, assessment, decision, and implementation (which includes follow-up monitoring). The assessment stage itself can take three possible routes based on the results of a brief initial survey as to the likely impact of the proposal. Class 1 proposals clearly will result in significant environmental impacts; Class 2 proposals may or may not have significant impacts, while Class 3 proposals will almost certainly have no significant impacts. This initial survey determines whether further assessment is needed (Council for the Environment 1989).

LEGAL MANDATE

Since many LDCs either have or will model their own environmental legislation on the United States model, it is useful to review the historical perspective of the nation's legal mandate. The environmental impact assessment (EIA) procedure and the accompanying environmental impact statement (EIS) process were legally promulgated in the National Environmental Policy Act (NEPA) of 1970. NEPA was a procedural reform to institutionalize environmental considerations into the federal planning and decision making process. The basic intent of NEPA was to require that environmental considerations be evaluated in relation to social, economic, and technological factors in policy, program, and project determinations. Specifically, all federal agencies were required to prepare a detailed EIS for actions that might affect environmental quality; environmental impacts, mitigating measures, and alternatives had to be considered in the EIS (Burton et al. 1983). A basic assumption of NEPA was that procedures (EIS) that generate better information will result in better decisions; however, this is not guaranteed. For instance, NEPA did not prohibit authorization of projects that have adverse impact, rather it was concerned with the procedural documentation of those impacts (Fairfax and Burton 1983).

A suite of additional federal legislation was passed that conceptually strengthened NEPA. Foremost among these was the Federal Water Pollution Control Act of 1972 (PL 92-500), which created the U.S. Environmental Protection Agency (USEPA) and set effluent limitations on industrial point source discharges based on availability and economics of control technology (Hocutt 1981). The stated intention of PL 92-500 was to "...restore and maintain the chemical, physical and biological integrity of the nation's waters." Other relevant information included the Clean Air Act, Toxic Substances Control Act, and Ocean Dumping Control Act. The Clean Water Act of 1977 and the Water Quality Act of 1987 amended PL 92-500 and broadened the regulations to monitor and improve water quality. Pollution was defined in the 1977 amendment as "... the manmade or man-induced alteration of the chemical, physical, biological, and radiological integrity of water." The Water Quality Act of 1987 directed the USEPA to "...develop ...methods for establishing and measuring water quality criteria for toxic pollutants on other basis than pollutant-by-pollutant criteria, including biological monitoring and assessment methods." In effect, this amendment signified a shift in the USEPA policy from pipe standard limitations to receiving system impact.

Equally important to the NEPA spirit, but with a perspective of expanding public involvement, was the Freedom of Information Act of 1974 that (1) assured public access to all public records except those under restricted classifications and (2) granted citizens the right to sue those federal agencies that wrongly withheld information. In this same vein, the Federal Advisory Committee Act of 1976 and the Government in the Sunshine Act sought to increase public involvement in the decision making process, ultimately requiring that proposed federal actions be publicly announced in the Federal Register (Fairfax and Burton 1983).

It is important to note (Burton et al. 1983) that the EIA/EIS process can be flawed and that the national presence of a legal mandate is not an indicator of reliable performance. To be effective, resources must be committed. There is a tendency for EIAs to be made in relation to identifiable and preferably quantitative impacts, sometimes without concern for long-term risk or future trends. In turn, the EIA process can be too project oriented, rather than taking into account programs of linked development, i.e., impact. Development that crosses over national jurisdictions poses special problems; the EIA/EIS must present an unbiased view of impacts and alternatives. Given these and other potential deficiencies, however, the EIA/EIS legally mandated procedure is a promise for considering alternatives to mitigate impact and sustain resources.

DEFINITIONS OF ENVIRONMENTAL ASSESSMENT

An environmental assessment is an activity designed to identify, predict, interpret, and communicate the socio-economic and ecological effects of a particular action affecting the environment, be it a policy, program, legislative proposal, engineering project, or industrial development. As such, ecosystem evaluations are an integral part of the assessment process. In this context, a strong data base on ecosystem structure and, where possible, function is desirable. A number of terms are often misused or misinterpreted. For this reason, it is useful to define some of the more commonly used environmental nomenclature. The first three definitions follow the U.S. Park Service (Conant et al. 1983).

The *resource inventory* is the initial stage in the evaluation of natural resources and their uses and refers to a catalogue of resources present. This inventory, or itemization of resources, may extend over any time period. Resource inventories identify natural resources in specific geographic areas and indicate their quantity and variety. Evaluation methods can include data sources such as maps, flora and fauna distribution lists, species checklists, geographic information system (GIS) analyses, and, perhaps, climatic reports, stream flow gauging station data, satellite imagery, and aerial photographs, depending upon available technology and personnel.

Environmental baseline surveys refer to information about the site, quality, and indicators of resource condition; these data are collected at the time the inventory is conducted. Baseline studies analyze and assess present characteristics and status of ecosystem processes such as productivity, nutrient cycling, seasonality in stream flow, temperature, faunal habitat requirements, presence of protected species, and a host of physicochemical parameters of water. Baseline data, with subsequent updates of ecosystem functioning, create a monitoring capability. Thus, baseline surveys are the basis of the environmental assessment process and are considered as "time zero" against which future comparisons will be made.

Resource monitoring refers to a subsequent assessment of the resources. Data must be collected in a standardized manner, consistent with the baseline survey, to allow analysis of trends resulting from resource use. Thus, monitoring programs measure ecosystem change over time against "time zero" baseline and are appropriately done at scheduled intervals to take into account daily, seasonal, or annual changes in pertinent biological or physicochemical conditions.

After a baseline is established and some intervention takes place in the environment, perturbations may occur in the ecosystem that can be considered environmental impacts. The ecosystem is then divided into *control* areas not affected by the intervention and *impacted* areas that are. Thus, the degree and extent of ecosystem damage can be ascertained and time to recovery monitored.

Related to environmental degradation of a water body is whether the impact represents a *point* or *non-point* discharge source. Point sources are more readily monitored, as they normally represent pipe effluents from industrial plants, sewerage facilities, or mining operations. Non-point sources are diffuse sources, perhaps as a consequence of runoff from highways, agricultural lands, pastures, forests, open mines, and urban or communal areas, that are likely to result in organic and inorganic loading of the catchment basin, including (but not restricted to) nutrients, pesticides, herbicides, fertilizers, and petrochemicals. In LDCs, the sooner a data base is established, the greater the utility it will have in the EIA procedure to ensure the long-term maintenance of the physical, chemical, and ecological integrity of the system.

Relative to the environmental assessment procedure, the USEPA has also attempted to foster the consistent usage of terms for regulatory impact monitoring programs (USEPA 1987). As discussed previously, USEPA has recognized that chemical-specific criteria or toxicity tests are insufficient in many instances for protecting ecosystem integrity. USEPA is now turning to biosurvey procedures and biocriteria to assess impact, especially in aquatic ecosystems. Some of the more commonly used terms are:

(1) Biosurvey - Qualitative or quantitative sampling of ambient communities in the short term;
(2) Biosurveillance - Continued systematic sampling of ambient communities over time;
(3) Biomonitoring - Surveillance to ensure that standards or effluent compliance are being met or to monitor impact;
(4) Bioassay - *In situ* or laboratory toxicity testing of organisms;
(5) Bioassessment - Broadly encompassing above terms;
(6) Biocriteria - Structural and functional condition of ambient community.

To summarize, resource inventory and baseline survey data are used in project planning and design. Information on resource abundance and vulnerability of functional processes dictates the magnitude and limits of a particular project. Resource monitoring is useful for tracking project success and making project adjustments as trends are identified. For instance, it may be identified after project completion that the natural assimilative capacity of the ecosystem has been over or under estimated. In the former case, effluent limitations would have to be strictly enforced, while the latter would allow for expansion of the particular project or increased multiple use of the system.

AFRICAN CASE HISTORIES

> The problems of conservation which face tropical countries are immense. The resources involved are intimately inter-related and inter-dependent, so that their solution is to be seen in a total approach and not on a single-project basis. This requires careful consideration and the need to seek advice. But the trouble is often that developments are not thought through to their logical conclusions before they are begun, and the ecologists are only brought in to clear up the mess after the project is completed and things begin to go wrong. In many tropical countries these problems are also beset with a mixture of ignorance, apathy and ruthless exploitation for personal gain.
>
> J. Yanney Ewusie (1980)

Case Study 1—The Practice of Apartheid in South Africa
The social philosophy of apartheid, based on a concept of perceived racial superiority and a desire to maintain ethnic purity, has long been regarded as a failure by the world community and is now recognized as such even by the South African government. The historical consequence of over 40 years under the apartheid system has been, on the one hand, an attempt to build up an infrastructure in terms of industry, technology, and quality of life approaching the standards set by the United States for a small sector of the population, while the largest portion of the population was comparatively ignored. The policy of concentrating Black populations (many of whom had grown up in urban surroundings) into small overcrowded "homelands" with little infrastructure and essentially rural economies has resulted in irreparable ecological damage to large parts of the country. These ecological disaster areas, however, are predictive of impact of rapid population growth in a third world context.

Interestingly, the main income of several homelands is generated from industry, not agriculture. For instance, over 90% of the average household income in Ciskei is derived from non-agricultural sources (Huntley et al. 1989), despite the fact that this is essentially a rural area and situated well away from South Africa's major industrial centers. The income is derived from both local employment in light industry and in the region's civil service, as well as from migrant labor in the major industrial centers.

Land is nonetheless heavily overstocked with cattle and goats, with overgrazing producing severe erosion. The invasion of unpalatable, arid-adapted vegetation that further reduces the land's stock-carrying capacity is another consequence of overgrazing. Indeed, if subsidy of income did not come from non-agricultural sources, the current population densities in most of South Africa's "homelands" would simply be untenable. It is possible, therefore, that the environment in such areas is more degraded because the economic consequences are not as severe as in regions more heavily dependent on agriculture.

Case Study 2—Desertification

The term *desertification* is most accurately used to describe the "conversion of productive land into wasteland by human mismanagement" (Timberlake 1985). The term is thus in one sense wider than simply the spread of deserts, since it could also apply in areas of high rainfall, but narrower in another sense, because it does not cover a purely natural spread of deserts. Africa, with its large arid and semi-arid areas, is particularly susceptible to this kind of mismanagement because of the low resilience of ecosystems in arid and semi-arid areas. Almost 7 million km^2 of sub-Saharan Africa are under direct threat of desertification (Mabutt 1983 as cited in Timberlake 1985).

Overgrazing, over-cultivation, deforestation, and poor irrigation are all causes of desertification. The eastward march of the arid-adapted Karoo vegetation in South Africa has been well-documented for over 30 years. In its turn, true desert is encroaching eastward into the Karoo despite South Africa's sophisticated agricultural infrastructure and expertise. This phenomenon is accelerated by white farmers who have exceeded the carrying capacity of the Karoo; there are 10 million sheep currently grazing in this semi-desert, which has an estimated capacity of 7 million (Huntley et al. 1989).

Timberlake (1985) discusses many examples of changes where traditional agricultural practices, developed by farmers from many centuries of experience, have been replaced by more "modern" but often inappropriate methods. The earliest of these is the notorious "Groundnut Scheme" where an elaborate infrastructure was established in the late 1940s to grow peanuts in an area where the mean rainfall was just sufficient to grow this crop. This mean rainfall included severely dry and wet years, however, resulting in the first of many spectacular failures of such schemes in Africa.

A giant rice cultivation programme on the Niger River in Mali was a similar failure, again the result of lack of local experience on the part of the expatriate experts promoting the scheme. The causes of these kinds of failures can be enumerated:

(1) Although clearly well intentioned, expatriate advisors are usually working on the assumption that if something works at home, it should also work elsewhere.
(2) These advisors often lack the right kind of information that should form the foundations of major agricultural projects. This applies particularly to information of a historical nature and is a problem that is

only now being addressed with the current interest in long-term data series.

(3) There is often a lack of communication between the experts and local farmers, i.e., those who actually know most about local conditions. The experts are far more likely to be in close contact with administrators. The reasons for this may include language difficulties and, of course, the education gap. The latter may also reduce the credibility and perceived usefulness of any information local farmers might be able to give.

(4) Experts are often brought in for projects that have never been used before in Africa. There is, consequently, a higher risk of encountering unforeseeable problems.

(5) Man-managed environmental disasters also occur in Europe and North America, and mankind's knowledge of ecological inter-relationships is still new and far from perfect.

Case Study 3—Watershed Development

Africa has been a recipient of grandiose schemes of water development since shortly after the turn of the century (Goldsmith and Hildyard 1984). For example, the 80-year development plan for the Senegal River (Mounier 1984) and the Volta River project had been conceptually planned in the 1920s (Graham 1984). Invariably, the need for such projects was identified by the developed countries (the colonizers) for the poorly developed African nation's (the colonized) "own good" (Goldsmith and Hildyard 1984). Development scenarios have generally been couched in lofty goals and ideals such as enhancing socio-economic development through increased agriculture productivity; development of electricity supplies; providing permanent and regular water supplies, linked to domestic and commercial uses, as well as irrigation; ending malnutrition; providing employment; and, hence, increasing living standards.

It is doubted that any multipurpose dam constructed in Africa has ever achieved its stated objectives (Goldsmith and Hildyard 1984). Analogous to the Cabora Bassa Dam in Mozambique where the Portuguese overestimated the resources of the region, while under-estimating the investment to exploit them, the social and environmental effects of most impoundments have been largely detrimental (Bolton 1984). At Cabora Bassa, over 25,000 people had to be resettled, a significant number, but still less than (a) the 60,000 persons along the Zambezi River when Lake Kariba was completed in 1959 between present day Zimbabwe and Zambia and (b) 80,000 people along the Volta River in Ghana. Mounier (1984) indicated that only 16,000+ persons were scheduled for relocation in the Senegal River Project, but over 300,000 would have to change their way of life.

As summarized by Graham (1984) for justifying the wholesale resettlement of Ghanians, the populace was increasingly referred to as "backward" and "primitive," and "...somehow, the idea emerged that, by virtue of their backwardness, the resettled deserved their fate." Graham (1984) contends that the Ghana government began to look at resettlement as a means of implementing "modernization theory."

The historical results of watershed development programs in Africa have in general been devastating. Common impacts have included destruction of biodiversity, forests, and traditional farms; social disruption, including separation of traditional family or ethnic groups, resettlement into unacceptable areas, land reallocation, and inundation of culturally important sites; destruction of traditional fisheries; desertification; salinization; alterations in water quality and quantity, including downstream impacts; accumulation of agro-chemicals; and ill health and disease (e.g., Graham 1984; Mounier 1984; Webb 1984). By way of example, Pollard (1984) documented the decline of a 3000-year old cotton based economy in Sudan along the Blue Nile's Gezira alluvial flood plain, which collapsed in 20 years under a "model for agricultural development."

LaVergne (1984) recognized the secondary effects of the Aswan High Dam to include the destruction of soil fertility; erosion of the Nile bed and bank degradation; retreat of the coastline of the Nile delta; overzealous use of agro-chemicals threatening human populations; collapse of buildings and foundations weakened by the increasing water table; loss of vast water volumes by evapo-transpiration; and the inability of the Nile to deposit fine silt on the floodplain annually, thereby decreasing its productivity. These and other considerations led LaVergne (1984) to recommend abandonment of High Aswan for the long-term storage of water, i.e., the benefits did not approach the disadvantages.

As Magadza (1986) related, large scale projects in developed countries are evaluated on their likelihood of reconciling often conflicting interests, but sociological and environmental impacts are usually secondary in importance to economic considerations in LDCs. Importantly, Magadza (1986) also noted that development plans in LDCs rarely include input from local inhabitants, but are usually implements by a central government often under pressure from foreign investors.

Case Study 4—Traditional Exploitation of Intertidal Resources in Transkei

The eastern seaboard of Africa has many documented cases of traditional exploitation of intertidal resources, such as the Bajun of southern Somalia and northern Kenya (Grottanelli 1955) and the amaThonga of southern Mozambique and northern Natal province (Pooley 1980). By far the best documented, however, is the exploitation of intertidal resources in the Transkei and the Cape Province of South Africa.

It is here that exploitation of intertidal resources, particularly in the form of intertidal mollusks on rocky shores, appears to have been going on continuously for well over 100,000 years. A by-product of such exploitation is the formation of middens (heaps of large shells) that abound all along the South African coast. The earliest of these have been estimated as dating from 120,000 years ago (Avery and Siegfried 1980). These earliest, Middle Stone Age middens show evidence of shellfish collection mainly from the upper part of the intertidal zone. This pattern continued until some 35,000 years ago, after which a gap occurs in records due to a substantial fall in sea level (over 100 m below present levels) corresponding with the last glacial period. The receding seashore was presumably followed by human exploiters, but their middens would lie submerged today (Avery and Siegfried 1980). The exploiters had developed culturally and were living in the Late Stone Age some 15,000 years ago. By this stage, lower areas of the intertidal zone were being exploited, as well as the subtidal fringe, which yielded larger animals. Even abalone were exploited, apparently with the aid of special levers made from the ribs of large antelopes. There is even evidence that accessible abalone stocks may have been periodically overexploited, forcing the people to switch to a mixture of smaller shellfish. Apparently there were two types of exploiters: nomadic or semi-nomadic peoples who moved to the seashore temporarily for varying periods of time and then moved inland again, and those who lived there permanently. Avery and Siegfried (1980) suggest that, in fact, these types merely represent extremes of a socioeconomic spectrum. These Late Stone Age peoples (forming part of the Khoisan racial grouping, which includes the Bushmen) were still present when the Dutch settlers arrived at the Cape in 1652 and gave them the name of Strandlopers (beach walkers).

Such exploitation continues today on the coast of the Transkei on the eastern coast of South Africa. The present day exploiters are comparative newcomers to the area, having probably moved to this region some 800 years ago. The practice of intertidal exploitation was probably passed on from the Khoisan peoples living there. Today, these resources are still extensively used, and, in the Transkei intertidal, mollusks form an important proportion of the indigenous coastal population's annual protein intake (Siegfried et al. 1985).

Intertidal shellfish are collected mostly by women using knives, crowbars, and pickaxes, with the greatest efforts at low water spring tide. The main species gathered in Transkei is the brown mussel *Perna perna* (Lasiak and Dye 1989), although limpets, rock oyster, and abalone are also collected (Hockey and Bosman 1986; Siegfried et al. 1985). Mussel gathering continues throughout the year and often results in laying bare large areas of rock as mussels are removed in clumps. The intensity of exploitation is believed to have increased substantially in recent years, mainly due to increased populations and drought related crop failures. As mussel stocks have declined, there has been an increase in exploitation of other mollusks, such as limpets (Siegfried et al. 1985).

Lasiak and Dye (1989) showed that the community structure of exploited areas remained altered three years after denudation, with bare rock and microalgae predominating in areas previously dominated by mussels. The possibility of the bare patches being successfully recolonized by other consumable species also appeared remote. Dye (1989, 1990) found that the rock oyster, *Saccostrea cucullata*, was exploited at a rate exceeding recruitment.

Siegfried et al. (1985), Hockey and Bosman (1986), and Lasiak and Dye (1989) all express concern as to the long-term success of this natural resource. While the rugged and exposed nature of the Transkei coast protects the exploited species from extinction, it seems clear that the level of exploitation has exceeded the sustainable yield level. The Transkei coast, unlike other fisheries where such situations are at least partly the result of improved technology, this state of affairs has been reached with a level of technology that is thousands of years old. The crisis is a direct result of human numbers.

Case Study 5—Crisis in the World's Smallest Floral Kingdom, the Cape Floristic Kingdom

The Cape Floristic Kingdom is the smallest in terms of area of the six major subdivisions of the world's flora. It lies entirely within the southern and western parts of the Cape Province of South Africa, but carries an enormous diversity of plant families and species. Although only occupying 0.04% of the world's land surface, 7316 surviving plant species have been described, of which 68% are endemic (Gibbs Russell 1987). Some areas have diversities rivalling those of tropical rain forests (Kruger and Taylor 1979). Thus, the Cape Peninsula, with an area of 47,000 hectares, has over 2200 indigenous species, about half as much as the whole of the eastern United States (Huntley et al. 1989; Moll 1987). This vegetation is normally referred to as "fynbos," a descriptive name of Dutch origin that refers to the fine-leaved low shrubs that predominate. Overall, it forms evergreen sclerophyllous heathlands and shrublands (Moll and Jarman 1984).

The small size of this floristic kingdom has, however, made it particularly vulnerable to disturbance. Covering an original 70,500 km^2 before 1650, only 47,000 km^2 remain, mostly in mountainous areas unsuitable for agriculture. Of the lowland component of the Cape Floristic Kingdom, only some 6% remain today (Moll 1987). The kingdom is characterized by a high degree of endemism, with most species having very small distribution ranges, and the number of plant species that have become extinct, particularly in the lowland areas, is unknown. Hall and Veldhuis (1985) listed over 1300 plant species as being endangered, vulnerable, or rare. Some 14%, in terms of area, is conserved, mostly in mountainous areas (Hilton-Taylor and Le Roux 1989).

Little is known about the invertebrate fauna associated with fynbos, but it appears that it, too, is very diverse, with leaf hoppers possibly making up a large percentage of the fauna. It seems likely that it too has suffered a considerable number of extinctions.

The causes for this present situation are several fold. Most obvious is the introduction of western style agriculture, which began with the Dutch who established a revictualing station in the mid-1600s for their ships sailing to the East Indies. Today, most of the low lying areas of the kingdom are under agriculture. Wheat and grazing cover the largest areas, followed by vines and fruit trees.

Equally obvious are the population centers and the networks of roads connecting these. Associated with this has been draining of wetlands and modification of water tables, affecting the natural vegetation further. Urbanization around Cape Town, in particular, has been proceeding at a very rapid rate in recent years, with the local press quoting figures of thousands of families per week moving into the greater Cape Town area.

Although the Cape Floral Kingdom is fire adapted, the frequency of fires has probably increased proportionately with the increase in human numbers. The natural succession cycle is in the region of 12 years, and Kruger (1979) estimated that the natural frequency of fires (mostly caused by lightning) ranged between 6 and 40 years, varying randomly between these limits, but probably averaging once every 20 years. Today, many areas are subjected to fires every 4-5 years, thus shortening the natural succession cycle of the vegetation.

Another major factor in the decline of the fynbos has been the introduction of a number of invasive alien species, in particular several Australian species of *Acacia* and *Hakea*, as well as some pine species. An analysis of threats to threatened plant species in the core area of the Cape Floral Kingdom by Hall and Veldhuis (1985) showed these invasive aliens to be the single greatest threat, followed by agriculture and genetic decline in critically small populations. The invasion of the largely woody species have had impacts on nutrient and hydrological cycles, as well as on the fire regime due to changing fuel properties (van Wilgen 1988). In addition, some of the *Acacia* species are adapted to high fire frequencies, making them more competitive than the natural vegetation in areas subject to frequent fires.

An interesting aspect is the suggestion by van Wilgen (1988) that the extirpation of 12 large mammal species from the region between 1700 and 1960 may have increased the susceptibility of the fynbos biome to alien invasion. Presumably, large herbivores would have helped check the spread of the aliens. The mammals mostly disappeared as a direct result of hunting, although loss of habitat due to competitive exclusion by humans probably also played a role in the case of the elephant (Parker and Graham 1989).

The crisis in the Cape Floristic Kingdom thus stems mostly from the activities of the Dutch colonizers and their successors. These imposed western farming methods and the trappings of western civilization drove many species to extinction, displaced indigenous species with both cultivated and invasive alien species, and imposed accelerated fire regimes on the indigenous vegetation. The fynbos vegetation had little practical value to the new settlers, and, indeed even today, its direct economic value is low. The large diversity does represent an enormous potential genetic resource, however, one which is continually being whittled away at the moment.

Case Study 6—Biodiversity and the African Rift Lakes

Biodiversity is defined as the variability and variation among organisms (which includes plants, animals, fungi, and monerans) as related to their ecology and habitats (Norse et al. 1986; OTA-USC 1987; Soule and Wilcox 1980) and is inclusive of genetic, population, species, and ecosystem diversities. Biological diversity, which provides basic resources for the life support system on our planet, is being reduced at an alarming rate throughout the world. This destruction directly undermines the capacity of the biota to respond to the changing environmental conditions that in recent years are mostly created by anthropogenic enterprises. Economic development activities are being accelerated throughout the world to meet the needs of an ever increasing human population, which has already reached the 5 billion mark. Such economic activities, coupled with rapid expansion of urban centers, destroy the habitats for plants and animals, thus causing massive extinction of species and reduction of biodiversity (Lovejoy 1980; Myers 1979; Wilson 1989).

The rift valley lakes of Africa harbor the most speciose ichthyofauna of any of the world's lakes. The rich faunas of these lakes are primarily attributable to the explosive radiation and speciation of the endemic haplochromine cichlids (Greenwood 1982; Regan 1921; Trewavas 1935). The fauna of the Great Lakes of Africa are threatened by over exploitation (McKaye and Stauffer 1983), species introductions (McKaye et al. 1985; see Case Study 7), possible chemical pollution due to oil exploration (NOVA, PBS Special "The World of Oil"), and the establishment of industries along their shorelines (McKaye and Stauffer 1987). Ecological and behavioral studies (Fryer 1959; Holzberg 1978; Marsh et al. 1981; McKaye et al. 1990; Ribbink et al. 1983) have been responsible in part for focusing attention on the large number of fish species in these lakes. These research efforts have been slowed and results often confused by the uncertain systematic status of some of the cichlids being examined. In Lake Malaŵi alone, only 35-40% of the estimated 1000 species that inhabit the lake are described. It is essential, therefore, to catalogue the species before they are lost, so as to guard against the extinction of species, which is occurring in Lake Victoria (Barel et al. 1985). The perpetuation of this fauna is extremely important when it is considered that most (93-97%) of these fishes are endemic to single lakes. Within a given lake, many species are endemic to certain islands (Stauffer 1988). Apart from the ethical arguments associated with the maintenance of biological diversity, there are some very practical reasons to protect the ichthyofauna of Lake Malaŵi. Fishes provide 70% of the animal protein consumed by Malaŵians. Furthermore, the rock-dwelling cichlids, "mbuna," are important components of the ornamental fish trade. In 1990, more than $250,000 in foreign currency was generated for Malaŵi via the export of these fishes. Moreover, recent work has demonstrated the usefulness of Malaŵi cichlids in controlling the human disease bilharzia (Chiotha and McKaye 1986; McKaye et al. 1986); and LoVullo et al. (in press) have urged the use of indigenous fishes for aquaculture purposes.

The preceding demonstrates the lack of information available on the fish fauna of Lake Malaŵi; however, it should be realized that there is even less knowledge of the invertebrates, protozoans, algae, vascular plants, etc. that inhabit these systems. In essence, the conservation and effective management of any ecosystem requires a basic knowledge and a comprehensive understanding of the species that occur in the system. Armed with data documenting the localized nature of many of these species, it may be possible to convince local governments of the need to protect their unique faunas.

Case Study 7—Introduced Species
The establishment of reproducing populations of non-native organisms have concerned ecologists and resource managers for decades (Stauffer 1984). Courtenay and Stauffer (1984) examined the worldwide benefits and consequences of the introduction of alien species, while Bruton and Merron (1985) and de Moor and Bruton (1988) provide the most comprehensive treatment for aquatic animals in southern Africa. In several cases, the purposeful introduction of fishes has been well intentioned and beneficial. Conversely, the release of exotic species may have a devastating impact on native species and may be irreversible.

Probably the most quoted example of the catastrophic effects of an introduced organism on native species in Africa is the introduction of the Nile perch, *Lates niloticus*, into Lake Victoria. The establishment of this large piscivore has caused the extinction of several native species (Barel et al. 1985; Harrison et al. 1989), many of which had not been described. The deleterious consequences of this introduction includes the decline of native food fishes from Lake Victoria (Ogutu-Ohwayo 1990). The problem is further exacerbated by the fact that since the Nile perch's flesh is oily, it is not a preferred food by the native people and must be smoked rather than sun-dried to be preserved (Barel et al. 1985). The need for fuel to smoke this fish is causing deforestation of several islands in Lake Victoria (Barel et al. 1985). Finally, since the Nile perch is a large fish, more expensive and stronger nets are needed to catch it than were required for the small haplochromine fishes, which the artisan fishermen formerly harvested (Barel et al. 1985).

Although not as dramatic as the establishment of the Nile perch, other introductions throughout Africa have had deleterious consequences on native fish faunas. Introduction of the European carp, *Cyprinus carpio*, to

Zimbabwe decreased fishery yields of the native African catfish *Clarias gariepinus* from man-made lakes, because it competed for chironomid larvae, which are the common food of both species (Costa-Pierce, personal communication). In Madagascar, the European carp's disturbance of the substrate destroyed native tilapia breeding areas and muddied lake waters, causing unsuccessful breeding of native fish species (Moreau 1981). The introduction of tilapia species into Lake Luhondo, Rwanda, caused the extirpation of two large cyprinids and a drastic decline in population numbers of a smaller cyprinid (de Vos et al. 1990). Additional examples of species introductions into southern Africa are discussed in Bruton and Merron (1985), and a worldwide documentation of alien fish introductions can be found in Welcomme (1988). To prevent the potential deleterious effects that such introductions can have on the native faunas, some difficult decisions and policy statements must be made. For example, the government of Malaŵi should be commended for its recent policy statement that bans the culture of the European carp in Malaŵi.

Conversely, Turner (1982) proposed that two fishes from Lake Tanganyika, *Stolothrissa tanganicae* and *Limnothrissa miodon*, be established in Lake Malaŵi. McKaye et al. (1985) predicted that should either of these fishes be established in the lake that they would directly compete with the indigenous haplochromine species. Furthermore, McKaye et al. (1985) hypothesized that these introductions could cause a collapse of the cichlid fisheries. Bruton (1990) states that since these introductions are essentially irreversible that:

> The highest priority should be to assist the governments of the riparian countries (Tanzania, Uganda, and Kenya) with monitoring and research programmes and to support their policies of non-introduction of further alien fishes into any of the African Great Lakes so as prevent the same cycle of events from occurring, for example, in Lakes Tanganyika and Malaŵi. The diverse animal and plant communities of the African Great Lakes are a heritage of all mankind and it is the duty of every country to play a role in their conservation. It is therefore proposed that an internationally funded research programme should be mounted on the African Great Lakes on the scale of the tropical forest biome project of the IUCN.

SUMMARY

It is not possible within these few pages to adequately discuss all the ramifications of environmental assessment issues faced by LDCs. The few case studies discussed present a number of contrasting pictures of the problems faced by Africa. For the future, the continent's rapidly growing population will be the focus of concern. The pressures exerted by this growth on the African environment have been building for some time. Writing almost 30 years ago, Hance (1964) estimated that 28% of the continent was overpopulated, with 38% of the population living in overpopulated areas.

Although Africa was the last continent to face the global population explosion, it is currently experiencing the highest growth rates. As is the case elsewhere, urbanization is also increasing, and Africa is experiencing a net depopulation of the rural areas. Urban populations are increasing at a rate of 5.1% annually, well in excess of the population growth rate (Huntley et al. 1989).

Intimately related to population growth, is "central" government's views on land use, land tenure, and land reallocation practices. Existing inheritance laws in many African countries, (including land fragmentation) and resource over utilization are in conflict with conservation goals of sustainable development (Mabogunje 1981). Another factor that can contribute to degrading conditions is government's "big brother" attitude. For instance, Beckman (1984) documented the rioting of Nigerian villagers who were neither consulted nor adequately compensated for their losses with the Bakalori Irrigation Project.

Perhaps the future in development activities, and mitigation of environmental impact, is to champion programs like ADMADE in Zambia and Operation Campfire in Zimbabwe. Such programs combine traditional land use with conservation of natural resources and limited transformation towards modernization, i.e., upgrading of living standards. Similarly, promotion of the concept of ecotourism (e.g., see Ceballos-Lascuráin 1990) has much promise for LDCs; ecological based tourism has been shown to constitute an important instrument for socio-economic development by attracting foreign exchange, and hence conserving natural and cultural heritage. Jointly, such programs can raise ecological awareness and promote environmental education among local and national populations, including central government. Akin to this is the prospect of conserving unique anthropological and renewable resources through the establishment of national parks (McNeely and Miller 1984) or transfrontier border parks for peace (McNeil 1990; Thorsell 1990).

REFERENCES

AVERY, G. and SIEGFRIED, W.R. (1980). 150,000 year tradition: Food gatherers along South Africa's seashore. Oceans 4:32-37.

BAREL, C.D.N., DORIT, R., GREENWOOD, P.H., FRYER, G., HUGHES, N., JACKSON, P.B.N., KAWANABE, H., LOWE-McCONNELL, R.H.J., NAGOSHI, M., RIBBINK, A.J., TREWAVES, E., WITTE, F., and YAMAOKA, K. (1985). Destruction of fisheries in Africa's lakes. Nature 315:19-20.

BECKMAN, B. (1984). Bakalori: Peasants versus state and industry in Nigeria. In: *The Social and Environmental Effects of Large Dams*, Vol. 2 (E. Goldsmith and N. Hildyard, eds), pp. 140-155. Wadebridge Ecological Centre, UK.

BOLTON, P. (1984). Mozambique's Cabora Bassa Project: An Environmental Assessment. In: *The Social and Environmental Effects of Large Dams*, Vol. 2 (E. Goldsmith and N. Hildyard, eds), pp. 156-167. Wadebridge Ecological Centre, UK.

BRUSSARD, P.F. (1991). The role of ecology in biological conservation. Ecol. Appl. 1(1):6-12.

BRUTON, M.N. (1990). The conservation of the fishes of Lake Victoria, Africa: An ecological perspective. Environ. Biol. Fish. 27:161-175.

BRUTON, B.N. and MERRON, S.V. (1985). Alien and translocated aquatic animals in southern Africa: A general introduction, checklist, and bibliography. South African National Science Program Report (CSIR), Pretoria 133:1-71.

BURTON, I., WILSON, J., and MUNN, R.E. (1983). Environmental impact assessment: National approaches and international needs. Environ. Monit. Assess. 3:133-150.

CAIRNS, J., JR. (1990). The emergence of global environmental awareness. J. Environ. Sci. (China) 2(2):1-17.

CALLICOTT, J.B. (1991). Conservation ethics and fisheries management. Fisheries 16(2):22-28.

CEBALLOS-LASCURÁIN, H. (1990). La Ruta Maya: A transfrontier ecocultural tourism circuit in the Yucatan Region. In: *Parks on the Borderline: Experience in Transfrontier Conservation* (J.W. Thorsell, ed), pp. 23-38. IUCN, Gland, Switzerland.

CHIOTHA, S.S. and McKAYE, K.R. (1986). Possible biological control of *Schistosomiasis* (Bilharzia) by Lake Malaŵi molluscivores. Luso J. Sci. Tech.

CONANT, F., ROGERS, P., BAUMGARDNER, M., McKELL, C., DASMANN, R., and REINING, P., eds. (1983). Resource Inventory and Baseline Study Methods for Developing Countries. American Association for the Advancement of Science. 539 pp.

COUNCIL FOR THE ENVIRONMENT. (1989). An Approach to a National Environmental Policy and Strategy for South Africa. Joan Lotter Publications, Pretoria, South Africa. 41 pp.

COURTENAY, W.R., JR. and STAUFFER, J.R., JR. (1984). Distribution, Biology, and Management of Exotic Fishes. Johns Hopkins University Press, Baltimore, MD. 430 pp.

DE MOOR, I.J. and BRUTON, M.N. (1988). Atlas of Alien Translocated Indigenous Aquatic Animals in Southern Africa, Report No. 144. South Africa National Science Program, Pretoria.

DE VOS, L., SNOEKS, J., and THYS VAN DEN AUDENAERDE, D. (1990). The effects of tilapia introductions in Lake Luhondo, Rwanda. Environ. Biol. Fish. 27:303-308.

DYE, A.H. (1989). Studies on the ecology of *Saccostrea cucullata* (Born, 1778) (Mollusca: Bivalvia) on the east coast of southern Africa. S. Afr. J. Zool. 24(2):110-115.

DYE, A.H. (1990). Episodic recruitment of the rock oyster *Saccostrea cucullata* (Born, 1778) on the Transkei coast. S. Afr. J. Zool. 25(3):185-187.

EWUSIE, J.Y. (1980). Elements of tropical ecology. Hienemann, London. 205 pp.

FAIRFAX, S. and BURTON. L. (1983). A decade of NEPA: Milestone or millstone? Fisheries 8(6):5-9.

FRYER, G. (1959). The trophic interrelationships and ecology of some littoral communities of Lake Nyasa with special reference to the fishes, and a discussion of the evolution of a group of rock-frequenting Cichlidae. Proc. Zool. Soc. London 132:153-282.

GIBBS RUSSELL, G.E. (1987). Preliminary floristic analysis of the major biomes in southern Africa. Bothalia 17:213-227.

GOLDSMITH, E. and HILDYARD, N., eds. (1984). The Social and Environmental Effects of Large Dams, Volumes 1, 2. Wadebridge Ecological Centre, UK.

GRAHAM, R. (1984). Ghana's Volta resettlement scheme. In: *The Social and Environmental Effects of Large Dams*, Vol. 2 (E. Goldsmith and N. Hildyard, eds), pp. 131-139. Wadebridge Ecological Centre, UK.

GREENWOOD, P.H. (1982). East African Cichlid Fishes: The Haplochromine Fishes of the East African Lakes. British Museum (Natural Historty), London.

GROTTANELLI, V.L. (1955). Pescatori dell'Oceano Indiano. Cremonese, Roma. 409 pp.

HALL, A.V. and VELDHUIS, H.A. (1985). South African Red Data Book: Plants - Fynbos and Karoo Biomes, Report No. 117. South African National Scientific Programmes, Foundation for Research Development, Pretoria. 160 pp.

HANCE, W.D. (1964). The Geography of Modern Africa. Columbia University Press, New York. 220 pp.

HARRISON, K., CRIMMEN, O., TRAVERS, R., MAIKWEKI, J., and MUTORO, D. (1989). Balancing the scales in Lake Victoria. Biologist 36:189-191.

HILTON-TAYLOR, C. and LE ROUX, A. (1989). Conservation status of the fynbos and karoo biomes. In: *Biotic Diversity in Southern Africa: Concepts and Conservation* (B. J. Huntley, ed), pp. 202-223. Oxford University Press, Cape Town.

HINDEN, S. (1990). Joan Bavaria's crusade for the environment. Washington Post 23Dec:H1,3.

HOCKEY, P.A.R. and BOSMAN, A.L. (1986). Man as an intertidal predator in Transkei: Disturbance, community convergence and the management of a natural food resource. Oikos 46:3-14.

HOCUTT, C.H. (1981). Fish as indicators of biological integrity. Fisheries 6(6):28-31.

HOLZBERG, S. (1978). A field and laboratory study of the behaviour and ecology of *Pseudotropheus zebra* (Boulenger) and endemic cichlid of Lake Malaŵi (Pisces: Cichlidae). Z. Zool. Syst. Evolut.-forsch. 16:171-187.

HUNTLEY, B., SIEGFRIED, R., and SUNTER, C. (1989). South African Environments into the 21st Century. Human and Rousseau and Tafelberg, Cape Town. 127 pp.

KRUGER, F.J. (1979). Fire. In: Fynbos Ecology: A Preliminary Synthesis, Report No. 40 (J. Day, W.R. Siegfried, G.N. Louw, and M.L. Jarman, eds), pp. 43-57. South African National Scientific Programmes, Cooperative Scientific Programmes, Pretoria.

KRUGER, F.J. and TAYLOR, H.C. (1979). Plant species diversity in Cape fynbos: Gamma and delta diversity. Vegetatio 41:85-93.

LASIAK, T. and DYE, A. (1989). The ecology of the brown mussel *Perna perna* in Transkei, southern Africa: Implications for the management of a traditional food resource. Biol. Conserv. 47:245-257.

LaVERGNE, M. (1984). The seven deadly sins of Egypt's Aswan High Dam. In: *The Social and Environmental Effects of Large Dams*, Vol. 2 (E. Goldsmith and N. Hildyard, eds), pp. 181-188. Wadebridge Ecological Centre, UK.

LOVULLO, T.J., STAUFFER, J.R., and MCKAYE, K.R. (In press). The diet and growth of a brood of *Bagrus meridionallis* Gunther (Siluriformes: Bagridae) in Lake Malawi, Africa. Copeia.

LOVEJOY, T.E. (1980). A projection of species extinctions. In: *Global 2000 Report to the President*, Vol. 2, Technical Report, pp. 328-332. Council on Environmental Quality and U.S. Department of State, U.S. Government Printing Office, Washington, DC.

MABOGUNJE, A. (1981). The Development Process: A Spatial Perspective. Holmes and Meier Publishers, Inc., NY.

MAGADZA, C.H.D. (1986). Conflicts of resource use on the Lake Kariba environs. Nature and Resources, UNESCO 22(4):2-12.

MANN, C. (1991). Lynn Margulis: Science's unruly Earth mother. Science 252:378-381.

MARSH, A.C., RIBBINK, A.J., and MARSH, B.A. (1981). Sibling species complexes in sympatric populations of *Petrotilapia trewavas* (Cichlidae, Lake Malaŵi). Zool. J. Linn. Soc. 71:253-264.

McKAYE, K.R. and STAUFFER, J.R., JR. (1983). Malaŵi Fisheries: An Assessment and Overview. USAID Project No. 698-0135-19, Lilongwe, Malaŵi. 72 pp.

McKAYE, K.R. and STAUFFER, J.R., JR. (1987). Foraging theory and fortuitous contributions to aquaculture and African Great Lakes management. News IAEF 10:47-50.

McKAYE, K.R., MAKWINJA, R.D., MENYANI, W.W., and MHONE, O.K. (1985). On the possible introduction of non-indigenous zooplankton-feeding fishes into Lake Malaŵi, Africa. Biol. Conserv. 33:289-307.

McKAYE, K.R., STAUFFER, J.R., JR., and LOUDA, S.M. (1986). Fish predation as a factor in the distribution of Lake Malaŵi gastropods. J. Exp. Biol. 45:279-289.

McKAYE, K.R., LOUDA, S.M., and STAUFFER, J.R., JR. (1990). An experimental field test of the importance of bower size to male reproductive success in a cichlid fish lek. Am. Nat. 135:597-613.

McNEELY, J.A. and MILLER, K.R., eds. (1984). National Parks, Conservation, and Development: The Role of Protected Areas in Sustaining Society. Smithsonian Institution Press, Washington, DC. 825 pp.

McNEIL, R.J. (1990). International parks for peace. In: *Parks on the Borderline: Experience in Transfrontier Conservation* (J.W. Thorsell, ed), pp. 23-38. IUCN, Gland, Switzerland.

MOLL, G. (1987). Table Mountain: A Natural Wonder. Wildlife Society of Southern Africa, Cape Town. 45 pp.

MOLL, E.J. and JARMAN, M.L. (1984). Clarification of the term fynbos. S. Afr. J. Sci. 80:351-352.

MOREAU, J. (1981). Le lac Alaotra a Madagascar. 50 ans d'amenagement des peches. Cah. Hydrobiol. ORSTOM 16:151-158.

MOUNIER, F. (1984). The Senegal River scheme. In: *The Social and Environmental Effects of Large Dams*, Vol. 1 (E. Goldsmith and N. Hildyard, eds), pp. 109-119. Wadebridge Ecological Centre, UK.

MYERS, N. (1979). The Sinking Ark: A New Look at the Problem of Disappearing Species. Pergamon Press, Oxford. 307 pp.

NORSE, E.A., ROSENBAUM, K.L., WILCOVE, D.S., WILCOX, B.A., ROMME, W.H., HOHNSTON, D.W., and STOUT, M.L. (1986). Conserving Biological Diversity in Our National Forests. The Wilderness Society, Washington, DC. 116 pp.

OGUTU-OHWAYO, R. (1990). The decline of the native fishes of lakes Victoria and Kyoga (East Africa) and the impact of introduced species, especially the Nile perch, *Lates niloticus*, and the Nile tilapia, *Oreochromis niloticus*. Environ. Biol. Fish. 27:81-96.

OTA/USC, Office of Technology Assessment, U.S. Congress. (1987). Technologies to Maintain Biological Diversity, OTA-F-330. Washington, DC. 334 pp.

PARKER, I.S.C. and GRAHAM, A.D. (1989). Elephant decline: Downward trends in African elephant distribution and numbers. Int. J. Environ. Stud. 35(A):13-26.

POLLARD, N. (1984). The Sudan's Gezira scheme: A study in failure. In: *The Social and Environmental Effects of Large Dams*, Vol. 2 (E. Goldsmith and N. Hildyard, eds), pp. 168-180. Wadebridge Ecological Centre, UK.

POOLEY, E.S. (1980). Some notes on the utilization of natural resources by the tribal people of Maputaland. In: *Studies on the Ecology of Maputaland* (M.N. Bruton and K.H. Cooper, eds), pp. 467-479. Rhodes University, Grahamstown and Wildlife Society of Southern Africa, Durban, South Africa. 560 pp.

REGAN, C.T. (1921). The cichlid fishes of Lake Nyassa. Zool. Soc. Lond. 21:675-727.

RIBBINK, A.J., MARSH, B.A., MARSH, A. C., RIBBINK, A.C., and SHARP, B.J. (1983). A preliminary survey of the cichlid fishes of rocky habitats in Lake Malaŵi. S. Afr. J. Zool. 18:149-310.

SIEGFRIED, W.R., HOCKEY, P.A.R., and CROWE, A.A. (1985). Exploitation and conservation of brown mussel stocks by coastal people of Transkei. Environ. Conserv. 12:303-307.

SOULE, M.E. and WILCOX, B.A., eds. (1980). Conservation Biology: An Ecological-evolutionary Perspective. Sinauer Associates, Sunderland, MD. 395 pp.

STAUFFER, J.R., JR. (1984). Colonization theory relative to introduced populations. In: *Distribution, Biology, and Management of Exotic Fishes* (W.R. Courtenay Jr. and J.R. Stauffer, Jr. eds), pp. 8-21. The Johns Hopkins University Press, Baltimore, MD.

STAUFFER, J.R., JR. (1988). Descriptions of three rock-dwelling cichlids (Teleostei: Cichlidae) from Lake Malaŵi, Africa. Copeia 1988:663-668.

THORSELL, J.W., ed. (1990). Parks on the Borderline: Experience in Transfrontier Conservation. IUCN, Gland, Switzerland.

TIMBERLAKE, L. 1985. Africa in Crisis: The Causes, the Cures of Environmental Bankruptcy. Earthscan, London. 232 pp.

TREWAVAS, E. (1935). A synopsis of the cichlid fishes of Lake Nyasa. Ann. Mag. Nat. Hist. 16:65-118.

TURNER, J.L. (1982). Lake flies, water fleas, and sardines. In: *Biological Studies on the Pelagic Ecosystem of Lake Malaŵi*, FI:DP/MLW/75/019. FAO Tech. Rep. 1:165-173.

USEPA (United States Environmental Protection Agency). (1987). Report of the National Workshop on Instream Biological Monitoring and Criteria. Lincolnwood, IL. 34 pp.

VAN WILGEN, B.W. (1988). Fynbos terrestrial ecosystems. In: *Long-term Data Series in Relation to Southern Africa: Renewable Natural Resources*, Report No. 157 (I.A.W. MacDonald and R.J.M. Crawford, eds), pp. 216-236. South African National Scientific Programmes, Foundation for Research Development, Pretoria.

WEBB, P.J.R. (1984). Of rice and men: The story behind Gambia's decision to dam its river. In: *The Social and Environmental Effects of Large Dams*, Vol. 1 (E. Goldsmith and N. Hildyard, eds), pp. 120-130. Wadebridge Ecological Centre, UK.

WELCOMME, R.L. (1988). International Introductions of Inland Aquatic Species. FAO Fisheries Technical Paper 294:1-318.

WILSON, E.O. (1989). Threats to biodiversity. Sci. Am. 261(3):108-117.

CHAPTER 6

MONITORING FOR GLOBAL CHANGE

John Cairns, Jr.
University Center for Environmental and Hazardous Materials Studies
Virginia Polytechnic Institute and State University
Blacksburg, Virginia

We like to believe that if we secure adequate data bearing on a scientific problem, then anybody with normal intelligence who takes the trouble to become acquainted with these data will necessarily arrive at the same conclusion regarding the problem in question. We like to speak of conclusions demonstrated, settled, proved and established. It appears, however, that no evidence is powerful enough to force acceptance of a conclusion that is emotionally distasteful.

Th. Dobzhansky

INTRODUCTION

Estimating the hazard of various substances to the biota is facilitated by having four categories of evidence: (1) screening or range-finding toxicity tests, (2) predictive toxicity tests, (3) validation or confirmation of the accuracy of the predictive tests, and (4) monitoring - surveillance to ensure that previously established quality control conditions are being met (Cairns 1980; Cairns et al. 1978). Although great strides have been made in monitoring for point source discharges or other localized perturbations of the environment, monitoring at the landscape level is just beginning and, at the global level, is still in the preliminary design stages. Although it may seem precipitous to attempt development of a monitoring system to detect global change, potential for such change has increased dramatically in the last few decades (Ehrlich and Ehrlich 1991). A system that requires continual updating could provide early warning of change that might permit prevention worth many orders of magnitude more money than the cost of the system itself. In short, we must learn by doing and treat monitoring for global change as an experimental activity despite the undoubted importance of detecting global change as soon as possible. The present generation of college students had either just been born or were just about to be born when the first Earth Day was celebrated in 1970. Before this environmental movement began, the United States had no Environmental

---

1. Corresponding Author: John Cairns, Jr., University Center for Environmental and Hazardous Materials Studies, Virginia Polytechnic Institute and State University, Blacksburg, VA 24061-0415.

2. Key Words: biological monitoring, environmental auditing, hazardous evaluation, predictive toxicity tests, validation of predictive models, global change, levels of biological organization.

Protection Agency, no Council on Environmental Quality, no National Environmental Policy Act, no Endangered Species Act, no SuperFund for Toxic Waste, and no strong National Clean Air or Water Acts. Unfortunately, many of the people suffering most from environmental degradation and over population simply do not have the time or resources to take substantive effective action. Those with the time and resources in the developed countries feel that life is good and will continue to be so and are not especially alarmed, although they may be disconcerted, by some of the signs of environmental deterioration that appear in the news media and, if we look closely, in some areas that were once regarded as pristine. Calling society's attention to the environmental peril or even potential peril is not a path to popularity. Aldo Leopold (1953) recognized this almost half a century ago:

> One of the penalties of an ecological education is that one lives alone in a world of wounds.....An ecologist must either harden his shell and make believe that the consequences of science are none of his business, or he must be the doctor who sees the marks of death in a community that believes itself well and does not want to be told otherwise.

WHY THE NEED IS URGENT

When Paul Ehrlich (1968) first published *The Population Bomb*, 3.5 billion people were on earth and the population was growing by approximately 70 million per year. When Paul and his wife Anne published an update, *The Population Explosion* (Ehrlich and Ehrlich 1990), the world had 5.4 billion people to support, not 3.5 billion; and, the population is expected to increase by roughly 95 million people in the same year. This growth of nearly 2 billion people from 1968 to 1990 has been supported by a worldwide expansion of food production. However, many of the gains have been achieved at the expense of non-renewable (at least, renewable only over millennia) supplies of soil, underground water, and other resources. Just several decades ago, we had no inkling of how swiftly tropical rain forests and other forests worldwide would be chopped down. In addition, the role of chlorofluorocarbons (CFCs) in destroying the ozone layer had not been discovered. The problem of acid rain had not reached public consciousness, nor had the possibility of global warming received the attention it is now getting. The annual production of synthetic organic chemicals in the United States alone increased ninefold between 1950 and 1985, from 24 billion to 225 billion pounds (Speth 1988). Of the slightly over 70,000 chemical compounds in fairly widespread commercial use today, approximately half are considered potentially or actually hazardous to health by the U.S. Environmental Protection Agency (Ehrlich and Ehrlich 1991). Solid waste is another worldwide problem that has acquired considerable mobility. For example, in about 10% of the European communities, approximately 25 million tons of waste generated each year is probably passed from one country to another for disposal (Lawrence and Wynne 1989). Much attention has been given to the Exxon Valdez oil spill in Prince William Sound,

Alaska, but smaller spills occur all the time. Approximately 10,000 smaller spills occurred in the year following the Exxon Valdez spill (Byrnes 1990).

The most compelling reason for global monitoring is that the developing countries naturally wish to improve the quality of life of their citizens. Slightly over two decades ago, Ehrlich and Holdren (1971) developed a simple equation to demonstrate that the impact (I) of any activity or population upon the environment is the product of its population size (P) multiplied by the per capita affluence (A) as measured by consumption, which in turn is multiplied by a measure of the damage done by the technologies (T) employed in supplying each unit of that consumption. The basic equation I = P x A x T shows dramatically that doubling a population size will double its impact on the environment, but, equally importantly, a small population can have an enormous impact upon the environment. A more recent discussion of the I = PAT equation can be found in Ehrlich and Ehrlich (1990). The equation also shows that a small population that is very affluent and employs a vast array of energy demanding technologies to support its lifestyle can be as damaging as a large number of much less affluent, low technology populations. If the populations in the developing countries have a relatively modest increase in affluence, particularly as expressed by energy consumption, the effects on the global climate, for example, might be substantive. Both the People's Republic of China (PRC) and India hope to double their per capita energy consumption in the first half of the next century (Ehrlich and Ehrlich 1989). At the present time, per capita consumption in the PRC is approximately 7% of the per capita energy use in the United States. Even if the PRC population remained at about 1.1 billion (which many people think it will not), increasing per capita consumption would introduce more additional carbon dioxide into the atmosphere than the United States is now releasing from coal burning. It may be a surprise to many people used to oil crises that coal still supplies almost a quarter of U.S. energy demand.

John Holdren, of the Energy and Resources Group at the University of California at Berkeley, has developed a well reasoned energy scenario for the planet. It is primarily based on maximizing the efficiency of energy use, which could well resolve the complex problem, a small portion of which was just described. In the Holdren scenario (Holdren 1990, in press), developing nations increase their per capita energy use by 2% per year between 1990 and 2025, doubling it from 1.0 to 2.0 kW (during this period their combined populations may increase by two-thirds). At the same time, developed nations reduce their per capita use by 2% annually through increased efficiency, reducing their per capita use from 7.5 to 3.8 kW (with marked increases in efficiency, per capita benefits might actually increase during this period). The population growth in the developed nations during this period is estimated to be less than 10% for the entire 35-year period from 1990 to 2025. Nevertheless, the environmental impact, given the rather high per capita energy use, will still be substantial. In the Holdren scenario, for the remainder of the next century, both developed and developing nations will converge on an average per capita energy use of 3 kW. A major uncertainty during this entire period will be the size of the world population peak which, in the Holdren scenario, is 10 billion people reached around 2100, after which the size begins to decline slowly. Holdren's somewhat optimistic scenario assumes both that population size can be limited to 10 billion and that no major disruption of society will occur such as destabilization due to epidemic diseases, such as AIDS, or dramatic increases in drug dependency and/or crime.

Energy, although probably the best single indicator of global environmental stress, is only one of a vast array of indicators. Deforestation is now occurring very rapidly. To first achieve a no-net-loss of ecological habitat and ecosystem services, dramatic changes in social action are necessary. If we wish to increase the repair or healing of the earth by beginning with restoration of damaged ecosystems, other dramatic ecological changes will also need to occur. Ehrlich and Ehrlich (1991) believe that such changes are essential for the survival of society as we know it. I agree and, more importantly, many leading mainstream scientists (E. O. Wilson, Peter Raven, Thomas Lovejoy) concur. During this period of dramatic, though gradual, change, where present trends are either arrested or reversed, it is important to know precisely how the ecosystems of the planet are responding. It is important to know, for example, if one hemisphere is responding at a different rate than another, if certain types of habitats or ecosystems are not responding at all while others are, and if rehabilitation of tropical rain forests is effective in the maintenance of the rainfall patterns necessary for their growth. All this will require a global monitoring system that will show both the aggregate impact of the changes and the regional or ecosystem type responses. Such a system is far from being in place. Such a monitoring system is not even in place in the United States, although the U.S. Environmental Protection Agency's Environmental Monitoring and Assessment Program (Hunsaker et al. 1990) is a step in the right direction. The basic purpose of the monitoring system, of course, will be to determine if the risk of anthropogenic activities is being reduced in natural ecosystems and, if so, to what degree. This chapter explores some of the strategies that might be helpful in developing a global monitoring program.

MONITORING AT DIFFERENT LEVELS OF BIOLOGICAL ORGANIZATION

If the integrity of natural systems is to be preserved at each level of biological organization, multiple breakpoints and thresholds at each level must be identified and methods must be developed for gathering the information. Until an understanding of the relationship between state and process variables becomes more robust, measurements should be made of both until the relationship is clear. Alternatively, the measurements might be defined as structural and functional at different levels of biological organization.

Three, not mutually exclusive, possibilities exist regarding this relationship: (1) process and state attributes are so intimately connected that it is impossible to change one without changing the other, and, therefore, monitoring one will enable robust predictions of changes in the other; (2) it is possible to diminish the functional capabilities (process variables) of organisms without killing them and, therefore, functional measurements (process variables) are more sensitive than structural; or (3) there is much functional redundancy in natural systems so that a multiplicity of species have overlapping functions and the loss of a few selected such species will not alter the functional attributes because the remaining species will fill the vacuum left. Thus, structural thresholds are more sensitive to anthropogenic stress than functional ones.

For some very simple ecosystems, such as those under extreme environmental conditions where functional redundancy is small and numbers of species are low, possibility (1) may be true. In other situations, either of the remaining possibilities may be valid, depending on which attributes are being compared. Given the high degree of uncertainty, prudence demands that measurements in each category at each level of organization be made until the ability to extrapolate from one attribute to another is more robust.

A number of methods exist for monitoring at the single species and community levels (e.g., Cairns et al. 1977) using both structural (e.g., Cairns 1986) and functional (e.g., Cairns and Pratt 1989) end points. Additionally, methods are available for automated biological monitoring (e.g., Cairns et al. 1970; Gruber and Diamond 1988). I cite my earlier work and that of a former colleague (Gruber) because I want to be critical of the limitations of this earlier work and do not want to criticize the research of others in the same area lest the criticism be misunderstood. This earlier research was on target in the sense of developing an information feedback loop to determine anthropogenic effects upon the environment with special emphasis on early warning systems that detected the onset of trouble before serious damage had occurred. There was even a systems level perspective (e.g., Cairns 1975), but this array of methods is inadequate even for the task of regional monitoring because some important information is not generated and erroneous decisions might be made as a consequence.

Future monitoring, particularly at the global scale, should involve the following as a bare minimum:

1. Species Level
   A. Structural - for example, tissue or organ damage
   B. Functional - respiratory rates or behavior
2. Community Level
   A. Structural - trophic relationship
   B. Functional - colonization rate or rate of detritus processing

3. Ecosystem Level
   A. Structural - trophic relationships characteristic of this particular ecosystem type in this locale
   B. Functional - nutrient spiralling or energy cycling
4. Landscape Level
   A. Structural - compatible with the landscape mosaic
   B. Functional - landscape used with appropriate duration and frequency by species that regularly use the larger mosaic of which this is a part.

All monitoring of systems requires attention to phenomena only displayed at certain temporal and spatial scales. The new but rapidly growing field of landscape ecology has already superbly documented this need (e.g., Dale et al. 1989; Delcourt and Delcourt 1988; Delcourt et al. 1983; Dyer and Vinogradov 1989; Dyer et al. 1988; Forman and Godron 1986; Meentemeyer and Box 1987; Risser et al. 1983). Allen and Starr (1982) espouse hierarchy theory to address scale problems.

Hierarchy theory recommends a focus on a specific level of interest utilizing an upper level, or coarser scale, as a constraint explaining the significance of study at the level of focus, and lower levels (finer scales) to explain processes that control the phenomena being examined (Risser 1987). Rosswall et al. (1988) also discuss approaches to integrating disparate spatial and temporal scales. It may be possible to scale up or down if relationships are linear and phenomena are uniformly defined for at least two different levels. In such cases, monitoring would be greatly simplified, but such situations may well be uncommon.

As noted by Risser et al. (1983), landscape ecology is not a distinct discipline or simply a branch of ecology, but rather is the synthetic intersection of many related disciplines that focus on the spatial-temporal pattern of the landscape. As noted in the excellent paper just cited, improvement in the conceptual base of landscape ecology will also assist in the communication of ideas between and among groups that now suffer from the absence of a common framework. Since effective global monitoring requires a landscape context, it will also catalyze a convergence of developing methodology and theory and accelerate improvements in existing methodologies. This perspective should be espoused. It is a pity that more methods are not available for at least illustrative comment in this chapter. The fact that methods for monitoring global change are in such an early stage in development should be a considerable cause for concern. It would be well if this concern were at least substantively reduced by the year 2000.

FALSE NEGATIVES AND FALSE POSITIVES

All monitoring systems may occasionally produce misleading information including false positives—signals that indicate harm is occurring when it is not, and false negatives—signals indicating that quality control conditions are acceptable when they are not. Once, while my students and I were using fish respiratory rhythms to detect appearances of deleterious materials in an effluent pipe, an abnormal alteration in respiratory rhythm indicated that something deleterious had appeared. Since the effluent had not changed in chemical or physical attributes being measured, a false positive was suspected. Fortunately, seismologists noted that a minor earthquake tremor had appeared on their seismograph at the identical time that the fish responded outside their normal pattern. Sometimes one or two fish in a series may deviate, but rarely the entire group. In this case, however, the entire group did produce a major alert signal (false positive). A false negative might result when the majority of the organisms in the natural system have a much lower response threshold than those being monitoring or the attributes not being monitored are more sensitive than those that are. Testing the monitoring system by simulating episodic events, etc. will improve confidence in the system and reduce the number of false negatives and false positives, but it is not wise to assume that this false information can be reduced to zero. Therefore, in addition to furnishing evidence at different levels of biological organization, an array of measurements at different levels of biological organization for both structural and functional attributes diminishes the likelihood of an erroneous judgment based on inaccurate information. If the cost of inappropriate decisions are included in the total monitoring costs, then the rather substantial array just discussed appears more cost effective. Whatever monitoring system is installed, attention to the false positive/false negative information situation will improve substantially the quality of the decisions being made on that information.

MAJOR GOALS FOR GLOBAL MONITORING

There are five primary goals global monitoring should support: (1) to maintain balanced biological communities, (2) to protect both the structural and functional integrity of ecosystems, (3) to facilitate no-net-loss of habitat by category or, if restoration becomes more widespread than it is now, a percentage increase of habitat by category, (4) to sustain the availability of customary ecosystem service, and (5) to maintain global biodiversity recorded by ecoregion. Types of monitoring to be carried out, the frequency of the monitoring, the level of detail, and the geographic area to be covered should all be highly correlated with the decisions being made. In this case, as in the Holdren scenario, it is important to know whether methods of energy use will be factored into environmental planning. Well beyond the lifetimes of those now living, the reduction in total and per capita use will have a profound effect on the planetary ecosystems. Although it is a *sine qua non* that ecologists must play a strong role in these activities, it would be arrogant to assume that they can be successful without major contributions from a large variety of other disciplines. In addition, progress will be greatly assisted by the use of decision analysis, which is commonly used in a number of other professions (Behn and Vaupel 1982; Keeney and Railla 1976). Although decision analysis has been used infrequently in the field of ecology, Macguire (1988) has shown how effective it can be.

USE OF INDICATOR, SENTINEL, OR SURROGATE SPECIES

Although the indicator species concept might be useful if sufficient data were available on the precise detailed ecological requirements of indigenous species (Kingston et al. 1983), this information base is not available for most species. In fact, Wilson (1988) makes a persuasive case that most of the species on earth have not yet even been named. The probable existence of abundant unnamed species means that their ecological requirements are unknown and consideration of them as indicator species would be ineffective. The point here is that in-depth knowledge is available for only a tiny fraction of the earth's species. Scientists maintain that species chosen as indicators are "representative," but is this justified when so many of the earth's species have not yet been named? Before global monitoring is begun, a fresh examination of these assumptions is in order. John Harte (personal communication) speculates that neotenic salamanders in the Mexican Cut Preserve of the Nature Conservancy on the western slopes of the Rocky Mountains just on the western side of Schofield Pass in Colorado have suffered a decline in reproductive success. Preliminary experiments indicate that this decline is due to a pulse of acidic water from snowmelt that occurs roughly at the same time the species spawns. Nevertheless, although 6 years of tests have been carried out and the salamanders have been under surveillance for a far longer period than that, scientifically justifying a statement that acid rain is inhibiting the spawning of the salamanders is still not possible. Circumstantial evidence indicates that this is probable, but making this statement in a professional peer-reviewed journal is not warranted. Due to the enormous information base necessary to support this hypothesis, the selection of "indicator" species must be made carefully. Not many of the earth's species can be monitored in sufficient depth to establish changes in the characteristics just named. In addition, such studies must be carried out over substantial periods of time.

DEVELOPING CRITERIA FOR SELECTION OF MONITORING AND SAMPLING SITES

Criteria for selecting sites, frequency of sampling, and level of detail required (together with a number of other factors) should all be determined by considering the ways in which the information will be used to make management decisions. Management is used in this connection with regard to environmental quality control in the sense that corrective or preventive action will be taken when necessary to ensure that natural systems stay within pre-determined quality control parameters. The most important conclusion likely to result from extensive surveys is that some information is redundant and that having information of one type will allow prediction of the outcome in other areas. Extensive surveys will also show which information is unique. Determination of information redundancy is exceedingly important since society simply will not pay to make every measurement that professionals know how to make. Additionally, even if this were to happen, over sampling, particularly in fragile ecosystems, might result in more damage than the other anthropogenic activities from which the ecosystem is being protected.

The second major criterion for site selection is determining the extent of variability of natural systems so that normal cyclic variation is not confused with a trend. This means that the sites would have to be sampled not only extensively but over a long period of time. Presumably, if the information redundancy investigations are carried out carefully, a significant reduction in the amount of effort necessary will result.

A third major criterion for carrying out extensive surveys is to enable decision makers to make a distinction between regional and global stresses. This distinction will not be an easy one because the effects will most likely be a series of gradients extending from local and easily identifiable effects to long-range effects where multiple sources are contributing to degradation.

An array of ecosystems representing a broad spectrum of ecosystems types should be chosen for these exercises. This is desirable as a means of identifying trends at both ecosystem and global levels. Each ecosystem should be chosen for its capability of providing evidence or information to the decision makers that cannot be obtained from other ecosystems. Since most ecologists are not accustomed to thinking in these terms (i.e., selecting information on the basis of its decision making value), initial stages of the development should be carefully designed through a standard decision analysis process.

DEVELOPING MONITORING PRIORITIES

Sometimes describing what a thing should not be is as effective as describing what it should be. A global monitoring system should not be a collection of ongoing or newly established monitoring systems primarily designed to meet local or regional needs. This does not mean that systems so designed are totally inappropriate to a global monitoring system, but rather that the latter should not be dependent on the hope that local monitoring needs will always be congruent with global monitoring needs. Similarly, those responsible for monitoring at a local or regional level should not attempt to utilize data gathered for global monitoring to fit local needs unless reasons to do so are persuasive. Since monitoring has been defined for many years (e.g., Hellawell 1978) as surveillance to ensure that previously established quality control conditions are being met, widespread agreement must exist on the quality control trend boundary conditions that will validate that ecological changes are or are not on predicted pathways. That is, stated simply, that certain ecological qualities are improving or, at the very least, remaining stable. Agreeing on these and establishing institutional responsibility for implementing the monitoring programs should not depend on individuals but rather on institutions. Additionally, while new and improved methodology may and should be used, it should be carefully coordinated for significant periods of time with the older methodologies so that the relationship between such measurements is clearly established before a particular methodology is abandoned in favor of a newer one. In short, confidence in the newer methodologies must be clearly established and the relationship with the older methodologies in difference in measurements and the like must also be clearly established before the older methodology is abandoned.

Coordinating monitoring activities and establishing appropriate responses to declining quality will be particularly difficult across political boundaries. Holdren's scenario for energy use and recent United Nations agreements on use of chlorofluorocarbons or global warming provide examples of the potential difficulties.

REDUNDANCY OF INFORMATION

Additionally, selection of too many sites rather than too few is advantageous so that a careful analysis of information redundancy can be made. This analysis need not be extensive but if, for example, five sites are giving the same information through normal seasonal cycles and the like, it is clearly not cost effective to maintain all five sites. Similarly, too many parameters should be selected to determine information redundancy among these parameters. For example, if changes in parameter X are a good predictator of changes in parameter Y, both measurements need not be continued. Above all, information not used regularly for decision making should be abandoned. The acid test of utility for long-term global monitoring is not whether someone is fond of the method used for the organism or the system but rather its importance in the decision making process. This latter will probably be the greatest difficulty to overcome because each specialist feels her/his methods are extremely important and should be made. Very often, these specialists are in positions of considerable power and, thus, able to facilitate the acceptance of their favorite method. Such situations will inevitably develop, but the requirement that the utility of the information be demonstrated in the decision making process will markedly reduce the "favorite method" problem.

CONCLUDING REMARKS

As this book was in its final stages in September, 1991, slightly over a year had elapsed since the invasion of Kuwait and its subsequent liberation, although the oil fires are still burning and the Gulf is still badly polluted. The whole series of events, beginning with the invasion of Kuwait, caught the world by surprise. The early part of 1991 was preoccupied with preparation for the war labeled Desert Storm, and the events that followed the liberation of Kuwait are still unfolding over a year after the invasion first took place. It seemed unlikely that 1991 would have a surprise that exceeded the one just described. Nevertheless, the unbelievable has happened as the coup that temporarily unseated Mikhail Gorbachev collapsed and led to a series of events that would have seemed impossible just months or even weeks before. The Communist party in the USSR has not disappeared, but it has suffered a very serious setback and may even have disintegrated. The Baltic states appear to be free, and the USSR National Parliament voted itself out of existence. Whatever ultimately transpires, the future will be dramatically different from the immediate past. Each small change was fiercely resisted by the persons in charge, often even Mr. Gorbachev. As recently as August 22, which was two weeks after the coup, he stated, "I am convinced that Socialism is correct. I am an adherent of Socialism." The important point is that resistance to change did not prevent it from occurring. Not all that long ago, rapid change occurred in Eastern Europe, then the Soviet Union, and next may be global ecological change. However, just because society has become modestly accustomed to one type of change (i.e., political) does not mean that it is prepared for other kinds of sudden dramatic changes (such as ecological change)! Unquestionably, events are still transpiring so quickly that further, and perhaps even more dramatic, revolutionary changes may have occurred by the time this book has been published. However, this is not the major point. The major point is that no one predicted either of these events, although a few claim to have done so. Neither can we do justice at present to the interpretation of recent events, and, more important, no one with caution or prudence will attempt to predict precisely how far the changes will go.

Clearly, major global changes can occur with amazing rapidity and may occur in ways that are totally unexpected in the political sphere, as did the events just described. There is no way to predict global change with any precision. What will occur cannot be predicted; neither can the timeframe or the spatial dimensions. This is indeed a time of enormous change, not only in the political order but also in population growth, in energy use, in destruction of forests, in the accumulation of persistent hazardous chemicals and radioactive wastes, of thinning of the ozone layer, and a variety of other events. It seems imprudent to assume that these events will occur simultaneously with no synergistic or multiplicative interactions among them. Furthermore, it would not be reasonable to assume that these changes will be linear and without accelerating feedback loops. We can, however, make predictive models, although they will likely not be robust, and validate them or falsify them with monitoring information. Monitoring information may give an early warning of changes in many cases and indicate the areas in which the changes appear most rapid and crucial. Monitoring will also give some false negatives and false positives discussed earlier. This should not discredit monitoring but rather be used in a constructive way to improve its quality. The era of most rapid change lies ahead. Establishment of a global monitoring system, whatever its weaknesses, is a prudent step. In this area, intellectual frontiers and problems of immense practicality intersect. We cannot just consider fragments of complex systems, as reductionist science has been prone to do for most of this century, but rather consider complex multivariate systems in an integrated rather than a fragmented fashion.

ACKNOWLEDGMENTS

I am indebted to Darla Donald for editorial assistance and to Teresa Moody for transcribing the dictation of this chapter on the word processor. I am also indebted to Bruce Wallace for calling the Dobzhansky quote to my attention. Comments of B. R. Niederlehner and James R. Pratt on early drafts of this manuscript are also greatly appreciated.

REFERENCES

ALLEN, T.F.H. and STARR, T.B. (1982). Hierarchy: Perspectives for Ecological Complexity. University of Chicago Press, Chicago, IL.

BEHN, R.D. and VAUPEL, J.W. (1982). Quick Analysis for Busy Decision Makers. Basic Books, NY.

BYRNES, P. (1990). The price of addiction. Wilderness, Summer Issue:3-4.

CAIRNS, J., JR. (1975). Criticial species, including man, within the biosphere. Naturwissenschaften 62(5):193-199.

CAIRNS, J., JR. (1980). Estimating hazard. BioScience 30(2):101-107.

CAIRNS, J., JR. (ed). (1986). Community Toxicity Testing, STP920. American Society for Testing and Materials, Philadelphia, PA. 350 pp.

CAIRNS, J., JR. and PRATT, J.R. (eds). (1989). Functional Testing of Aquatic Biota for Estimating Hazards of Chemicals, STP988. American Society for Testing and Materials, Philadelphia, PA. 242 pp.

CAIRNS, J., JR., DICKSON, K.L., SPARKS, R.E., and WALLER, W.T. (1970). A preliminary report on rapid biological information systems for water pollution control. J. Water Pollut. Control Fed. 42(5):685-703.

CAIRNS, J., JR., DICKSON, K.L., and WESTLAKE, G.F. (eds). (1977). Biological Monitoring of Water and Effluent Quality, STP607. American Society for Testing and Materials, Philadelphia, PA. 246 pp.

CAIRNS, J., JR., DICKSON, K.L., and MAKI, A.W. (eds). (1978). Estimating the Hazard of Chemical Substances to Aquatic Life, STP657. American Society for Testing and Materials, Philadelphia, PA.

DALE, V.H., GARDNER, R.H., and TURNER, M.G. (eds). (1989). Predicting across scales: Theory development and testing. Proceedings of a Workshop at Oak Ridge, TN, November, 1988. Landscape Ecol. 3(3/4).

DELCOURT, H.R. and DELCOURT, P.A. (1988). Quarternary landscape ecology: Relevant scales in space and time. Landscape Ecol. 2:23-44.

DELCOURT, H.R., DELCOURT, P.A., and WEBB, T., III. (1983). Dynamic plant ecology: The spectrum of vegetation change in space and time. Quat. Sci. Rev. 1:153-175.

DYER, M.I. and VINOGRADOV, B.V. (1989). The role of biosphere reserve in landscape and ecosystem studies. Nat. Resour. 26(1):19-28.

DYER, M.I., DI CASTRI, F., and HANSEN, A.J. (1988). Geosphere-biosphere observatories: Their definition and design for studying global change. Biol. Int., Special issue 16.

EHRLICH, P.R. (1968). The Population Bomb. Ballantine, NY.

EHRLICH, P.R. and HOLDREN, J. (1971). The impact of population growth. Science 171:1212-1217.

EHRLICH, P.R. and EHRLICH, A.H. (1989). How the rich can save the poor and themselves: Lessons from global warming. In: *Proceedings of the International Conference on Global Warming and Climate Change: Perspectives from*

Developing Countries (S. Gupta and R. Pachauri, eds), pp. 287-294. Tata Energy Research Institute, New Delhi, India.

EHRLICH, P.R. and EHRLICH, A.H. (1990). The Population Explosion. Simon and Schuster, NY. 320 pp.

EHRLICH, P.R. and EHRLICH, A.H. (1991). Healing the Planet: Strategies for Solving the Environmental Crisis. Addison-Wesley Publishing Co., NY. 353 pp.

FORMAN, R.T.T. and GODRON, M. (1986). Landscape Ecology. John Wiley & Sons, NY.

GRUBER, D.S. and DIAMOND, J.M. (eds). (1988). Automated Biomonitoring. Ellis Horwood Limited, Chichester, England. 208 pp.

HELLAWELL, J.M. (1978). Biological Surveillance of Rivers. Water Research Centre, Stevenage, England.

HOLDREN, J. (1990). Energy in transition. Sci. Am. September Issue:156-163.

HOLDREN, J. (In press). The transition to costlier energy. In: *Energy Transitions* (L. Schipper and S. Meyers, eds). Stockholm Environmental Institute, Stockholm, Sweden.

HUNSAKER, C.T., CARPENTER, D.E., and MESSER, J.J. (eds). (1990). Environmental Monitoring and Assessment Program: Ecological Indicators, EPA/600/3-90/060. National Technical Information Service, Springfield, VA.

KEENEY, R.L. and RAIFFA, H. (1976). Decisions with multiple objectives: Preference and value tradeoffs. John Wiley & Sons, NY.

KINGSTON, J.C., LOWE, R.L., STOERMER, E.F., and LADEWSKI, T.B. (1983). Spatial and temporal distribution of benthic diatoms in northern Lake Michigan. Ecology 64(6):1566-1580.

LAWRENCE, D. and WYNNE, B. (1989). Transporting waste in the European community: A free market? Environment 31(6):12-35.

LEOPOLD, A. (1953). Round River. Oxford University Press, NY. (Quoted from *A Land County Almanac with Essays on Conservation from Round River*, Ballantine Books, NY, 1970, pp. 197).

MAGUIRE, L.A. (1988). Decision analysis: An integrated approach to ecosystem exploitation and rehabilitation decision. In: *Rehabilitating Damaged Ecosystems* (J. Cairns, Jr., ed), pp. 105-122. CRC Press, Boca Raton, FL.

MEENTEMAYER, V. and BOX, E.O. (1987). Scale effects in landscape studies. In: *Landscape Heterogeneity and Disturbance* (M.B. Turner, ed), pp. 15-34. Springer-Verlag, NY.

RISSER, P.G. (1987). Landscape ecology: State of the art. In: *Landscape Heterogeneity and Disturbance* (M.B. Turner, ed), pp. 1-14. Springer-Verlag, NY.

RISSER, P.G., KARR, J.R., and FORMAN, R.T.T. (1983). Landscape Ecology: Directions and Approaches, Sp. Publ. 2. Illinois Natural History Survey, Champaign, IL. 18 pp.

ROSSWALL, T., WOODMANSEE, R.G., and RISSER, R.G. (1988). SCOPE 35, Scales and Global Change: Spatial and Temporal Variability in Biospheric and Geospheric Processes. John Wily & Sons, Chichester, England.

SPETH, G. (1988). Environmental Pollution, Earth '88: Changing Geographic Perspectives. National Geographic Society, Washington, DC.

WILSON, E.O. (1988). Biodiversity. National Academy Press, Washington, DC. 521 pp.

CHAPTER 7

THE RISK OF CRYING WOLF

Gordon Tullock[1] and Gordon Brady[2]

INTRODUCTION

Everyone knows the story of the boy who cried wolf, but the fable is discussed here in more detail than usual. When the boy cried wolf, although no wolf was there, he inflicted significant cost on the village in that the villagers dropped whatever they were doing in order to drive off the non-existent wolf. The fable ignores these costs and simply points out that the boy lost credibility. Of course, the actual cost to the villagers of interrupting their work could have been considerable.

THE ANALOGY OF CRYING WOLF

If the analogy is extended further, the possibility can be considered that the boy deliberately calls a false alert in order to inflict costs on the community. However, this discussion considered the case where the boy simply has poor eyesight and mistakenly thinks he sees wolves when he actually does not. Under these circumstances, he would be innocent of any intention to cause damage but, nevertheless, would impose considerable cost. The first kind of cost is inflicted on the villagers by their going out and chasing the wolf. The second kind of cost results if the villagers conclude that, whether deliberately or accidentally, the boy gives false alarms, in which case they do not respond and the wolves have free reign.

In addition, the boy with poor eyesight does not bear the full cost of his actions. If the boy makes a mistake, the cost of his action is borne by the villagers who respond by driving off the wolf. On the other hand, if the wolf is there, the incidence of the cost is on the owners of the sheep.

The optimal situation is one in which the boy calls wolf when and only when there is a wolf. The villagers would turn out and drive the wolf away, with the result that no sheep are lost. Although this is optimal, related costs are incurred. Nevertheless, these costs are lower than those of any other alternative. The other two alternatives (that the villagers do not turn

---

1. Corresponding Author: Gordon Tullock, Karl Eller Professor of Economics and Political Science, Department of Economics, College of Business and Public Administration, University of Arizona, Tucson, Arizona 85721.

2. Professor and Director of Environmental Studies, Sweet Briar College, Sweet Briar, Virginia 24595.

3. Key Words: cost of environmental decisions, environmental safety alarms, health safety alarms, Malthusian error, resource depletion.

out to drive off the wolf or that they turn out to drive off a nonexistent wolf from time to time) are both inferior. The pros and cons of these two alternatives depend on the details of each specific situation.

The relevance of the optimal situation for crying wolf to environmental and energy concerns is obvious. Society gains by driving real environmental wolves away. On the other hand, society loses each time it turns out to repel an environmental wolf if the wolf is the result of poor eyesight on the part of the environmentalists or even a deliberate false claim on their part. The latter are probably rare, but both the ALAR and the poison scare for Chilean grapes are examples. In both cases, it seems likely that the people who raised the alarm were more interested in the social consequences of the disturbance than with the more or less nonexistent dangers that they fomented.

The basic danger in such cases is that continuous alarms when no wolf is actually present will eventually reduce the credibility of the environmentalists and hence their potential to affect sound environmental policy. If environmentalists continually claim severe environmental damage when it is not there, eventually the citizenry will begin to discount their warnings. Consequently, when a real wolf comes, the sheep will be lost.

Environmentalists regularly raise alarms about environmental damage. Energy advocates have voiced concerns about the depletion of oil and its predictable increase in price. Such alarms are sometimes ignored, but often a great amount of resources are expended in dealing with these alarms.

Publicity generated recently by *Diet for a Poisoned Planet* (Steinman 1990) is an example of the risks of crying wolf. The author rates about 100 foods according to levels of alleged toxic residues. Foods are described as being "drenched" in pesticides and "saturated" with preservatives. Fearing consumer reaction, the National Food Processors Association retained a public relations firm to counter allegations made by the book, and food producers reacted in a similar fashion. For example, the California Raisin Advisory Board allocated $558,000 of its marketing funds to address charges that raisins contain various toxic residues (Ferguson 1990) [For a discussion of health and safety alarms, see "Those Terrifying Cows," *The Wall Street Journal*, 1991a. In this particular case, opponents of BST (bovine somatotropin), which is capable of increasing milk production by 25% with the same feed, mount two arguments: (a) cheaper milk will force smaller dairy farms to close and (b) BST renders cows more susceptible to infection and thus endangers the safety of the country's milk supply. Both claims have been proven false.] Likewise, the ALAR scare has reportedly cost apple growers over $200 million (Ray and Guzzo 1990).

Of course, if the risk is real and significant, appropriate action must be taken. On the other hand, if the risk is not real, it should be ignored. However, in many cases, only after substantial costs have been imposed on industries (apples) or countries (Chile) does the alarm prove unfounded.

ALAR, a chemical used as a growth regulator on a number of agricultural products is not a carcinogen as claimed by the Natural Resources Defense Council. In the case of ALAR, a human would have to eat 28,000 pounds per day for 70 years in order to produce tumors similar to those suffered by mice exposed to mega doses of ALAR (Ray and Guzzo 1990). On the other hand, mice fed one-half the maximum amount (in human equivalents, 14,000 pounds per day for 70 years) produced no tumors. Evidence is mounting that the risks of exposure to radon and asbestos have been severely exaggerated (Brody 1991; Gladwell 1990; Koshland 1989; McInnis 1990; Mossman et al. 1990).

It is, of course, reassuring that accurate information eventually emerges. However, the information may reach the public too late to avoid damaging innocent commercial interests. Lags in the development of scientific information may occur for several reasons. First, science may simply be slow in catching up with policy. Or second, a more plausible explanation, for cases such as asbestos, is that those adversely affected by the regulations applied political pressure to obtain an objective review. However, organizing coalitions to oppose erroneous political decisions takes time. In the case of asbestos, pressure from school districts may have been translated through their congressman to force USEPA to back off the asbestos removal requirements.

THE RISE AND FALL OF ENVIRONMENTAL, HEALTH, AND SAFETY ALARMS

The rise and fall of such issues is a troubling social problem because alarmists bear only a small part of the costs they impose. For example, Random House, the publisher of *Diet for a Poisoned Planet*, will pay only the cost of producing the book plus a 10-week promotional program. Of course, at some point the publisher may face lawsuits over the claims in the book. However, Random House will no doubt reap benefits from the higher sales of the book due to the publicity generated by the controversy.

There are, of course, cases in which environmental damage occurred without the environmentalists crying wolf. For many years the U.S. Bureau of Land Management and the U.S. Forest Service engaged in "chaining," a method of clearing land that involves dragging a 600-foot anchor chain with links weighing 90 pounds between two "crawler" tractors. This method of removing pinon pine and juniper trees was used on hundreds of thousands of acres in the Southwest during the 1960s and 1970s. The effect of uprooting by chaining has been severe damage to lightly wooded areas in the western United States (Anderson and Leal 1991; Lanner 1981; National Center for Policy Analysis 1991). However, for a long time, very little objection came from environmentalists.

Another example of government destruction of the environment is found in the Forest Service subsidies for clearing trees on the eastern slopes of the Rocky Mountains. Regrowth will be slow if it occurs at all. Such marginally productive land would never be cut by private sector lumbermen because of the expense of transporting the wood to the mill. As a consequence of such subsidies, the Forest Service encourages pointless injury on the environment (there are political reasons why the Forest Service would do this).

Only recently have environmentalists become aware of these problems. In general, most environmental organizations have chosen to ignore such problems. This is particularly depressing because, in both chaining and the subsidized cutting of uneconomic trees, there is no resource cost in preventing environmental damage; all that is necessary is that the government merely stop wasting money.

It has been almost 200 years since Malthus first predicted environmental disaster and widespread starvation (Malthus 1817). The prediction was, of course, falsified. Nevertheless, these predictions continue to be made. Fears of population growth prompted Paul Ehrlich, some 20 years ago, to predict an even more severe environmental and natural resource disaster than Malthus. In addition, Ehrlich gave his prediction the range of 10 years (Ehrlich 1968; Ehrlich and Ehrlich 1990). In his later work, Malthus was fully aware of the possibility of birth control, but he, as a clergyman, may have disapproved even though he expected it to happen. However, Malthus did not anticipate the remarkable technical improvements in agriculture.

On the other hand, latter day Malthusians made the same error, even after a period in which immense improvements had been made in agricultural technology due to the green revolution (Ehrlich 1968; Meadows et al. 1974). They failed to notice, as any economist would, that a very large part of all vegetable matter is fed to animals that are subsequently eaten. This means that the population of the developed world could nearly double without any increase in total food production if humans simply reduced the eating of meat. Malthusians also failed to notice the vast areas currently under jungle cover that are suitable for growing wet rice.

Environmentalists may, of course, fail to note some risks. Electromagnetic fields (EMF), while little acknowledged by environmentalists, may be the "asbestos of the 1990s" (Lehman 1991). The strength of the invisible magnetic force increases with the amount of electric current running through a wire. High voltage transmission stations and lines create EMFs that travel through most physical matter without losing strength, although the intensity declines rapidly with distance from the source. About 40 scientific studies conducted over the past 15 years in homes and on electrical workers link electromagnetic radioactivity to childhood leukemia, brain tumors, birth defects, breast cancer, and other health problems. The electric utility industry is understandably resistant to opening this issue since millions would have to be spent to reroute, rewire, bury or shield electrical cables (Morgan 1990; Pool 1990; Shulman 1990).

RESOURCE DEPLETION

Resource depletion is another issue that is subject to deep-rooted fears. Fears of resource depletion—"energy crises" and "timber crises"—have occurred in the past and the Neo-Malthusians have many antecedents. For example, in 1865 William Stanley Jevons published "The Coal Question: An Inquiry Concerning the Progress of Nations and the Probable Exhaustion of our Coal Mines" (Jevons 1865). (For a popular discussion of Jevons' views, see Tierney 1990.) In order to demonstrate the inadequacy of British coal reserves, Jevons' graph of population and coal consumption showed a direct relationship. In later lectures, however, Jevons argued that it was not the absolute amount of consumption that he found alarming, but the rate at which consumption increased year by year. He noted that England used one and one-half times as much coal in 1865 as in 1854 (Jevons 1871).

The current view of population as a problem has evolved from Malthus' fear of population growth, a concern that has appeared intermittently since his time. At present, it is fashionable for ordinary intellectuals to worry about population as a menace and for specialists in population demography or economics to downplay it (Kelley 1988; Simon 1990a).

The dimensions of this phobia are demonstrated in a well publicized bet between ecologist Paul Ehrlich and economist Julian Simon. Respectively, they lead two intellectual schools—called the Malthusians (or doomsters) and the Cornucopians (also called "boomsters"). Ehrlich argues that the earth's growing population is running out of resources. Simon, on the other hand, argues that babies do not cause poverty, that human beings create more than they destroy, and that population growth is an important component of economic growth. Simon's message is that raw materials and energy are getting less scarce, the world's food supply is improving, pollution in the United States has been decreasing, population growth has long-term benefits (though added people are a burden in the short run), and the United States benefits from more immigrants.

In 1980, Simon issued a challenge to the neo-Malthusians. Simon offered to let anyone select five commodities—grain, oil, coal, timber, or metals. If the price, adjusted for inflation, rose over 10 years, Simon would pay $10,000 and his opponents would pay the same if the price fell. Ehrlich and two colleagues accepted the offer and bet that the price of five metals—chrome, copper, nickel, tin, and tungsten—would increase over a 10-year period. When the bet came due in 1990, the five metals showed an unadjusted decline of 38% (Tierney 1990; *Wall Street Journal* 1990). Ehrlich and his friends sent a check, but without a letter of enclosure. Simon offered them an opportunity to be $20,000 richer for the next 10 years, and they have not responded. (Simon's offer is open to anyone—if any readers feel that the price of raw materials is going up over the next 10 years and would like to pick up an easy $20,000, they should write to Simon.) Simon was clearly the winner, but he received little media attention for his predictions while Ehrlich remained immensely popular among environmentalists. On Earth Day in 1990, Ehrlich spoke to several thousand people; Simon spoke before a group of 16 economists (Tierney 1990).

The point, of course, is that human beings come into the world equipped with a stomach, two hands, and a brain. The question is whether the stomach *or* the two hands *and* the mind are more important. Historically, the latter seems to be the case. Developed countries, for example, are now trying to manage what they regard as an excess of food. These countries have large populations but have less land under cultivation than the acreage that previously provided a much less adequate diet to much smaller populations 50 to 100 years ago.

Many environmentalists seem to regard human beings as pollution in themselves. Nothing is easier than reducing the world's population. However, most environmentalists would be horrified at the quick and inexpensive methods available to accomplish this goal. Instead, they want either to prevent its growth or to slow it by controlling births. The contrary concern that the population of some developed countries is declining does not appear to be a major concern; Hungary, Germany, and Denmark are experiencing slow population decline, and France, Britain, and the United States are expected to experience slow population growth after the year 2000 (Simon 1990b). Not long ago, a number of countries adopted pronatalist policies because of this concern. Whether that will happen in the West is an open question. As Sam Goldwin put it, "Prediction is difficult. Especially the future."

The invention of agriculture in 6000 or 7000 B.C. in the Middle East and more recently in the Western continents is the largest single environmental change that humankind inflicted on the world. For example, the Mesopotamian Basin is one of the most fertile areas of the world when it is under irrigation, but is naturally a desert. Egypt is similar.

The changes now being imposed by human civilization are minor in comparison with those that occurred in the Mesopotamian Basin and Egypt. Indeed, in many cases, modern developments are moving the environment closer to its earlier natural condition. For example, the eastern states of the United States have immense areas that were farmed as late as World War I and that are now reverting to forests. This is occurring because agriculture has become more intensive and requires less real estate than in the past.

The problem with present day agriculture is not the inadequate production of food for the world, but rather that agricultural production has been sharply restricted by government policy. In developed countries, efforts to keep the price of agricultural products above the market clearing levels are enforced partly by restricting the acreage that is farmed. This lowers total world food supply, although it is not clear by how much. In Europe, farmers who want government subsidies and artificial restrictions on production have formed an alliance with the Greens who want land taken out of production. Ehrlich and other Neo-Malthusians support taking agricultural lands out of production for environmental reasons; the ghost of Malthus must be astonished (Ehrlich and Ehrlich 1990).

In the developing world, on the other hand, the bitterly poor farmers are subjected to prices that are held artificially low in order to benefit the urban population. This makes the poorest people in these societies even poorer while only the upper classes and the urban population benefit. Such a situation also sharply reduces total output. As a consequence of maintaining low prices, farmers return to subsistence farming instead of producing food for market. It is likely that government attempts to subsidize the urban population in this way are one of the basic causes of widespread famines at present in Africa.

Fears of the Loss of a Species

Another example of "crying wolf" is the case of the endangerment of African elephants. Worst case scenarios suggest the extinction of elephants in East and Central Africa possibly as early as 1995. From 1979 to 1989, the population of Central African elephants is estimated to have declined from 497,400 to 274,800, while the East African population fell from 546,650 to 154,720. Game parks estimate the population to have fallen by 56%, while outside the parks elephants are estimated to have declined by 78%. Given these rates, elephants should be extinct by 2005. Burning ivory and contraband wildlife products have become a means to demonstrate a country's firm resolve to address the issue.

The opposite approach of managing elephants as resources has been much more successful, but has received less publicity. Zimbabwe and several southern African countries have given ownership of the elephants to nearby villages. The property rights structure allows the villagers control over the use of elephants, including possible agreements under which private hunters are permitted to shoot an elephant. Some villages received as much as $30,000 per elephant.

As a consequence of the property rights structure, the villagers, instead of viewing elephants as pests, regard them as a valuable asset and the population of elephants is rapidly increasing. In fact, elephants may be increasing at the rate of 5% annually in Zimbabwe. Also, the incentive to conserve elephants is exactly the same as the rancher who protects and manages his cattle. The introduction of property rights in the management of almost any species that is endangered provides a purely commercial reason for protection. This is more effective than having a bureaucrat offer protection, because bureaucrats in the developing world can be bribed easily.

Fears of Global Climate Change

Pollution, population, and resource depletion are by no means the most outstanding examples of "crying wolf." About 15 years ago, concerns were voiced about a new ice age. Recent discussions have been offered on the results of various nuclear winter computations on climate and the greenhouse effect (Crutzen and Birks 1982; Sagan and Turco 1991) and on global cooling (Stevens 1990). While fears of an ice age are still present, the major concern of the 1990s is global warming, although predictions of high temperatures keep receding into the future and getting lower. In any event, cooling the earth to former levels will not be extrememly difficult. Boeing would be delighted to provide the aircraft necessary to place dust in the upper atmosphere to increase the aldebo with resultant temperature decrease.

Fears of a gradually warming planet have captured the attention of the public (Singer 1990; *Wall Street Journal* 1991b). A British documentary entitled "The Greenhouse Conspiracy" compared the evidence for global warming with the positions of the "doomsayers" (*Wall Street Journal* 1991b). They charged that a completely fraudulent case had been used to call for political action costing billions of dollars. The U.S. Public Broadcasting System (PBS) refused to air the program despite the ease with which it aired "After the Warming," a 2-hour PBS documentary that purported to describe the world in 2050 after the damages of rising temperatures and oceans resulting from a global warming catastrophe (Miniter 1990).

On the other side, a United Nations scientific report argues that, if nothing is done, the global mean temperature could rise 5.4°F by the end of the 21st century (*New York Times* 1990; *Washington Post* 1990). Furthermore, the report states that "ocean water would expand and ice stored at the poles could melt, raising the level of the sea by 25.6 inches ... enough to submerge the Maldives and inundate the coastal plains of Bangladesh and the Netherlands" (*New York Times* 1990; *Washington Post* 1990).

The idea that a rise in sea level of 25.6 inches would "inundate the coastal plains of .. the Netherlands" is of course absurd. The coastal plains of the Netherlands are already over 25.6 inches below sea level. The Dutch, with several hundred years of hydraulic engineering experience, would simply raise the dikes.

Indeed, the Netherlands regards the possibility of global warming and a rise in sea levels as a major economic opportunity. No doubt they feel that Dutch hydraulic engineers would be in great demand in places like Malaŵi and Bangladesh. Of course, various places in the United States and, indeed, many other countries might decide that their coastal lowlands are worth preserving by importing a few Dutchmen.

In theory, global warming will be caused by the buildup of CO_2 and other greenhouse gases in the atmosphere. Data, however, cast considerable doubt on this prediction. According to climate models, the effects should appear first in the northern hemisphere. In fact, temperature has not changed in the United States over the last 100 years. Indeed, most of the rise in temperature in the southern hemisphere occurred before the buildup of CO_2 in the atmosphere (Singer 1990).

Greenhouse models have also proven faulty in other ways. According to these models, warming would be more intensive in the polar regions than the tropical regions. The reverse has been true—the polar regions warmed prior to the CO_2 buildup, and then cooled by an equal amount.

The cost of reducing CO_2 to prevent global warming will be enormous. Granted the fact that much of the world is still in poverty, the proposal to prevent generation of more greenhouse gases such as CO_2 seems particularly hardhearted. If (and this seems unlikely) it were necessary to let the ocean rise to improve living standards in India, China, and Africa, many would regard this as a good bargain.

If current trends for carbon-based energy continue, approximately half the CO_2 emissions will come from developing countries by the middle of the next century. For example, China plans to double its use of coal by the year 2000, while India plans to triple its fossil use. If this occurs, coal use in these countries alone will exceed that of all members of the Organization for Economic Cooperation and Development.

The obvious way to reduce carbon emissions is, of course, to switch to nuclear power. Indeed, without the opposition of environmentalists, society would now be much farther along that road. Environmentalists could back away from their current anti-nuclear power position by merely claiming that new plants are safer (which, of course, they are). Both technological progress and changes in the political climate with regard to nuclear power are likely.

In the early 1980s, acid rain was an issue that generated equal, if not greater, environmental concern. It was the focal point of efforts to amend the U.S. Clean Air Act in 1981 and was subject to much skepticism and fears of high cost solutions. Congress established the National Acid Precipitation Program (NAPAP; PL 96-294, Title VII) in 1981 to address the question. NAPAP was coordinated by an interagency task force consisting of representatives of 12 federal agencies, four national laboratories, and four presidential appointees. Over the next nine years, NAPAP succeeded in getting Congress to appropriate almost $600 million for careful scientific study (Kulp 1990). The results of the NAPAP scientific study did not support most of the fears; it also failed to disprove most of them either. However, it indicated that knowledge of acid deposition was poor and that further research was necessary before action. This was not the course chosen by the government. Congress chose to ignore the NAPAP report and required massive SO_2 reductions by the largest power plants in the United States. However, a small ray of hope may be in the choice of Congress to rely on emission trading for these reductions.

Media treatment of environmental wolves is not always over alarming. For example, in February 1991, CBS "60 Minutes" probed the Natural Resource Defense Council's claims on the damage of acid rain. The program highlighted the findings of the NAPAP, which found "no evidence of a general or unusual decline of forests in the U. S. and Canada due to acid rain."

THE RISK OF CRYING WOLF

There has always been a great deal of "crying wolf," but little on the "risk" of "crying wolf." The reason for the risk is that, as in the myth, people begin to disbelieve. Environmental problems thus far have been genuine although they have not been extreme. One enduring example is the English downs that are the result of late Stone Age agriculture which successfully changed the actual environment in those areas. The areas most likely will never return to their pristine state.

Risks of "crying wolf" surface over false issues also. Often these are small for the perpetrators. In many cases, the government becomes the focal point for its failure to act. On the other hand, government will be criticized for excessive control. Both sides of the equation incur costs. Excessive control leads to costs unjustified by results. People lose faith in government when large expenditures and dislocations occur with few benefits. Groups who continually "cry wolf" over time lose support.

Moreover, serious problems, such as fear of cancer or a gradually warming planet, are apt to carry significant emotional baggage that eliminate effective solutions in favor of policies that more quickly appeal to emotions, are easily visualized, or just simply "sound good." When a decision is made in a highly politicized environment, the losers feel disappointment and frustration and also an indignation that associates evil with those who prevailed and with the process that allowed them to do so.

Concern for the environment is a luxury of good economic times that occur in rich countries. The concern peaks just after economic booms peak (1973 and 1989 in the United States) and recedes when recession is imminent (*The Economist* 1990a). Public opinion on the environment has undergone major changes in the past 10 years. In June 1989, a CBS/*New York Times* poll in the United States showed that almost 80% of those surveyed "agree that protecting the environment is so important that requirements and standards cannot be too high, and continuing environmental improvements must be made regardless of cost" (*New York Times* 1989). This percentage is double the number holding that view in 1981. In the 18 months from December 1987 to June 1989, Britain witnessed a dramatic increase [5% to 35%] in those who "say the environment is one of the most important issues facing many of the voters in the June 1988 election for the European Parliament."

However, after the November 1990 election, Germany had only 8 quasi-green members with 4.8% of the vote. In the 1987 election, Greens received 8.3% of the vote and 42 seats in the West German government (*The Economist* 1990a,b; *International Environment Reporter* 1990a, 1991). Those, of course, were not East Germans, but no doubt the jobs/environment tradeoff played a prominent role in the decision calculus of the average voter. Certainly, jobs and the regional economic effects were important factors in the rejection of "Big Green" in California and similar high priced environmental initiatives in New York, Washington, and Oregon.

The German Greens opposed the unification of East and West Germany and suffered from their own internal anarchy. The California Greens proposed programs that voters were persuaded would cost dollars and jobs. The Greens learned that costs matter and that if they continue to behave as if costs are irrelevant, their influence will continue to decline. In the January 20, 1991 local elections in Germany, the Green coalition (with the Social Democrats) was very successful. While the Greens hold only 10 seats in the state of Hesse, Germany's largest state, the "Red-Green alliance" now controls 56 of Hesse's 110 parlimentary seats (*International Environment Reporter* 1990b).

In November 1990, the American Greens lost the "Big Green" referendum in California. The referendum became a battle between environmentalists and agricultural interests opposing restrictions on pesticides (*The Economist* 1990a). Activists learned that people weigh costs in their voting calculus. Second, the industry advocates were successful in fomenting dissonance about Big Green. It became a battle where special interest groups could generate the more believable numbers on jobs and costs. Finally, the Greens recognized the importance of institutions in transmitting public preferences as a policy input. The Greens are more successful in Washington in pushing projects with high costs and low benefits. Washington will continue its importance in the battle to avoid majority rule collective decisions.

The problem of the boy who called wolf should be kept in mind. At present, American voters have demonstrated, through "Big Green" and other environmental initiatives, a lack of interest in environmental precautions. The public cannot claim to have made significant studies of all of the environmental proposals that have been on ballots and American voters do not know whether their judgment was better than that of the environmentalists who put the measures before them.

In many instances, however, environmentalists clearly cry wolf—by alleging dangers that either are not in existence or are drastic exaggerations of genuine dangers. The world is constantly changing, and many of these changes have only recently been measurable. Thus, both the ozone hole over the Antarctic (which was, it turns out, first noted in 1956) and global climate concerns have only recently become measurable. Geological studies show that global temperatures have changed very sharply at various times in the past (Singer 1990). If temperatures are changing now, the causes may be man-made or some rather mysterious events that caused temperature changes in the past.

In these circumstances, scientific research is necessary; however, it is too optimistic to expect that this scientific research will bear immediate fruit. Certainly, mankind is not imposing changes on the environment that are as great as the neolithic revolution that brought in agriculture. Mankind has learned to live with that revolution, and there is no reason to believe that humans will not adapt to future changes.

To reiterate, the risk of crying wolf is both the costs of not taking action when it is necessary and the costs of taking action when it is not. One of the authors of this chapter lives in Tucson, which has a moderate air pollution problem. This problem is probably regarded as trivial by the poor people (Mexicans) who live 100 miles south of Tucson. However, Tucson also has a university and a group of intellectuals who want to support antipollution programs.

On the other hand, the city government and, for that matter, the state and federal governments that subsidize the civic government have distinct limits on the amount of money they are willing to spend. The result in Tucson is a set of programs that are more ceremony than actuality.

The municipal bus line in Tucson is particularly irritating. Based on a subsidy provided partly by the city but mainly by the federal government, there is a bus line that moves empty buses around the city. The main function of this bus line is to permit local environmentalists to think they are doing something without expending many funds. In addition to the environmentalists whose interest is ideological, two minor special interest groups support the bus system. One of these groups is composed of the people who operate the buses and who would be fired if the buses stopped. The other is that small number of people who actually ride the buses.

The buses do, of course, carry some passengers, and it is possible that the net effect is a reduction in pollution. However, a heavy bus with only a few passengers is likely to be a net contributor to pollution over the alternative chosen by the nonbus riding public. We know of no scientific studies of the matter. Citizens of Tucson would probably be willing to pay for this study based on the very localized nature of the pollution in Tucson. There is little reason to expect anyone else to be interested. One example of such a situation is the debate over rail transit for Honolulu. Evidence cited for Honolulu showed that the price for each car removed from traffic is $30,000 per year. Surprisingly, the local League of Women Voters came out against the rail, arguing that the ridership would be too low to justify the expense. The Chamber of Commerce supported the mayor of the city, although rejecting a special tax to help pay for the rail system (Ferguson 1991). The citizens of Tucson are, however, willing to send congressmen to Washington who then trade votes on various matters for money to be spent for local projects. The Central Arizona Project (a public works project initiated in the mid-1950s that reroutes water from the Colorado River to various parts of Arizona) is a far more negative factor from the standpoint of the United States government, which funds it, than is the bus line. Nevertheless, Tucson citizens favor having the bus line and the Central Arizona Project as long as someone else pays for them.

Environmental considerations enter into many political decisions. Congressman Udall had an alliance with environmentalists under which he voted to restrict growth in Alaska and the environmentalists, in turn, did not object too loudly to the funds he obtained for Arizona. Of course, this is quite common in environmental politics.

Many people are aware of the deal between the environmentalists and the Eastern coal producing interests to require full scrubbing of coal regardless of sulfur content. This decision led to an increase in the value of Eastern coal mines, employment of members of the mine workers union, and an increase of air pollution in the Northeast (Ackerman and Hassler 1981; Yandle 1989).

CONCLUSIONS

As a concluding thought for improving the responsibility of those making public alerts, we offer one modest proposal. We suggest that those making such claims and/or predictions be required to post a bond. If the alarms prove false, the proceeds would be distributed randomly in the economy. An alternative approach would allow civil suits for unfounded claims. The Colorado legislature has a bill pending that would allow civil suits against those who unfairly "disparage" food products, i.e., make "reckless and unfounded claims" about food. A sponsor of the bill provided an example of a possible suit: a health food store in Boulder recently advertised that beef "should be sold in drug stores" because it contains residue from the antibiotics given to cattle (*New York Times* 1991). If, on the other hand, the predictions are proven true, the prophets would have the notoriety (free advertising) of winning the bet while the public would be protected from the costs of at least some false alarms.

This is exactly what Ehrlich and Simon did, but on a grander scale. The $10,000 stakes were not high enough to provide the kind of public attention that would have occurred if the bet had been higher, e.g., $1 million.

REFERENCES

ACKERMAN, B.A. and HASSLER, W.T. (1981). Clean Coal/Dirty Air, or How the Clean Air Act Became a Multibillion-Dollar Bail-out for the High Sulfur Coal Producers and What Should be Done About It. Yale University Press, New Haven, CT.

ANDERSON, T.L. and LEAL, D.R. (1991). Free Market Environmentalism. Pacific Research Institute for Public Policy, San Francisco, CA. 191 pp.

BRODY, J.E. (1991). Some scientists say concern over radon is overblown by E.P.A. New York Times 18January:A-1.

CRUTZEN, P.J. and BIRKS, J.W. (1982). The atmosphere after a nuclear war: Twilight at noon. Ambio 11:114-125.

THE ECONOMIST. (1990a). Green storm falling. 8December:14.

THE ECONOMIST. (1990b). Variegated winners, red-green losers. 8December:56.

EHRLICH, P.R. (1968). The Population Bomb. Ballantine Books, NY.

EHRLICH, P.R. and EHRLICH, A.H. (1990). The Population Explosion. W. H. Freeman, San Francisco, CA.

FERGUSON, A. (1990). Better eat nothing—just to be safe. The Wall Street Journal 7December:B-1, col. 2.

FERGUSON, T.W. (1991). Hawaiian jewelry maker assays flaws in rail link. The Wall Street Journal 29Janaury:A-19.

GLADWELL, M. (1990). Greenpeace digs deep into dioxin debate. Washington Post 30November:A-27.

INTERNATIONAL ENVIRONMENT REPORTER. (1990a). Greens in west fall short of votes. 5December:500.

INTERNATIONAL ENVIRONMENT REPORTER. (1990b). After defeat in federal elections. 30January:44.

INTERNATIONAL ENVIRONMENT REPORTER (1991). Green group moves closer. 13March:147.

JEVONS, W.S. (1865). The coal question: An inquiry concerning the progress of nations and the probable exhaustion of our coal mines. Macmillan's Magazine 63:143.

JEVONS, W. S. (1871). Lecture IX. In: *Science Lectures for the People*, pp. 107-118. Manchester, England.

KELLEY, A. (1988). Economic consequences of population change in the third world. J. Econ. Lit. 26:1685-1728.

KOSHLAND, D.W. (1989). Scare of the week. Science 244:9.

KULP, J.L. (1990). Acid rain: Causes, effects, and control. Regulation, Winter:41-51.

LANNER, R.M. (1981). Chained to the bottom. In: *Bureaucracy vs. the Environment* (J. Baden and R.L. Stroup, eds), pp. 154-169. University of Michigan Press, Ann Arbor, MI.

LEHMAN, J.H. (1991). Homeowners confront risks of electromagnetic fields. The Washington Post 1June:E-1.

MALTHUS, T.R. (1817). An Essay on the Principle of Populations; or A View of its Past and Present Effects on Human Happiness; with An Inquiry into Our Prospects Respecting Future Removal or Mitigation of the Evils which it Occasions. John Murray, London.

MCINNIS, D. (1990). EPA changes course on asbestos removal. Human Events 20October:11.

MEADOWS, D.H., MEADOWS, D.L., RANDERS, J., and BEHRENS, W.W.

(1974). The Limits to Growth: A Report for the Club of Rome's Project on the Predicament of Mankind. Potomac Associates, New American Library, NY.
MINITER, R. (1990). PBS chills global warming debate. CEI Update 12:1.
MORGAN, G. (1990). Flying blind: The making of EMF policy. Science 250:23-25.
MOSSMAN, B.T., BIGNON, J., CORN, M., SEATON, A., AND GEE, J.B.L. (1990). Asbestos: Scientific developments and implications for public policy. Science 247:294-301.
NATIONAL CENTER FOR POLICY ANALYSIS. (1991). Progressive Environmentalism: A Pro-Human, Pro-Science, Pro-Free Enterprise Agenda for Change. Dallas, TX. 18 pp.
NEW YORK TIMES. (1989). Plight of the planet. 18August:A-31.
NEW YORK TIMES. (1990). Scientists warn of dangers in a warming earth. 26May:A-6.
NEW YORK TIMES. (1991). Meant for the stomach, really taken to heart. 17February:30, col. 1
POOL, R. (1990). Is there an EMF-cancer connection? Science 249:1096-1098.
RAY, D.L. and GUZZO, L. (1990). Trashing the Planet: How Science Can Help Us Deal with Acid Rain, Depletion of the Ozone, and Nuclear Wastes (Among Other Things). Regnery Gateway, Washington, DC.
SAGAN, C. and TURCO, R. (1991). A Path Where No Man Thought: Nuclear Winter and the End of the Arms Race. Random House, NY.
SHULMAN, S. (1990). Cancer risks seen in electro-magnetic fields. Nature 345:463.
SIMON, J.L. (1990a). The unreported revolution in population economics. Public Interest, Fall:89-101.
SIMON, J.L. (1990b). Population Matters: People, Resources, Environment and Immigration. Transaction Publishers, New Brunswick and London.
SINGER, S.F. (1990). Global climate change: Facts and fiction. World Climate Change Report, November:19-23.
STEINMAN, D. (1990). Diet for a Poisoned Planet. Random House, NY.
STEVENS, W.K. (1990). In the ebb and flow of ancient glaciers, clues to a new ice age: Greenhouse effect could delay the onset of the cold, glaciologist say. New York Times, 16January:C-1.
TIERNEY, J. (1990). Betting the planet. New York Times Magazine 2December:75,78.
THE WALL STREET JOURNAL. (1990). Simon's safe bet. Asides:A-18.
THE WALL STREET JOURNAL. (1991a). Those terrifying cows. 7January:A-14, col. 1.
THE WALL STREET JOURNAL. (1991b). Warming up to the facts. 11January:A-10, col. 1.
WASHINGTON POST. (1990). U.N. report predicts dire warming. 26May:1.
YANDLE, B. (1989). The Political Limits of Environmental Regulation. Quorum Books, NY.

CHAPTER 8

ECOLOGICAL RISKS ASSOCIATED WITH THE EXTINCTION OF SPECIES

James R. Pratt* and John Cairns, Jr.**
*School of Forest Resources and Graduate Program in Ecology
Pennsylvania State University, University Park, Pennsylvania
**University Center for Environmental and Hazardous Materials Studies and Department of Biology
Virginia Polytechnic Institute and State University, Blacksburg, Virginia

REDUCTION IN ADAPTIVE ABILITY OF COMMUNITIES TO CLIMATIC AND OTHER ENVIRONMENTAL CHANGES

A large array of species may fulfill a similar or even identical role in a complex natural community (e.g., Patrick 1977; Pratt and Cairns 1985). The advantage to the community and, therefore, to the ecosystem is that, when environmental conditions change, species X with a better fit to the new conditions may replace species Y that was best fitted to the previous environmental conditions. This orderly transition ensures the ability of natural communities to function on a sustained basis during normal environmental changes. The array of species available for each particular function (Fig. 1) under a wide spectrum of environmental conditions is enormously beneficial to sustained and unimpaired functional performance of communities and ecosystems. The loss of a few species from a substantial array means that the fit of the biota to a unique set of environmental conditions is likely to be poorer with a smaller array than with a larger array. Loss of species may ultimately mean that, for certain conditions, no species in a particular functional category is available. The community and, therefore, the ecosystem are placed into severe disequilibrium because of the loss of one functional group (or perhaps even several simultaneously), and this does not bode well for a complex, multivariate system with strong interdependencies. In some instances, in harsh or marginal environments such as those at the high altitudes of the poles and in other rigorous conditions, natural communities are simplified since the array of species capable of surviving and functioning under these conditions is markedly reduced. Under such circumstances, the loss of even a few species may mean irreparable loss of an entire functional category, and disruption may occur abruptly rather than gradually as in the case of a more complex system with a large array of species.

---

1. Corresponding Author: James R. Pratt, School of Forest Resources, 7 Ferguson Building, Pennsylvania State University, University Park, PA 16802-4301.

2. Key Words: hazard evaluation, microbial communities, toxicity, ecological effects.

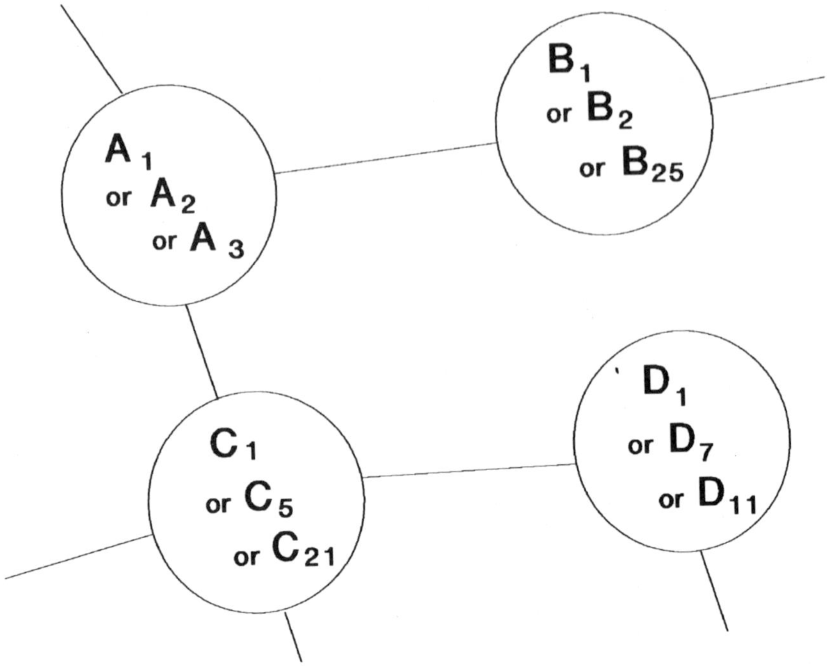

Figure 1. An illustration of community functional redundancy. Several alternative species may fill the same, or a similar, functional role (or "realized niche") in a community.

In human societies, the loss of a single service, such as the power failure in the Northeast some years ago, can cause severe disruption of other services and considerable disequilibrium within the society. While the analogy to complex natural systems is far from perfect and while humans have more difficulty recognizing disruptions in natural systems as swiftly and clearly as in their own society, there are important parallels. Unfortunately, society's knowledge of the functional roles of most of the earth's species remains unknown. This is not surprising in view of the fact that it appears highly probable that a substantial majority of the earth's species remains unnamed and unstudied (e.g., Wilson 1988). Disruptions will have little early warning and will be abundantly clear only after severe ecological effects have occurred because predicting these disruptions, or even tracking loss of species, is difficult since the inventory of species is so inadequate. As a consequence, risks will be poorly estimated and will not give significant advanced warning. The ecosystem risk situation is not dissimilar to the risks of sudden exposure to unfavorable and often violent bad weather of ships at sea and the like before the wireless and other means of communication provided warnings.

RISKS TO THE GLOBAL FOOD SUPPLY OF BOTH HUMANS AND OTHER ORGANISMS

A substantial number of calories consumed by humans are provided by seven or eight cereal grains. The fact that these are truly domesticated plants and have lost many of the characteristics essential for survival in the wild (including resistance to disease, pests, and the like) has been well documented. Loss of species in certain categories, such as plants that provide cereal grains, deprive human society of alternative genetic information that may be valuable during periods of climatic or other global changes. For example, if global warming occurs, even to the extent of several degrees over a decade, agricultural practices might well be required to undergo major shifts, particularly if the temperature change is accompanied by alterations in rainfall patterns. Under these circumstances, the availability of species with genetic information suitable to the new conditions will help human society enormously. It is almost platitudinous to say that a collapse of human society is likely to have even greater disruptive effects upon certain natural systems than a stable human society. The loss of valuable food stuffs to organisms in natural ecosystems, whether cereal grains or other species, is likely to have comparable effects on populations of organisms dependent upon them. These might well be as severe to these species as they might be to the human species. Again, information is sufficient for an awareness of the risks, but not enough information is available to predict the risks or even to restore damage when it occurs.

LOSS OF ECOSYSTEM CAPABILITY FOR SELF-MAINTENANCE

Natural ecosystems are self-maintaining in the sense that they retain structural and functional characteristics representative of that locale or ecoregion without any management or intervention from humans. Self-maintenance is important because unmanaged systems provide a variety of ecosystem services at no expense to human society. These services include macro- and microclimate regulation, transformation of at least some societal waste including sewage, storage of carbon that otherwise would become excessive in the atmosphere as carbon dioxide, and production of oxygen (to mention a few of many services). Just as disequilibrium conditions occur when a regulator disappears (such as a predator and the prey population exceeds its food supply or other necessary resources, often in the process creating substantial ecological damage), a complex multivariate system such as an ecosystem depends on a very complicated series of mutualistic, competitive, and other interactions that have evolved over long periods of time (e.g., Janzen 1988). Just as human society can be thrown abruptly into disequilibrium by an event such as a cessation of delivery of electricity, so can a complex ecosystem when one of the primary components (in this case, one or more species) is removed. When ecosystems are thrown into disequilibrium as a consequence of the situations just described, it is at least theoretically possible that human management might restore both structure and function through various techniques of restoration ecology. However, in some cases, restoration can be expensive and management may be necessary in perpetuity. As an example, Nonsuch Island is being restored to a semblance of its pre-settlement condition by David Wingate (1990). This is a complex situation that deserves more attention than is possible in this discussion, but the Wingate research is a good summary of the situation. Basically, one of the major problems is the fact that some of the lost species (e.g., the yellow-crown night heron) have disappeared from the Bermuda Islands and had to be imported from other localities where they still existed. One wonders whether the race of yellow-crown night heron inhabiting Nonsuch Island at one time and the imported race might not be physiologically or behaviorally distinct and, if so, what implication this might have. A more important problem is that even the rather heartening restoration that has been possible to date would probably disappear if Wingate and some of his staff did not continually remove the exotics that invade from other islands of the Bermuda system. This occurs because species of birds feed on the main island of Bermuda and many roost on Nonsuch, defecating seeds from exotic plants existing elsewhere in the island system. Some of these flourish on Nonsuch and would displace the native species if they were not continually removed. Either a temporary cessation of removal or a diminishing in intensity of removal could cause serious displacements of native species. This is a fine, but not unique, illustration of the dangers to ecosystems of loss of self-maintaining qualities through loss of species. In order to be truly effective, the entire Bermuda system would have to undergo the kinds of rehabilitation characteristic of Nonsuch, an event that seems unlikely at the moment. As a consequence, there is probably no terrestrial ecosystem similar to pre-settlement ecosystems in Bermuda that is capable of self-maintenance.

It is quite possible, in fact virtually certain, that some of the species lost will be key regulators. These could be predators that keep prey species populations under control by consuming them or competitors that keep another species under control by excluding them from certain territories of resources. Elimination of the control species then results in expansion of the prey population.

PROBABILITY OF DETECTING LOSS OF SPECIES

Structural changes in ecosystems are more likely to occur than functional changes in response to anthropogenic stress unless the changes are of an "atom bomb" nature. For those ecosystems with little or no functional redundancy, such as those in severe climates or ecosystems in which communities are greatly simplified due to other naturally occurring stresses (e.g., the pools in which brine shrimp occur that are so salty that not very many species can survive), the reduction in species richness is likely to be one of the first responses to anthropogenic stress. The question arises then of how likely will such changes be detected. Wilson (1988) has indicated that approximately 17,000 systematists are required just to complete an inventory of species in tropical rain forests. Persons with such skills are graduated by universities over night. Even more unsettling is the fact that systematists were quite common in university departments of biology, zoology, botany, and microbiology a generation ago, but some entire departments of biological sciences today may not have a single systematist on the faculty (that is, a faculty member may have taxonomic skills but not have primary research and publications in the field of systematics). Furthermore, research funding likely to lead to the graduation of additional systematists is exceedingly rare, and museums and other institutions where systematics normally flourish are having severe budgetary difficulties. Wilson's interesting volume *Biodiversity* (Wilson 1988) superbly describes the problem of estimating species loss on the planet. Only crude estimates of biotic impoverishment can be made because the planetary inventory of species is grossly inadequate. Estimates of the total number of species on the planet range from below 30 million to approximately 50 million. Contributors to Wilson's book express concern that species are being lost annually without society's knowledge because of an inadequate inventory coupled with inadequate surveillance and monitoring of global biodiversity. The National Academy of Sciences and the Smithsonian Institution, which co-sponsored the symposium that led to the book *Biodiversity*, represent mainstream science rather than highly emotional environmental activist fringe groups. If species richness and other structural attributes are sensitive indicators of anthropogenic stress, the inability to measure these, except in the grossest terms, severely hampers any capabilities of predicting and detecting ecosystem risk. Not only will the predictive models be weak because the evidence base is inadequate, but the validation of these predictive models will be severely hampered by the lack of adequate personnel to track species richness on a global basis.

The book *Biodiversity* focuses mostly on species that do not have a cosmopolitan distribution. Although not explicitly stated, presumably the reason for this is that species with a cosmopolitan distribution appear much more resistant to extinction than species that are found only in areas where appropriate ecological conditions exist, even if only on a temporary basis. Thus, the Hawaiian flora and fauna both have lost substantial numbers of species that are found only in the Hawaiian Islands, and a very significant number of species in this locale are endangered or seriously threatened. This story of endangered species with a limited distribution can be repeated over and over again throughout the planet. Even in large land masses, such as the United States, species have very, very limited distributions (such as the desert pupfish, which is likely to become extinct). The extinction of these species, although important for a particular locale, may not be important individually outside of their restricted habitat area. Cairns (1991a) speculates that species with a cosmopolitan distribution can also become extinct. The reason for this is that many species with a cosmopolitan distribution are, in a sense, perpetual fugitive species moving from one ecological island to another as conditions permit. In an earlier paper, Cairns (1991b) discusses the consequences of a cosmopolitan distribution. Briefly, to maintain a cosmopolitan distribution, no species or group of species should become established in a particular area so well that they can exclude other invading and potentially colonizing species. A second feature is that dissemules (any life history stage suitable for transport over considerable distances) must be adequate to ensure that when favorable ecological conditions develop in a particular patch, colonists will reach them. Finally, the article speculating on the possibility of extinction of species with a cosmopolitan distribution calls for an adequate number of ecological "stepping stones" so that all the species are maintained somewhere at all points in time and that the number of such "stepping stones" at any one point in time is sufficient to produce dissemules that will have a high probability of reaching other ecological patches as they develop temporarily in time and space. The significant aspect of this is that a critical number of "stepping stones" are necessary to maintain a cosmopolitan distribution, although this would certainly be extremely difficult and perhaps impossible to test experimentally. One body of experimental evidence shows the influence of distance from an epicenter (a source of colonizing species), but numerous other factors, such as the durability of the dissemules, length of time they can survive transport, and the like, are also important. Cairns (1991b) calls attention to the fact that some species, such as *Stokesia vernalis*, are reported only rarely. The infrequency of such reports is almost certainly influenced by the number of observers. While a precise figure is difficult to obtain, the number of persons capable of identifying protists in a complex, highly perishable community is relatively small. Furthermore, most of these persons have other primary activities and such identifications are not their primary activity. Thus, a species with a cosmopolitan distribution could become extinct, and decades could pass before this extinction was noted for reasons already described. If species with a cosmopolitan distribution play any significant ecological role (and it is difficult to imagine that they have existed for so long without developing a capability of doing so), the loss of a significant number of such species might have considerably greater ecological consequences globally than the loss of endemic species at a particular locale.

Functional attributes of ecosystems furnish important information about their condition (e.g., Cairns and Pratt 1986; Pratt and Cairns 1985). At one time, it seemed moderately probable that measuring functional attributes of complex multivariate systems might markedly reduce the need for taxonomic information. Although this probability should not be entirely excluded, its occurrence may be much less likely than believed previously. H. L. Mencken stated the situation succinctly: "For every complex problem there is a solution, simple, direct, and wrong." If systematists have a greater role in predicting ecosystem risk than they have had thus far, a no-net-loss policy, at least, should be implemented for professionals in this category. In doing so, it would be prudent to recognize the considerable educational and training period required before a systematist can function adequately in an ecosystem risk determination. Better still would be to produce enough systematists to furnish a robust estimate of how many species share the planet. Such an inventory is essential in the event that the United Nations or various member states decide to either reduce the rate of biotic impoverishment or establish a no-not-loss policy. Enough scientists would be available to provide evidence on whether the policy is a success or failure. At present, this is not possible.

THE POSSIBLE RELATIONSHIPS BETWEEN STRUCTURE AND FUNCTION

Historically, community ecologists strongly based in taxonomy and autecology have viewed systems as collections of species that interact. Under this view, the identity of the individual taxa carries the important information. Ecosystem ecologists view ecosystems as processors of matter and energy, and the construction of both qualitative and quantitative models of ecosystem dynamics has required amalgamating compartments of the ecosystem (that is, processing units). Under this view, system function is measurable and predictable given the flows between compartments (Fig. 2). Many investigators have married the concepts of structure and function by categorizing the roles that various species play in processing matter (Cummins and Klug 1979; Karr et al. 1986). An ecosystem ecologist might view the identity of the species comprising a particular compartment as a detail so that the loss of a rare species may be unimportant unless it has an impact on ecological function. As such, the loss of a rare species might be unimportant if the species is a member of a large array of taxa with equivalent function (e.g., diatom algae). On the other hand, the loss of an individual species such as a keystone predator might significantly alter ecological function. The application of community ecology to assessing environmental damage has relied on the use of taxonomic information, which is high in information content.

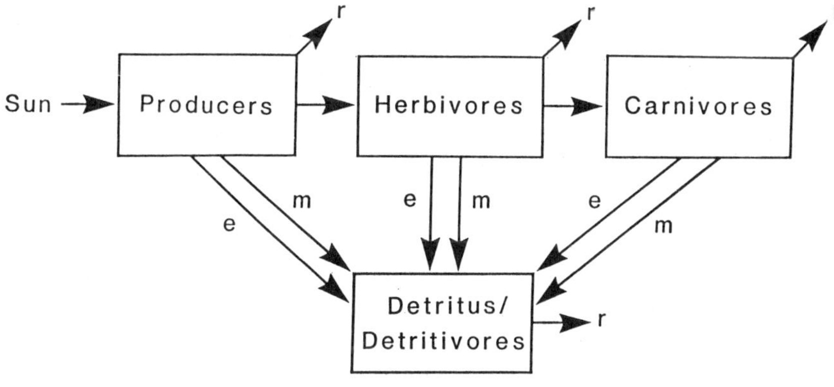

Figure 2. Standard compartment model of an autotrophically driven ecosystem.

The prediction of ecological risk is based on a presumption that ecologists are capable of measuring change above background noise. A debate continues about the effectiveness and sensitivity of various measures of ecological change. Simply stated, three extreme positions can be taken on the question of the sensitivity of structural and functional changes (Cairns and Pratt 1986). First, structure may change before function, that is, the structural complexities of communities allow taxonomic composition and dominance to maintain ecological function (Levine 1989; Fig. 1). Second, function may change before structure, that is, the broad tolerances of species for environmental conditions may allow them to adjust their physiological functioning (and, therefore, ecological function) without materially affecting community structure. Finally, structure and function may be so intimately linked that change in structure produces an immediate change in function and vice versa.

Resolving the structure-function debate is important because, from a practical viewpoint, time will be wasted in measuring ecological end points that have little probability of detecting effects and predicting risks. An immediate concern is statistical. Which measures have sufficiently low variability that they can reliably be used to detect a signal in response to some anthropogenic stress? Statistically-based decisions carry the weight of objectivity (but see Berger and Berry 1988) in deciding when an effect has occurred. However, there are two components to the detectability of stress responses. Some ecological measures may be inherently variable. Alternatively, sufficient precision in some measurements of detectability may be impossible to make (see Pratt and Bowers 1992; Green 1989). A further complication in field sampling is that common inferential statistics may be inappropriate to resolving a question of hypothesized stress (Hurlbert 1985).

PREDICTED EFFECTS OF STRESS

Several authors have predicted the effects of stress in ecosystems (Table 1), but few studies have quantitatively tested these predictions. The most notable whole ecosystem manipulations have been serious disruptions, such as the watershed manipulation studies at Hubbard Brook (New Hampshire; Bormann and Likens 1972) and similar studies at the Coweeta Hydrologic Laboratory (North Carolina; e.g., Johnson and Swank 1973). Similar large scale manipulations have been conducted in the Experimental Lakes Area (ELA) of northern Ontario, Canada (Schindler 1987). Because these studies are large scale manipulations that require significant long-term commitments for sampling, they usually lack replication.

Recent reports reviewing examples of stress effects (Barrett and Rosenberg 1980; Levin et al. 1989; Sheehan et al. 1984) provide little evidence of the sensitivity of ecological variables to stress. The extensive review by Sheehan and colleagues provides evidence of ecosystem stress effects but little guidance for the selection of sensitive measures. Significant resources are spent

TABLE 1
Summary of several authors' predictions of ecosystem responses to stress

Effects on	Odum (1985)	Schindler (1987)	Schaeffer et al. (1988)
Structures	*r*-strategists increase Size of organisms decreases Life span of organisms decreases Food chains shorten Species diversity decreases and dominance increases	The most vulnerable species are organisms with short life cycles Sensitive indicators include life-table parameters, short-lived species, and changing community structure	Failing numbers of native species Changing standing crop biomass Overall regressive succession Changes in mineral macro-nutrient stocks
Processes	Community respiration increases P/R<>1 P/B, R/B increase Exported primary production increases Nutrient turnover increases Nutrient loss increases	Decomposition is a function of primary production and is unaffected by pollutants Nutrient cycling is not disrupted by toxicants	Changing gross or net primary production Change relative energy flow to grazing and decomposer food chains Changes in mechanisms of, and capacity for, dampening undesirable oscillations

measuring chemical contaminants and physical stresses such as changes and hydrologic regime; however, these measures fail to identify linkages between stressors and ecosystem damage. Many measures are available, and some ecologists have expressed hope that particular measures can be used to evaluate ecosystem health in a manner similar to that used in medical diagnosis (Schaeffer et al. 1988). Continuing improvement in the use of standard methods in ecological analysis is needed, and linkages must be drawn between particular measures and the mechanisms of stress in natural communities.

An alternative to the measurement of ecological structure and process in natural ecosystems is the use of artificial ecosystems such as microcosms and mesocosms (Giesy 1980; Odum 1985). Such studies have the advantage of replication and experimental control and should be viewed as optimum conditions under which to measure the variability and sensitivity of ecological measures. Only recently have investigators considered the statistical implications of measurement variability in the design of experiments for detecting the effects of stress (Conquest 1983; Green 1989, Pratt and Bowers 1992).

Population level approaches traditionally used in ecotoxicology are capable of estimating the levels of chemical stress that produce adverse effects in biological material; however, testing focused on single species cannot predict the kinds of effects that would be expected at the community and ecosystem level (Cairns 1983). Single species toxicity tests can overestimate and underestimate chemical toxicity and, therefore, ecological effects. Effects are underestimated when a pollutant bioaccumulates or when a pollutant eliminates the key species. Effects are overestimated when laboratory conditions fail to realistically simulate the fate of chemicals in natural systems. For example, many potentially toxic organic chemicals are subject to microbial degradation, but clean laboratory systems lack the degradative microflora. Such problems are further compounded when degradation products are also toxic, so that estimates of disappearance of apparent compound and degradation tests cannot be melded effectively with toxicity test results in predicting the environmental effects of a particular compound.

Consequences of Species Loss

Several examples are presented here that examine the effects of stress on biotic diversity. Many of the examples are based on results of laboratory microcosm experiments using natural microbial communities that share many of the characteristics of ecosystems. These experiments were conducted following a standard protocol (Pratt and Bowers 1990), and the results are used to compare adverse ecological effects on process and structure.

Ecosystem Structure

Changes in standing stock and biotic composition have long been regarded as the primary evidence of environmental impact. Sensitive individuals and species are eliminated by the stress and do not return until stress is removed (assuming a species source is available for recolonization). Under toxic stress, surviving species may exhibit broad tolerance, rapid reproduction, or resistance to the toxicant. In each of these cases, evidence of stress would be observed in species composition changes from the unstressed condition and, usually, the number of species would be reduced. Such changes may lead to alterations in community function, the shortening of food chains, or to a reversal of normal successional trends. The reversal of successional trends includes community regression to dominance by small taxa with high reproductive rates (Cairns et al. 1980).

Reduced species richness and altered standing crop are common results of stress. For example, in laboratory microcosms, the number and kinds of

TABLE 2
Sensitivity of measured responses to toxic effects of selected chemicals in laboratory microcosms. Table values are lowest observed effect concentrations (LOECs), the lowest tested toxicant concentration producing a response significantly different from the control; USEPA criterion levels are shown for a hardness of 100 ppm

	Toxicant Concentration, μg/L			
Variable	Copper	Zinc	Atrazine	Chlorine
Species richness	9.9	89	3.2	6.1
Biomass				
Protein	9.9	4.2	3.2	100
Chlorophyll	19.9	10.7	3.2	2.1
Net primary production	90	4.2	3.2	25
APA[1]	NS	89	100	6.1
USEPA criterion	12	110	—	11

[1]Alkaline phosphatase activity.

species change quickly with response to stress. Changes in biological structure include changes in standing crop, changes in the number of species, and changes in community composition (e.g., Tables 2, 3). Historically, many species were deemed valuable as indicators of environmental conditions (Kolkwitz and Marsson 1909). More recently, the importance of changes in community structure has been given added meaning by the development of inferential methods for detecting changes in community similarity (Smith et al. 1990). These newer methods improve upon descriptive techniques such as cluster analysis and ordination.

TABLE 3
Responses showing sensitivity to toxicant action in laboratory ecosystems. Responses are listed in order of relative sensitivity. The number of responses is the number of experiments showing any significant response to a toxic chemical. The median coefficient of variation (CV) for each measurement is also given

Variable	# Of Tests	# Of Responses	%	CV%
Species richness	14	11	79	9.6
Chlorophyll biomass	12	10	83	16-27
Dissolved oxygen	13	6	46	4.53
APA[1]	11	4	36	18.4
pH	11	3	27	2.22
Protein biomass	12	3	25	23.6

[1]Alkaline phosphatase activity.

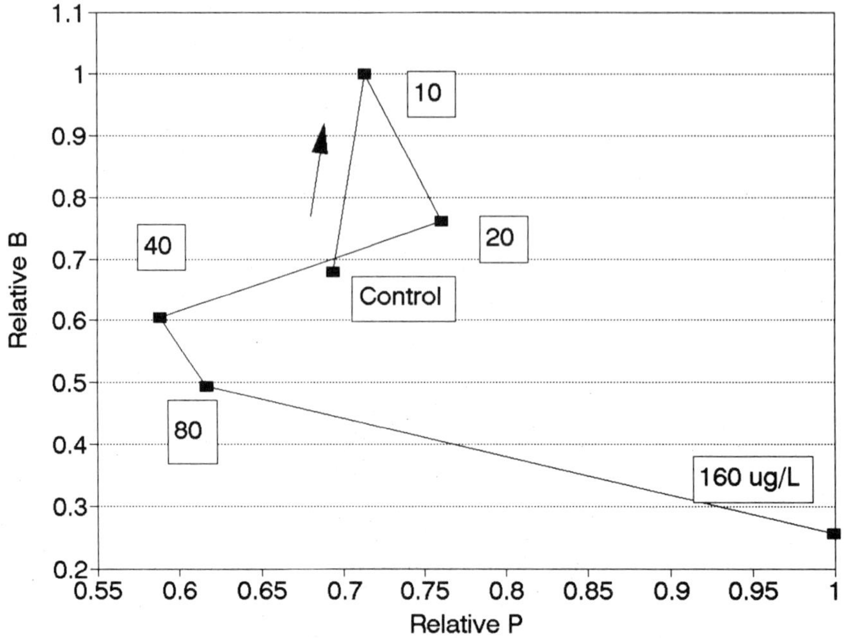

Figure 3. Relative biomass (B) and primary production (P) in microcosms dosed with copper. Plotted data show the trajectory of B and P change. Copper concentrations are shown adjacent to each point. The scale (0-1) is relative to the largest B or P observed in all treatments.

Field evaluations of stress effects have focused on changes in biological structure. In aquatic ecosytems, changes in species richness and composition often precede the ability to detect changes using nontaxonomic measures of structures and processes (Fig. 3), even when levels of stress are extremely high.

Ecosystem Processes
Important ecological processes such as primary production, community respiration, and nutrient cycling are expected to change under stress. For example, Odum (1985) hypothesized that increased maintenance (respiration) would occur under stress and that ratios of production to biomass (P/B) and respiration to biomass (R/B) would increase because of imbalances in P/R. Such energetic changes are measurable in laboratory ecosystems at environmentally realistic toxicant levels, probably because confounding factors such as physical changes in the natural environment are less problematic. For example, diurnal cycles of pH or oxygen in microcosms have been shown to be related to stress (Franco et al. 1984; Shannon et al. 1986; Stay et al. 1985; van Voris et al. 1978). Microcosm experiments on copper effects showed alteration of diurnal photosynthesis patterns (Fig. 4) with increasing stress. However, in lake phytoplankton

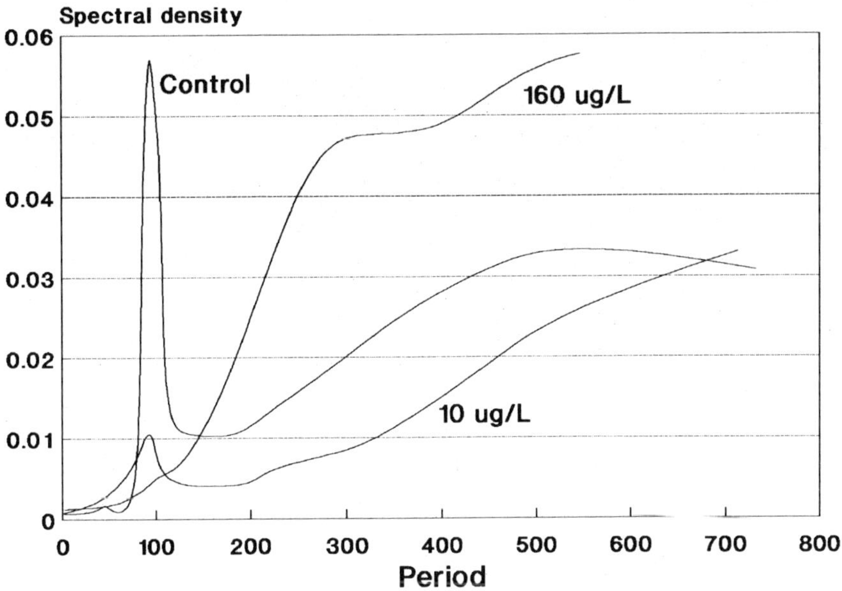

Figure 4. Changes in functional complexity measured as the power spectrum of pH recorded every 15 min in microcosms dosed with copper.

stressed with copper, P/B increased with increasing stress (Wurtsbaugh and Horne 1982). In both cases, production in standing crop decreased with increasing toxic dose, so changes in individual measures may be more meaningful than the ratio of production to biomass.

In field experiments, de Noyelles and Kettle (1985) observed shifts in production in replicated ponds. We have observed periphyton P/B changes in replicated streams with selenium (Fig. 5; Pratt and Bowers 1990). Low concentrations of selenium stimulated algal growth. However, Schindler (1987) reports no changes in planktonic production energetics and experimental lakes either under nutrient or acid stress. Similarly, experiments on chlorpyrifos in outdoor streams showed no evidence of energetic shifts even though invertebrate communities were significantly altered (Eaton et al. 1986).

Toxicants affect the ability of ecosystems to retain and efficiently process nutrients. For example, in response to added zinc (Pratt et al. 1987), microbial communities lost phosphate and showed large increases in alkaline phosphatase activity (a measure of organic phosphorus recovery; Fig. 6). Odum et al. (1979) call such ecosystems "leaky," and evidence of the lack of macronutrient retention clearly demonstrates stress effects on nutrient cycles. Related effects of a toxicant (copper) have been demonstrated by Taub et al. (1986) for nitrate assimilation.

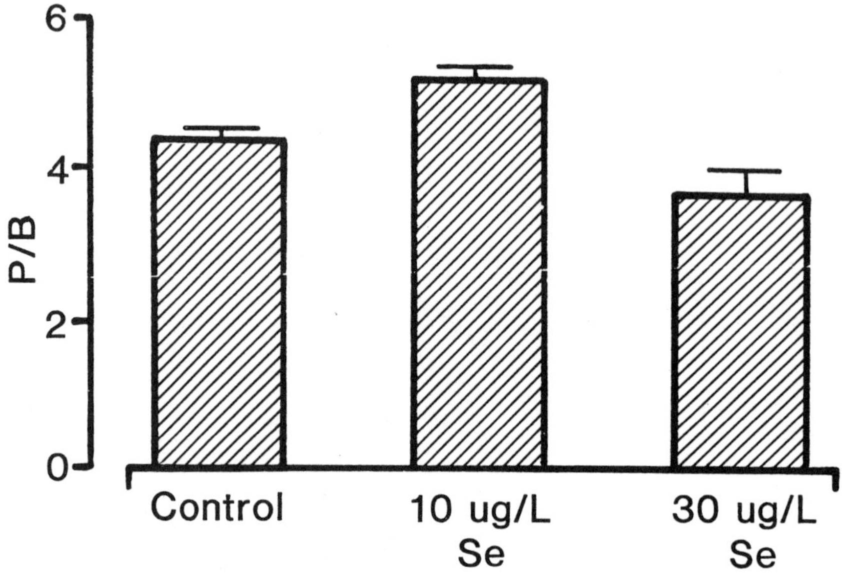

Figure 5. Changes in primary production (P) to biomass (B) ratios in outdoor stream mesocosms dosed with selenium.

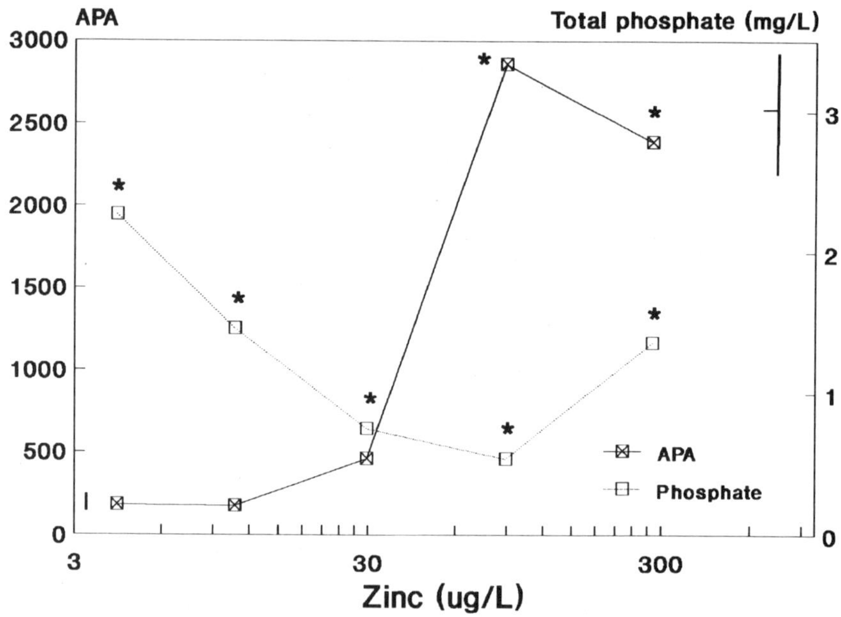

Figure 6. Effect of zinc on total phosphate and alkaline phosphatase activity (APA) in laboratory microcosms. APA units are nmoles p-nitrophenol/mg protein/hr.

Field experiments do not regularly show effects of toxicants on nutrient cycles unless perturbations are great. Schindler (1987) reports that neither added nutrients nor acidification affected nitrogen or phosphorus cycling. In a recent study of impacts of sewage wastewater containing ammonia and chlorine, we observed no effect on alkaline phosphatase activity despite readily observable alterations in community structure (Fig. 3). As noted above, Eaton et al. (1986) observed no effects on microbial nutrient cycling under chlorpyrifos stress, even though the most vulnerable organisms (invertebrates) were severely affected.

ECOSYSTEM LEVEL TRENDS

Results of some experiments have been used to evaluate the nature and sequence of changes that occur in model ecosystems stressed with toxic chemicals. A variety of community structure and process measurements were made. Additionally, summary data on the statistical properties of various measures examined the power of various measures to detect changes in ecological structure and function.

Effects of Heavy Metals
A variety of metals appear in industrial and municipal waste streams and have the ability to adversely impact aquatic ecosystems. Unlike many organic contaminants that may be specific biocidal compounds, heavy metals often have multiple modes of action affecting cytosolic and membrane bound enzyme systems and thereby producing numerous effects at a variety of concentrations. The experiments reported here are experiments using freshwater communities exposed to levels of metals that are expected to be chronically toxic to standard test species (fishes and invertebrates).

Copper is used as a biocidal compound for controlling algae and snails, and it also appears in waste streams from copper-forming point sources and from municipal waste discharges that contain substantial inputs from copper and brass plumbing fixtures. The most immediate effects of copper on communities is a severe reduction in species richness (Fig. 7). Additionally, species composition changes significantly (Table 4) as copper effects are manifested by the removal of species. Significant effects also occur on standing crop biomass as evidenced by severe reductions in chlorophyll and protein content of communities (Fig. 8). Effects on community structure measures are repeatable using communities from different ecosystems (Table 5), suggesting that copper is broadly toxic to species from a variety of ecosystems.

Effects on process measures are more problematic even though copper is expected to adversely affect important enzyme systems. For example, while copper effects are discernible as changes in the activity of alkaline phosphatase enzyme systems and in diurnal patterns of production and respiration, effects are most pronounced at higher levels of stress than levels that affected community structure.

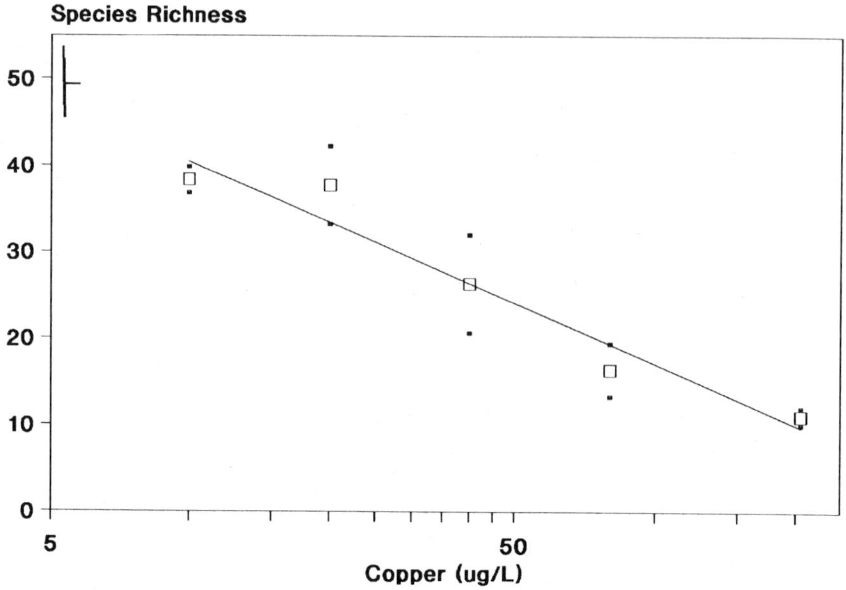

Figure 7. Effect of copper on species richness in laboratory microcosms.

TABLE 4
Results of similarity analysis of copper treatments within treatment similarity based on Jaccard's measure. Lambda (λ) values indicate the degree of difference between controls and treatments. Mean within and between similarity and p-values are based on randomized permutations of similarity coefficients. Species listed increased similarity among treatments when removed

Treatment	Algae Within Treatment Similarity	λ	Protozoa Within Treatment Similarity	λ
Control	0.59	1.00	0.43	1.00
10	0.48	0.93	0.37	0.81
20	0.63	0.81	0.48	0.70
40	0.67	0.76	0.34	0.76
80	0.70	0.77	0.33	0.44
160	0.66	0.83	0.27	0.35
Mean within similarity	0.62		0.37	
Mean between similarity	0.56		0.22	
P	<0.001		<0.001	
Indicator Algae			Indicator Protozoa	
Achnanthes sp.			*Balladyna elongatum*	
Cyclotella sp.			*Chlorogonium elongatum*	
Neidium binode			*Chlorogonium* sp.	
Nitzschia amphibia			*Chroomonas caudata*	
Surirella sp.			*Loxophyllum* sp.	

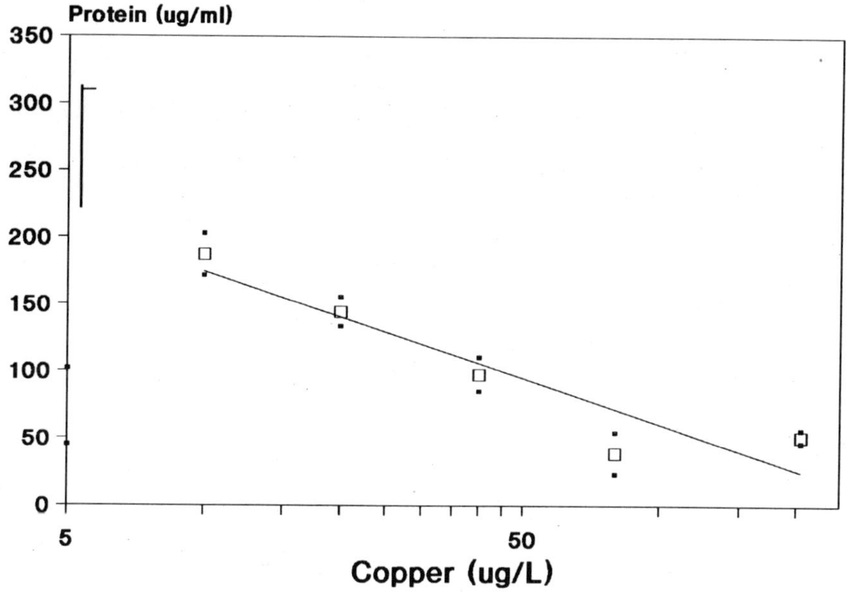

Figure 8. Effect of copper on standing crop biomass in laboratory microcosms.

TABLE 5
Summary of copper toxicity by response class in microcosms. Table values are μg Cu/L. Letters in parentheses identify the most sensitive variable in each response class: P = Protein, C = Chlorophyll *a*, A = Alkaline phosphatase activity, DO = Dissolved oxygen, ATP = Adenosine triphosphate biomass. The water quality criteria are calculated for dilution water hardness. ND = Not done

Test/ Date	WQC	MATC or LOEC Species Richness	Biomass	Function
Pennsylvania				
Nov88	19.5	19.9	9.9 (P)	40 (DO)
Nov89	19.5	ND	113 (C)	40 (DO)
Feb90	19.5	20[1]	73.3 (C)	N.S.
Apr90	19.5	9.1[1]	53.9 (C)	25.8 (DO,A)
Michigan				
Aug89[2]	15.7	18.9	82.1 (C)	18.9 (DO)
May90[3]	19.5	66.2	12.1[1] (C)	17.1 (A)
Aug90[3]	19.5	95.8	95.8 (C)	N.S.
Maryland				
Mar89	22.1	13.9	6.8 (P,C)	4.7[1] (DO)
Virginia				
Feb86	8.2	9.2	6.6 (C,ATP)	ND

[1]LOEC.
[2]Test conducted with MI dilution water.
[3]Test conducted with Pennsylvania State University dilution water.

Like copper, zinc is a ubiquitous waste material. Zinc appears in many industrial processes ranging from galvanizing to trace levels that appear in coal fly ash. Zinc effects include reduction in species number and significant changes in standing crop as measured by total phosphate, chlorophyll (as fluorescence), and dry weight contents of communities (Table 6). These effects occur at levels well below water quality criteria derived from invertebrate and fish testing.

The effects of zinc on process measures are delayed in terms of zinc concentration when compared to changes in community structure. For example, microcosms dosed with zinc lost phosphate as communities simplified, but the effects of zinc on the alkaline phosphatase enzyme system that recovers organic phosphates in nutrient limited communities increased dramatically only at the highest zinc doses. That is, structural effects were established before detectable changes in community function occurred. In fact, because the expression of alkaline phosphatase activity is based on protein biomass, it should be clear that gross activity (the quotient of alkaline phosphatase activity and protein biomass) is comparatively constant in the face of significant stress (Fig. 6). Similarly, effects on net production as measured by pH or dissolved oxygen changes in microcosms are less affected than are structural variables.

Biocides

Chlorine is a widely used disinfectant that is added to virtually every point source discharge to streams. The oxidizing powers of hypochlorite ion are utilized to remove bacteria from wastewaters. The effects of chlorine on

TABLE 6

Nontaxonomic responses of microbial communities to zinc. *In vivo* fluor = *in vivo* estimate of chlorophyll fluorescence; FU = fluorometric units

Treatment (μg Zn/L)	Total Phosphate (mg/L)	Dry Weight (mg/L)	*In Vivo* Fluor (FU)
Control	3.06	179.7	507
(<2.0)	(0.42)	(39.5)	(38)
4.2	2.27[1]	139.3[1]	477
	(0.30)	(12.3)	(49)
10.7	1.46[1]	89.7[1]	290[1]
	(0.26)	(10.7)	(46)
29.8	0.75[1]	36.7[1]	88[1]
(0.12)	(2.3)	(12)	
89.2	0.54[1]	25.3[1]	18[1]
	(0.08)	(2.3)	(4)
279.8	1.36[1]	39.0[1]	12[1]
	(0.58)	(8.2)	(2)

[1]Significantly different from control at $p < 0.05$. Data are means (standard deviation).

microbial communities might be expected to be substantial since most cells are exposed directly to the toxic material, and organic material is vulnerable to oxidation. Chlorine affects community structure significantly by limiting the production of biomass, decreasing the richness of communities, and altering community structure at levels of chlorine exposure comparable to or less than levels that adversely affect fishes (USEPA 1986; Table 7). Again, process measures are less effective in detecting the effects of chlorine (Table 8). For example, even though oxygen concentrations drop in microcosms in response to chlorine toxicity, these changes are detectable at levels no greater than those affecting community structure.

Atrazine is a specific photosynthetic inhibitor that affects the passage of electrons from the Hill reaction to photosystem II. The effects of atrazine on community structure are more striking than effects of many other toxicants. At low levels of atrazine, community species richness actually increases (Table 9) and community biomass also changes. Communities stressed with atrazine elaborate chlorophyll biomass, presumably as a response to the limited effectiveness of native chlorophyll. Concomitant with these effects, community biomass as measured by protein also increases, but processes are unaffected. This effect is particularly striking because communities actually have more photosynthetic material (chlorophyll) but are unable to produce significant dissolved oxygen gains because of the activity of atrazine. As noted above, the effects on community structure precede other changes in functional capacity such as the ability to transport organic phosphorus. Alkaline phosphatase activity increases dramatically but only after significant stress has already occurred.

TABLE 7
Effects of chlorine on structure of microbial communities in laboratory microcosms. Seq = equilibrium species number

Treatment	Seq	Chl a (mg/L)	ATP Biomass (mg/L)	Protein (mg/L)
Control	40.7	10.2	300.7	129.3
	(2.2)	(4.0)	(63.7)	(5.0)
2.1 μg/L	39.1	5.3*[1]	217.3	111.3
	(2.3)	(1.9)	(102.0)	(14.3)
6.1	31.0*	7.4	301.0	159.2
	(2.5)	(1.2)	(64.0)	(20.4)
25	30.4*	4.0*	228.7	120.5
	(3.7)	(1.2)	(83.1)	(8.8)
100	25.7*	3.5*	97.7	43.0*
	(5.2)	(2.3)	(35.1)	(15.4)
308	18.7	0.6*	22.7*	12.0*
	(19.0)	(0.3)	(12.6)	(8.0)

[1]Means significantly different from control ($p < 0.05$) are marked with an asterisk. Data are means of triplicates (standard deviation).

TABLE 8

Effects of chlorine on processes in laboratory microcosms. Data are means of triplicates (standard deviation). Means significantly different from control are marked with an asterisk. The units for alkaline phosphatase activity (APA) are nmol p-nitrophenol/mg protein/hr

Treatment	APA	Dissolved Oxygen
Control	1157 (301)	11.7 (0.30)
2.1 μg/L	1218 (185)	11.5 (0.17)
6.1	323* (83)	11.6 (0.20)
25	186* (89)	11.1* (0.06)
100	67* (11)	10.2 (0.00)
308	Trace[1]	10.1* (0.06)

[1]Activity was detected but was below the lowest standard.

TABLE 9

Effects of atrazine on protozoan colonization estimates based on MacArthur-Wilson model. Estimates of equilibrium species number and colonization rate based on fit to model equation using nonlinear regression. Numbers in parentheses are asymptotic standard errors. Values significantly different from control are marked with an asterisk

Treatment	Seq[1]	G[2]
Control	33.2 (2.34)	0.349 (0.104)
Solvent control	34.4 (1.92)	0.309 (0.068)
3.2	42.4* (2.65)	0.167* (0.030)
10.0	52.6* (5.06)	0.122* (0.028)
32.0	61.7* (12.26)	0.081* (0.031)
110	34.8 (3.05)	0.138* (0.031)
337	24.1* (1.18)	0.266 (0.048)

[1]Equilibrium species number.
[2]Colonization rate.

Statistical Power
Experiments with laboratory ecosystems have demonstrated changes in community structure that appear to occur earlier and are more likely to be detected than in system level processes. In summarizing a number of experiments, it became clear that the variability associated with enumerating the number of taxa was generally lower than most (but not all) other measures that had been used. In fact, we endeavored to find nontaxonomic measures that had acceptable prediction and detection power so that some time could be saved in conducting experiments. The number of manhours required to enumerate taxa is significantly greater (by a factor of 2 or 3) than those required to measure a suite of other variables. Two factors seem to account for the importance of measures of community structure (taxonomy, standing crop) when compared to other measures. First, the measurement variability associated with counting is usually low. Measurement variability associated with extracting and determining protein levels were lower than those for chlorophyll determination. These measures are less variable than measures of process where procedures are less well developed and standardized (Table 3). That is, the effectiveness of community structure may be solely based on statistical properties of the measures: measures with low variability are more likely to detect effects than those with greater variability. A second important factor to be considered is that some measures do not respond to toxic stress. For example, measurement of dissolved oxygen levels or pH also produces comparatively invariant results when compared with other process measures such as activities of nutrient transporting enzymes (Table 3), but these measures do not show consistent abilities to detect adverse effects even at high levels of stress (Table 3). This second factor is more biologically interesting than the statistical properties of the measures. Failure to detect changes in oxygen content, net primary production, or community respiration suggests that these processes and other processes such as enzyme activities of whole communities do not change as sensitively as does the taxonomic composition. This is surprising because changes in standing crop biomass should be reflected in changes in community primary production or respiration. That is, if biomass is being eliminated and needs to be decomposed, oxygen levels might be expected to fall. Alternatively, if biomass is not being formed, then oxygen levels might not increase. In either case, oxygen levels in the laboratory ecosystem should be sensitive to these community changes, but they are not.

Community processes seem to be maintained in the face of significant stress. A coherent explanation of the reason for this failure to detect process level change is provided by Levine (1989). She notes that most processes are substrate limited, and the effects of toxic materials rarely altered the supply of substrates. The changing community composition continues to exploit substrates in limited supply (for example alkaline phosphatase activity) and maintains rates of process in the face of significantly changing taxonomic composition. In other words, most species are not operating at full capacity because they are limited by some limiting nutrient. As species are eliminated from the community, competition for the limiting nutrient(s) is reduced and the functional capacity of the remaining taxa is increased. The net effect of these changes is the maintenance of community function in the face of significant changes in the standing crop and composition of the community. In other words, as Odum (1990) predicted, functional measures are robust.

In summary, structural measures of community change are more effective indicators of stress than process measures. The loss of species and the reductions in community standing crop have consequences that are only measurable at greater degrees of stress because the resistant portion of the community that remains is capable of exploiting the additional available resources now being used by fewer competitors. Coupled with this underlying biological explanation is the strictly statistical phenomenon that counting species and measuring biomass has comparatively low variability, and so the potential for detecting community change is improved by reliance on structural measures.

Our ability to detect changes in composition that are not coupled to the loss of species is more problematic. Methods are only now becoming available (e.g., Pratt and Smith 1991; Smith et al. 1990) that allow investigators to measure using inferential techniques. The changes in community similarity or dissimilarity are measured by common indexes such as coefficients of community or distances in multidimensional space.

The Relation Between Structure and Function

The functional composition of communities can be estimated by understanding the underlying biology of the component species. For example, a variety of functional group classifications allow investigators to infer from the taxonomic composition of communities and the importance of energy sources affecting community structure. These relationships are clearest for freshwater invertebrate communities (e.g., Cummins and Klug 1979; Merritt and Cummins 1984). Additional regional analyses are important in understanding the factors that structure communities and allow investigators to infer community and ecosystem function. For example, Karr and colleagues (1986) developed a structural and functional classification of fishes that must be adapted to ecoregional differences (see Plafkin et al. 1989; Whittier et al. 1988). We classified protozoans according to functional feeding groups (Pratt and Cairns 1985) and found that communities affected by a variety of ecological stresses shifted into dominance of various energy gathering modes. Therefore, it is possible to infer function from structure, while the reverse is unlikely to be true.

REFERENCES

BARRETT, G.W. and ROSENBERG, R. (1980). Stress Effects on Natural Ecosystems. John Wiley & Sons, NY. 305 pp.

BERGER, J.O. and BERRY, D.A. (1988). Statistical analysis and the illusion of objectivity. Am. Sci. 76:159-165.

BORMANN, F.H. and LIKENS, G.E. (1972). The nutrient cycles of an ecosystem. Sci. Am. 220:92-101.

CAIRNS, J., JR. (1983). The case for simultaneous testing at different levels of biological organization. In: *Aquatic Toxicology and Hazard Assessment: Sixth Symposium*, STP802 (W.E. Bishop, R.D. Cardwell, and B.B. Heidolph, eds), pp. 111-127. American Society for Testing and Materials, Philadelphia, PA.

CAIRNS, J., JR. (1991a). Can microbial species with a cosmopolitan distribution become extinct? Spec. Sci. Tech.

CAIRNS, J., JR. (1991b). Probable consequences of a cosmopolitan distribution. Spec. Sci. Tech. 14(1):41-50.

CAIRNS, J., JR. and PRATT, J.R. (1986). On the relation between structural and functional analyses of ecosystems. Environ. Toxicol. Chem. 5:785-786.

CAIRNS, J., JR., HART, K.M., and HENEBRY, M.S. (1980). The effects of a sublethal dose of copper sulfate on the colonization rate of freshwater protozoan communities. Am. Midl. Nat. 104:93-101.

CONQUEST, L.L. (1983). Assessing the statistical effectiveness of ecological experiments: Utility of the coefficient of variation. Int. J. Environ. Stud. 20:209-221.

CUMMINS, K.W. and KLUG, M.G. (1979). Feeding ecology of stream invertebrates. Ann. Rev. Ecol. Syst. 10:147-172.

DE NOYELLES, F., JR. and KETTLE, W.D. (1985). Experimental ponds for evaluating the bioassay predictions. In: *Validation and Predictability of Laboratory Methods for the Assessment of the Fate and Effects of Environmental Contaminants in Aquatic Ecosystems*, STP865 (T.P. Boyle, ed), pp. 91-103. American Society for Testing and Materials, Philadelphia, PA.

EATON, J., ARTHUR, J., HERMANUTZ, R., KIEFER, R., MUELLER, L., ANDERSON, R., ERICKSON, R., NORDLING, B., RODGERS, J., and PRITCHARD, H. (1986). Biological Effects of Dursban in Outdoor Experimental Streams, pp. 1-58. U.S. Environmental Protection Agency, Washington, DC.

FRANCO, P.J., GIDDINGS, J.M., HERBES, S.E., HOOK, L.A., NEWBOLD, J.D., ROY, W.K., SOUTHWORTH, G.R., and STEWART A.J. (1984). Effects of chronic exposure to coal-derived oil on freshwater ecosystems: I. microcosms. Environ. Toxicol. Chem. 3:447-463.

GIESY, J.P. (1980). Microcosms in Ecological Research. National Technical Information Center, Springfield, VA. 1110 pp.

GREEN, R.H. (1989). Power analysis and practical strategies for environmental monitoring. Environ. Res. 50:195-205.

HURLBERT, S.H. (1985). Pseudoreplication and the design of ecological field experiments. Ecol. Monogr. 54:187-211.

JANZEN, D.H. (1988). Guanacaste National Park: Tropical, ecological, and cultural restoration. In: *Rehabilitation of Damaged Ecosystems* (J. Cairns, Jr., ed), pp. 143-192. CRC Press, Boca Raton, FL.

JOHNSON, P.L. and SWANK, W.T. (1973). Studies of cation budgets in the Southern Appalachians on four experimental watersheds of contrasting vegetation. Ecology 54:70-80.

KARR, J.R., FAUSCH, K.D., ANGERMEIER, P.L., YANT, P.R., and SCHLOSSER, I.J. (1986). Assessing biological integrity in running waters: A method and its rationale, pp. 1-28, Sp. Publ. 5. Illinois Natural History Survey, Urbana, IL.

KOLKWITZ, R. and MARSSON, M. (1909). Ökologie der tierischen Saprobien. Beiträge zur Lehre von des biologischen Gewasserbeurteilung. Int. Rev. gesamten Hydrobiol. Hydrograph. 2:126-152.

LEVIN, S.A., HARWELL, M.A., KELLY, J.R., and KIMBALL, K.D. (1989). Ecotoxicology, Problems and Approaches. Springer-Verlag, NY. 547 pp.

LEVINE, S.N. (1989). Theoretical and methodological reasons for variability in the responses of aquatic ecosystem processes to chemical stresses. In: *Ecotoxicology: Problems and Approaches* (S.A. Levin, M.A. Harwell, J.R. Kelly, and K.D. Kimball, eds), pp. 145-180. Springer-Verlag, NY.

MERRITT, R.W. and CUMMINS, K.W. (1984). An Introduction to the Aquatic Insects of North America, 2nd. Ed. Kendall/Hunt Publishing Company, Dubuque, IA. 722 pp.

ODUM, E.P. (1985). Trends expected in stressed ecosystems. BioScience 35:419-422.

ODUM, E.P. (1990). Field experimental tests of ecosystem-level hypotheses. TREE 5:204-205.

ODUM, E.P., FINN, J.T., and FRANZ, E.H. (1979). Perturbation theory and the subsidy-stress gradient. BioScience 29:349-352.

PATRICK, R. (1971). Diatom communities. In: *The Structure and Function of Fresh-water Microbial Communities* (J. Cairns, Jr., ed), pp. 151-164. Research Monograph 3, Virginia Polytechnic Institute and State University, Blacksburg, VA.

PLAFKIN, J.L., BARBOUR, M.T., PORTER, K.D., GROSS, S.K., and HUGHES, R.M. (1989). Rapid Bioassessment Protocols for Use in Streams and Rivers: Benthic Macroinvertebrates and Fish, EPA/444/4-89-001. U.S. Environmental Protection Agency, Assessment and Watershed Protection Division, Washington, DC.

PRATT, J.R. and BOWERS, N.J. (1990). A microcosm procedure for estimating ecological effects of chemicals and mixtures. Tox. Assess. 5:189-205.

PRATT, J.R. and BOWERS, N.J. (1992). Variability of community metrics: Detecting changes in structure and function. Environ. Toxicol. Chem. 11:451-457.

PRATT, J.R. and CAIRNS, J., JR. (1985). Functional groups in the protozoa: Roles in differing ecosystems. J. Protozool. 32:415-423.

PRATT, J.R. and SMITH, E.P. (1991). Significance of change in community structure: A new method for testing differences. In: *Biological Criteria: Research and Regulation*, EPA 440/5-91-005, pp. 1-14. Office of Water, U.S. Environmental Protection Agency, Washington, DC.

PRATT, J.R., NIEDERLEHNER, B.R., BOWERS, N.J., and CAIRNS, J., JR. (1987). Effects of zinc on freshwater microbial communities. In: *International Conference on Heavy Metals in the Environment*, Vol. 2 (S.E. Lindberg and T.S. Hutchinson, eds), pp. 324-326. CEP Consultants Ltd., Edinburgh.

SCHAEFFER, D., HERRICKS, E.E., and KERSTER, H.W. (1988). Ecosystem health I: Measuring ecosystem health. Environ. Manage. 12:445-455.

SCHINDLER, D.W. (1987). Detecting ecosystem response to anthropogenic stress. Can. J. Fish. Aquat. Sci. 44(Suppl. 1):6-25.

SHANNON, L.J., HARASS, M.C., YOUNT, J.D., and WALBRIDGE, C.T. (1986). A comparison of mixed flask culture and standardized laboratory model ecosystems for toxicity testing. In: *Community Toxicity Testing*, STP920 (J. Cairns, Jr., ed), pp. 135-157. American Society for Testing and Materials, Philadelphia, PA.

SHEEHAN, P.J., MILLER, D.R., BUTLER, G.C., and BOURDEA, P. (1984). Effects of Pollutants at the Ecosystem Level. John Wiley & Sons, NY. 443 pp.

SMITH, E.P., PONTACSH, K.W., and CAIRNS, J., JR. (1990). Community similarity and the analysis of multispecies environmental data: A unified statistical approach. Water Res. 24:507-514.

STAY, F.S., LARSEN, D.P., KATKO, A., and ROHM, C.M. (1985). Effects of atrazine on community-level responses in Taub microcosms. In: *Validation and Prediction of Laboratory Methods for Assessing the Fate and Effect of Contaminants in Aquatic Ecosystems*, STP 865 (T.P. Boyle, ed), pp. 75-90. American Society for Testing and Materials, Philadelphia, PA.

TAUB, F.B., KINDIG, A.C., and CONQUEST, L.L. (1986). Preliminary results of interlaboratory testing of a standardized aquatic microcosm. In: *Community Toxicity Testing*, STP920 (J. Cairns, J., Jr., ed), pp. 93-120. American Society for Testing and Materials, Philadelphia, PA.

U.S. ENVIRONMENTAL PROTECTION AGENCY. (1986). Quality Criteria for Water, EPA 440/5-86-001. Washington, DC.

VAN VORIS, P., O'NEILL, R.V., SHUGART, H.H., and EMANUEL, W.R. (1978). Functional Complexity and Ecosystems Stability: An Experimental Approach. Environmental Sciences Division Publication No. 1123, Oak Ridge National Laboratory, Oak Ridge, TN. 120 pp.

WHITTIER, T.R., HUGHES, R.M., and LARSEN, D.P. (1988). Correspondence between ecoregions and spatial patterns in stream ecosystems in Oregon. Can. J. Fish. Aquat. Sci. 45:1264-1278.

WILSON, E.O. (1988). Biodiversity. National Academy Press, Washington, DC. 521 pp.

WINGATE, D.B. (1990). The restoration of Nonsuch Island as a living museum of Bermuda's precolonial terrestrial biome. In: *The Earth in Transition: Patterns and Processes of Biotic Impoverishment* (G.M. Woodwell, ed), pp. 133-150. Cambridge University Press, Cambridge.

WURTSBAUGH, W.A. and HORNE, A.J. (1982). Effects of copper on nitrogen fixation and growth of blue-green algae in natural plankton associations. Can. J. Fish. Aquat. Sci. 39:1636-1641.

CHAPTER 9

RISKS OF ALTERING THE HYDROLOGIC REGIME OF LARGE RIVERS

Richard E. Sparks
Illinois Natural History Survey
River Research Laboratory of the Forbes Biological Station
Havana, Illinois

Water regimes in virtually all the large rivers of the world are either intentionally regulated by dams and other structures or inadvertently affected by deforestation, agriculture, and urbanization. The environmental effects of altering water regimes too often appear as unwelcome surprises because: (1) the dynamic nature and interaction of the physical and biological components of river ecosystems are not understood and (2) some of the effects are long term and long range. Risks are more likely to be considered and reduced if projects are broadly conceived in the context of managing *a complex, dynamic system that fulfills many natural functions (e.g., transport of water, sediment, and nutrients and maintenance of deltas and estuaries) and human services, many of which have not been included in traditional analyses of costs, benefits, and risks in narrowly conceived economic development projects. Disciplines that should be represented on planning teams, in addition to economists and engineers, include fluvial geomorphologists, hydrologists, rural sociologists, fisheries ecologists, and systems ecologists with experience on large rivers. Epidemiologists should be included when waterborne diseases are likely to be a problem. The risks of losing attributes of the river-floodplain system (the floodpulse, floodplain habitats) that sustain fish and wildlife populations and local economies (e.g., floodplain agriculture and aquaculture and recreational, commercial, and subsistence fisheries) should be evaluated by considering what it would cost to replace (if replacement is even feasible) or maintain these services. Risk analysis should extend to economies that may be distant but directly affected by the local project, as in the Nile and Mississippi deltas where coastlines eroded and fisheries declined due to trapping of sediments and nutrients behind upstream dams. In view of the accumulated worldwide experience with effects of water resource developments and watershed manipulations, there is little excuse now for failing to evaluate risks during planning stages of projects, including projects that mitigate damage already present.*

---

1. Corresponding Author: Richard E. Sparks, Illinois Natural History Survey, River Research Laboratory of the Forbes Biological Station, P. O. Box 599, Havana, IL 62644.
2. Key Words: disturbance, ecosystem, flood, floodplain, large rivers, water regime.

INTRODUCTION AND SCOPE

The water regime and the geological substrate in rivers and streams set the physical stage for the drama of life. On a time scale of years to centuries, floods shape the morphometry of streams, rivers, and deltas. Within the span of a season, flow controls physical attributes of the habitat (habitat volume, current velocity) for aquatic organisms. At the micro scale of a few millimeters, there is a world of difference to a midge between laminar flow over smooth surfaces of a rock and vortices induced by surface irregularities (Osborne and Herricks 1987). Although all these scales are important, this chapter focuses on water regimes measured on temporal scales of hours to decades and on the spatial scale of reaches of mainstem rivers, with particular reference to the Mississippi and its major tributaries.

The water regimes in the largest rivers of the world are already managed by man, or soon will be. By the year 2000, 66% of the world's total streamflow will be controlled by dams (Croome et al. 1976). The United States has over 2.5 million dams, so it is virtually impossible for water to flow any great distance from the land to the sea without being regulated (National Research Council 1992). Hydropower use exceeds the sustained yield of water in the United States by a third, because water is used repetitively as it moves through river systems. Offstream uses (irrigation, domestic water supply, cooling) amount to 61% of sustained yield, a typical situation in the developed countries of the world (National Research Council 1992).

The few relatively undisturbed rivers in the tropic and arctic zones are being developed rapidly or inadvertently affected by land use practices. Extensive deforestation in the Amazon Basin is expected to affect the water regime of the Amazon River (Gentry and Lopez-Parodi 1980). Both the Soviet Federation and Canada are developing water resource projects on their north-ward-flowing rivers (Baxter 1977; Thurston 1991). Thus, the question to be addressed is not whether to manage water regimes, but *how* to manage to avoid or minimize environmental risks.

WHAT IS THE WATER REGIME?

To a hydrologist, the water regime is fundamentally the temporal pattern of water flow or discharge in the river, commonly measured in cubic meters per second and averaged over hourly, daily, or longer intervals (Fig. 1). Most people are aware of great regional differences in water regime—from the arid Southwest where arroyos that are dry most of the year become torrents during rare cloudbursts to the steady flows of spring-fed "runs" in Florida to the spring freshets reliably produced every year by snowmelt in the Rockies. These water regimes differ greatly in their predictability, a term Colwell (1974) defined as consisting of two components: constancy and contingency. In these examples, the flow of the Florida run in a certain month of the year is highly predictable because it is so constant from month

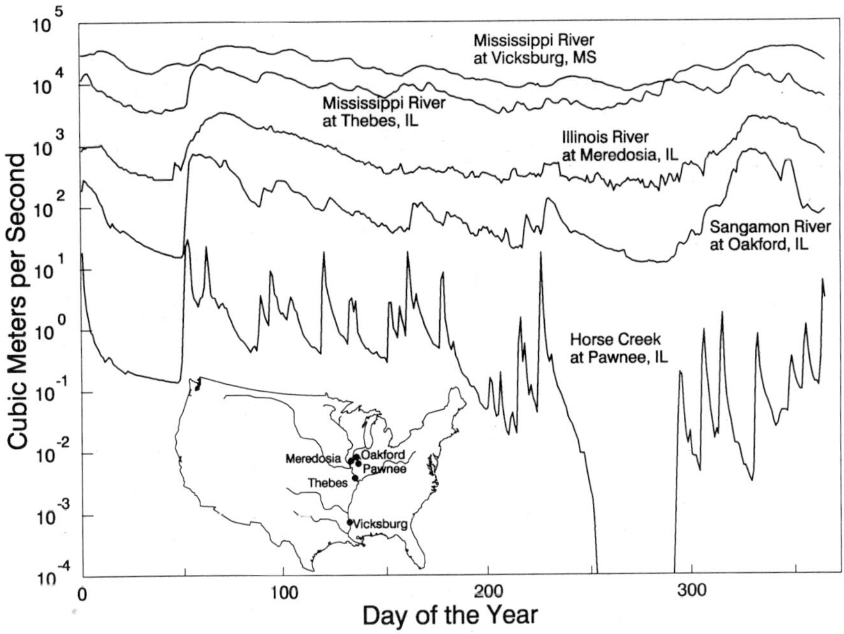

Figure 1. Variation in discharge pattern with position in the Mississippi drainage network. Plotted values are daily averages for 1985, from United States Geological Survey District Office records in Urbana, IL, and Vicksburg, MS. Horse Creek is a first order, intermittent tributary. The Mississippi River near Vicksburg is 12th order. Discharge becomes smoothed with distance downstream and the range between peak flow and low flow diminishes downstream.

to month. Predictability would be close to the maximum value of 1.00, with most of the predictability attributable to the high degree of constancy. The monthly flow of the mountain stream is also highly predictable, not because the flow is constant from month to month but because the flow is highly contingent on events (winter freeze and spring snowmelt) that usually occur about the same time each year. In contrast, the predictability of the water regime in the southwestern stream is low, perhaps close to zero, because the stream has neither a constant base flow (constancy is low) nor a strongly seasonal influx (contingency is also low).

Effect of Stream Order on Water Regime

The discharge pattern changes progressively with stream order in a drainage basin, usually becoming more predictable downstream because both constancy and contingency increase. Discharge hydrographs for 1985 are given in Figure 1 for the Mississippi River and various tributaries ranging downward in size to a first order, intermittent stream, Horse Creek at Pawnee, Illinois. Stream order in this discussion refers to the Strahler-Horton (Strahler 1952, 1957) system of ordering a drainage hierarchy. The smallest streams that have no tributaries are classified as first order. Two first-order streams join to form a second order stream. A second-order stream does not increase to third order by the addition of a first-order stream, but only when joined by another second order. The classification process proceeds similarly downstream to the mouth of the mainstem; e.g., the Mississippi is a 12th-order river at its mouth.

In the upper Midwest, the discharge in first-order streams may vary over five orders of magnitude, including zero, during periods of low rainfall in the summer (e.g., Horse Creek in Fig. 1). The discharge pattern is spiky because the stream floods with every local thunderstorm. Although the pattern undoubtedly has become more extreme because of deforestation, drainage of wetlands, and channelization associated with row crop agriculture in this region of Illinois, the fundamental pattern is attributable to the stochastic nature of local rainfall and relatively low water retention in these small, upland drainages.

In progressively higher stream orders, the peaks become smoother and the range of variation less. In sharp contrast to Horse Creek, the range between minimum flow and peak flow in the Mississippi at Vicksburg was much less than one order of magnitude in 1985, more on the order of 3-5x (Fig. 1). The two sharp peaks in Horse Creek at days 50 and 60 become one large, rounded peak downstream in the Sangamon River, and successive rounded peaks become part of the spring flood further downstream in the Illinois River (Fig. 1). The higher order streams integrate flows over progressively larger basins and, therefore, respond more to regional weather patterns than to local thunderstorms. In addition, the lowland rivers tend to have lower slopes and broader floodplains, which together increase the water retention capacity, even though approximately half this capacity has been lost through leveeing of the floodplains, primarily for agriculture (Mills et al. 1966).

Other Factors Affecting Water Regime

Although this brief discussion has focused on climate and stream order, other factors also affect flow regime. The geological history of the basin determines the parent material and the morphology of the basin, floodplain, and channels. Two rivers that have similar total annual discharges may have quite different water regimes because of differences in their floodplains. The Rock River in northern Illinois had approximately the same annual discharge as the Illinois River before the man-made diversions of Lake Michigan water into the Illinois River. However, its productivity in terms of fish and wildlife was much lower because it lacked the protracted flood and the large floodplain and backwaters that served as nurseries and feeding areas in the Illinois River. The broad floodplain of the Illinois River is a geological legacy from a much larger ancestral river. The size and low slope of the floodplain provide a large water retention capacity that creates a protracted flood from a relatively small discharge.

The water regime at a particular reach depends upon: (1) upstream factors such as the vegetative cover and soil of the drainage basin (which govern interception storage and infiltration capacity for rainfall), tributary slopes, lengths, and channel capacities, (2) local interchanges with floodplains and groundwater, and (3) downstream characteristics such as backwater effects and tidal fluctuations. Season of the year is important also, because vegetation and temperature determine rates of water loss due to evapotranspiration. The type and density of vegetation and woody debris in the channel, riparian zone, and floodplain influence stage and discharge by impeding flow. Although most people think immediately of impressive structures such as dams when flow alterations are mentioned, it is important to remember that widespread land use changes associated with agriculture and silviculture also have dramatic effects on flow regimes.

WHAT IS THE IDEAL WATER REGIME?

Having defined flow regime and predictability of flow, the next question becomes: what constitutes the "best" flow regime. The answer depends on the point of view of the uses of the stream or river. This is a familiar question for water resource engineers who specify the ideal flow regime for a particular human use and then design water management systems that will match the water supply to the water need. Complications arise when multiple human uses require contradictory water regimes, for example, storage of water on the upper Missouri River for recreation vs release of water to the lower Missouri to maintain minimum depths for navigation or lateral wetlands for fish and wildlife (Hesse et al. 1982). Even these problems can be addressed by techniques such as adding enough storage capacity for both irrigation and navigation flows and operating the entire system of dams so that no one use is maximized, but multiple uses are optimized.

Ecologists have difficulty describing the ideal flow regime because natural systems, particularly large river-floodplain systems, are so complex. Consider the difficulties of applying the engineering approach of specifying a water regime for a particular use, in this case the requirements for a particular species. This approach has been taken by fisheries and wildlife biologists for a few, highly-valued species (Lambert and Handley 1984). The first complication is that different life history stages of a single species usually have quite different requirements. Eggs of channel catfish are hatched in underbank cavities (muskrat burrows and root masses of riparian trees) whose accessibility depends on river stage. In the upper Mississippi River, adult catfish feed in backwaters when river stages permit access from the channel and overwinter in a dormant condition in the main channel where they use rocks or other obstructions to protect themselves from the current (Hawkinson and Grunwald 1979; Lubinski 1984; Talbot 1982, 1984; Todd et al. 1989). The ideal water regime for channel catfish might be high discharge in spring and summer, so that river stages are high enough for them to access spawning and feeding areas, but low discharge in winter so that they are not swept out of their wintering areas in the main channel. This species-specific information usually is obtained by the empirical approach of noting where each life history stage occurs in a stream or river, then measuring the physical characteristics (e.g., depth, velocity, substrate) of that site. Water regime then can be adjusted to provide habitat with the suitable characteristics at the appropriate time in the life cycle of the species. The difficulty in applying this approach to a large river-floodplain system is that there are not just a few highly valued sport or commercial fishes, but as many as 20. If the objective is broadened to include conservation of fish biodiversity, then the interest is in all fish species, particularly the rarer ones whose precise habitat requirements may be unknown because they are rarely observed. In the mainstem Mississippi, there are 260 true freshwater species of fishes and an additional 46 if saltwater or anadromous species are included (Fremling et al. 1989). If water birds, mammals, invertebrates, and plants that are vital components of the river-floodplain ecosystem are added, then the list of "user groups," each with their specific requirements, potentially numbers in the thousands.

Another difficulty with the species use approach is that water regimes that maximize habitat suitability for one group of species may be detrimental for another, so that imposition of one water regime year after year would cause the diminution or disappearance of some species. For example, water regimes could be manipulated to prevent "summerkill" in floodplain lakes and pools, a condition where fish and other aquatic organisms die as some of the shallower lakes contract because the water heats up from the sun and oxygen levels are depleted by the concentrated mass of aquatic animals and decaying organic matter. However, a water regime that eliminated summerkill would adversely affect nesting colonies of herons and egrets that are located near such lakes precisely because the summer drawdown concentrates fish that adult herons feed to their fledglings.

Natural Water Regimes

An entirely different approach to defining the ideal regime is based on evolutionary theories that apply to both the physical and biological components of the river system. The morphology of an undisturbed river system is a product of its water and sediment loads and is generally regarded as being in a state of dynamic equilibrium, at least on a time scale of decades. Within a given year, periods of scouring may be followed by deposition, and individual features such as bends may migrate, but the *pattern* (meandering, braided) is relatively constant from year to year. Alterations in the sediment or water regimes cause the channel to adjust until a new equilibrium is achieved.

Just as the morphology of the river has adapted to its water and sediment loads, so have organisms adapted to the physical characteristics of their natural environments; hence, the way to preserve whole assemblages of species is to preserve the habitat, including the water regime. If the characteristics of the natural regime that maintain both the physical and biological characteristics of fluvial systems can be identified, then it should be possible to avoid or minimize the negative impacts of flow alterations.

The problem now becomes one of identifying the natural regime. Streams or small rivers that lie wholly within undisturbed watersheds, or watersheds that have revegetated to a predisturbance condition, can be presumed to have natural flow regimes, at least within the constraints of the current weather patterns. The water regime in these streams can serve as a reference standard for other, more disturbed fluvial systems of similar size in the same ecoregion (sensu Omernik 1987; see also Hughes et al. 1990 and National Research Council 1992). The relationships between types of water regime and types of biological communities in reference streams indicate what aspects of flow (e.g., constancy, contingency, intermittency) are important to organisms. One then can deduce how water regimes in other streams may have changed, based on the presence or absence of certain species.

There are virtually no undisturbed large rivers, because there are virtually no pristine large basins. Rivers in developing countries may be less altered than rivers in the industrialized nations of the temperate zone, but even the mainstem Amazon River at Iquitos shows a pattern of increased flood heights and decreased low flows since 1970. Gentry and Lopez-Parodi (1980) attribute this to deforestation in the upper Amazon watershed in Peru and Ecuador. All large rivers in the United States have had their flow regimes altered to some degree. In a few cases, historical discharge or stage data that predate the major alterations are available. As in the smaller rivers and streams, the life history characteristics of some species also may provide clues to the predisturbance water regime. Examples using these types of information to determine natural flow regimes are discussed next. One other approach to reconstructing the predisturbance water regime is to apply hydrological simulation models to watersheds that are "revegetated" according to broad scale presettlement vegetation patterns (Vorosmarty et al. 1986). A more precise modeling approach would be to revegetate watersheds according to soil types (soils are heritages of former vegetation patterns).

Regimes in Relatively Undisturbed Streams and Small Rivers

Poff and Ward (1989) classified the flow regimes of 78 relatively undisturbed (e.g., no dams or major channel alterations) streams and mid-sized rivers throughout the United States according to criteria that most strongly correlated with the structure and function of the biological communities in the streams. The characteristics they found to be most important were degree of intermittency of the stream, predictability of flow, and the frequency and predictability of flooding (Fig. 2)—variations on the general theme of constancy and contingency of flow.

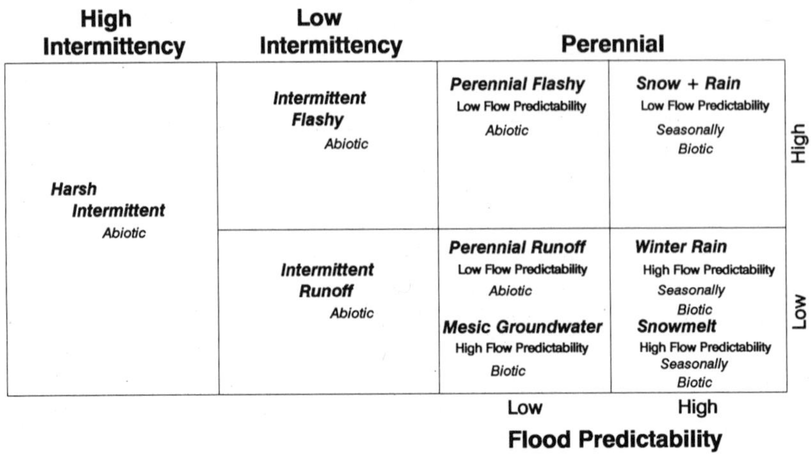

Figure 2. Classification of flow regimes in streams and small rivers, according to factors important to aquatic organisms. Abiotic factors (temperature, flow regime) are more important than biotic factors (predation, competition) in structuring biological communities in the harsher, more unpredictable environments (adapted from Poff and Ward 1989).

Even the harshest of flow regimes (those with a high degree of intermittency) has some forms of aquatic life, although life largely is controlled by abiotic factors (Fig. 2). Only species that have short life cycles, refuge-seeking behavior, drought-resistant resting stages, or rapid recolonizing abilities occupy the drastically fluctuating environments. As the predictability of the water regime increases, Poff and Ward (1989) found that biotic interactions assume more importance and communities include species that can compete efficiently for resources or prey upon other species. In groundwater-fed streams with the most constant water regimes, biotic interactions are important year round. Biotic interactions are seasonally important during periods of stable flow in perennial streams where floods occur predictably. Floods in these smaller streams and rivers presumably are equivalent in their effects to cessation of flow, in that they are abiotic disturbances that organisms can only survive or avoid through life history adaptations.

Regimes in Large Rivers

In contrast to the smaller fluvial ecosystems studied by Poff and Ward (1989), large river-floodplain systems typically have highly predictable floods (in terms of seasonal timing) of sufficient duration (in relation to time required for completion of critical life stages) that the flood is not a disturbance but an expected event, the *floodpulse*, which organisms have adapted to exploit, rather than to avoid (Junk et al. 1989). The average period of inundation of the floodplain forest on the upper Mississippi River at Burlington, Iowa, is 22 days (Swanson and Sparks 1990). Far downstream in the Mississippi delta, the flood duration on the Atchafalaya distributary of the Mississippi River is 160 days (Dr. C. F. Bryan, personal communication, June 15, 1990; Louisiana Cooperative Fish and Wildlife Research Unit, School of Forestry, Wildlife, and Fisheries, Louisiana State University, Baton Rouge, Louisiana). In the Amazon, the full cycle of flood advance, cresting, and recession requires 8 to 9 months at Obidos, so the flood is a longer lasting event than the low flow (Welcomme 1985)!

In rivers such as the Mississippi that are more disturbed than the mainstem Amazon or the rivers that Poff and Ward (1989) investigated, the natural flood regime sometimes can be reconstructed from historical records that predate major disturbances. For example, almost continuous daily records of water-surface elevation (stage) at Burlington, Iowa, have been kept on the upper Mississippi River, since 1878—34 years before the dam building era that began in 1913 (Grubaugh and Anderson 1988). Likewise, discharge has been determined daily since 1878 at Keokuk, Iowa, approximately 60 km downstream of Burlington. The stage hydrograph during the predam era is bimodal, with a major flood peak in the spring and a minor one in autumn (Fig. 3). Both peaks are largely attributable to seasonal changes in the weather patterns over the midwest region, with clashes between warm, moist Gulf air and colder northern air masses causing widespread precipitation (Wendland 1987). The spring flood usually is augmented by snowmelt and iceout in the northern part of the drainage. The autumn flood is augmented by leaf fall and cooler temperatures that together reduce evapotranspiration rates. When the temperature drops quickly and ice formation begins in early winter, the flow may decrease by 283 m^3/s (10,000 ft^3/s) in a few days to the stable winter low flow (Rock Island District Corps of Engineers 1991).

Water Regime and Biota

Careful study of how organisms use the river-floodplain ecosystem can reveal what aspects of the water regime are important to them. Many Amazonian fishes use the floodplains for spawning and feeding; some have anatomical and behavioral specializations for consuming the fruits and seeds of certain trees and are important dispersal agents for plants (Goulding 1980); physiological adaptations include the deposition of fat during the feeding period on the floodplains to prepare for the energy demands of migration and spawning during the low water period (Junk 1985).

A variety of species took advantage of the bimodal water regime characteristic of the predam era in the upper Mississippi River (Fig. 3). Fishes that use the spring flood for access to spawning and nursery sites on the floodplain and in its associated lakes include: crappies (*Pomoxis*), yellow perch (*Perca flavescens*), suckers (Catostomidae), sunfishes (*Lepomis*), gizzard shad (*Dorosoma cepedianum*), silver chub (*Hybopsis storeriana*), and white bass (*Morone chrysops*) (Holland 1986; Kwak 1988; Sheaffer and Nickum 1986). Waterbirds (diving ducks, dabbling ducks and geese, shorebirds, wading birds) use the spring and fall floods for access to the supply of seeds, roots, invertebrates, and fish that are produced during the

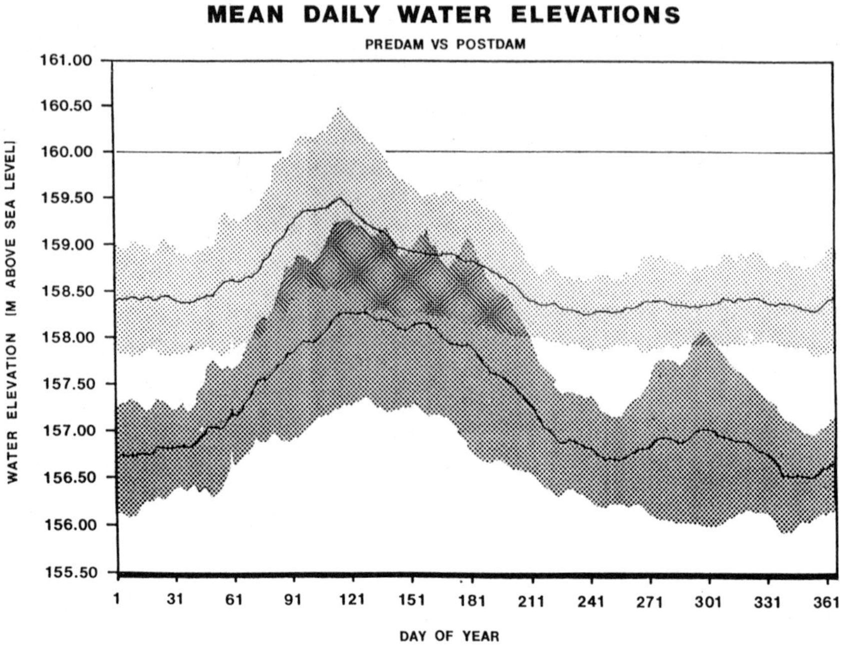

Figure 3. Average daily water surface elevations of the upper Mississippi River at Burlington, IO. Dark shading indicates the predam period, 1878-1912 (N = 34), and light shading the postdam period, 1913-1987 (N = 74). Shading indicates ±1 SD from the mean. The dam increased the water elevation and shortened the duration of the spring flood peak and the autumnal flood (Grubaugh and Anderson 1988).

growing season; their spring and fall migrations may be adjusted to the annual timing of these floods, as well as to other factors such as weather on the breeding and wintering grounds. The low flow regime is just as important as the flood regime. Submersed vegetation and moist soil plants benefit from low, stable water levels during the summer growing season. The timing (in relation to migration), rate of flood rise, and the difference in flood heights in spring and autumn are also important. In autumn, the annual food crop is made available in stages to successive waves of migrants by rising water levels. Groups of birds have preferred feeding depths, as the common names imply: shorebirds typically feed on mudflats or in very shallow water; the feeding depth of dabblers is governed by the length of their necks; and the feeding depth of wading herons and egrets depends on the length of their legs. If water levels were stable in spring or autumn instead of rising slowly, all the food at the preferred feeding depth would be exhausted by the first migrants. Since no additional food is produced in winter, it is probably important that the spring rise reach into areas of the floodplain that are not accessible during the lower fall flood. If the spring and autumn floods reached the same stages, the consumers would be going over the same feeding grounds twice.

Shifts in the seasonal timing and duration of the spring flood are important, as some theoretical examples will show. Fish in the temperate zone benefit most from a smooth (slow rise, slow fall, moderate duration) flood that is closely coupled with the annual temperature cycle (Fig. 4A). The fish will not be ready to spawn if the flood peak arrives when the water is too cold. If the fish do spawn late in the flood cycle when the water is warm, the retreating water will strand eggs and young (Fig. 4B). However, the early flood that is suboptimal for fish will benefit aquatic and moist soil plants by providing a longer growing season with low, stable water levels. Dabbling ducks in turn benefit from the increased primary production, especially if a fall flood makes it accessible (Fig. 4B).

Most species using the upper Mississippi River are opportunists. Fishes take advantage of favorable spring floods through their high reproductive potentials; most are r-selected species (sensu Pianka 1970). In farm ponds and reservoirs with stable water levels, over-population and stunting of these fishes often are problems because high rates of reproduction are followed by too-successful avoidance of predators when the young fish hide in dense littoral plant beds. In contrast, the falling stages of the flood draw prey out of cover and concentrate them, so more biomass moves up food

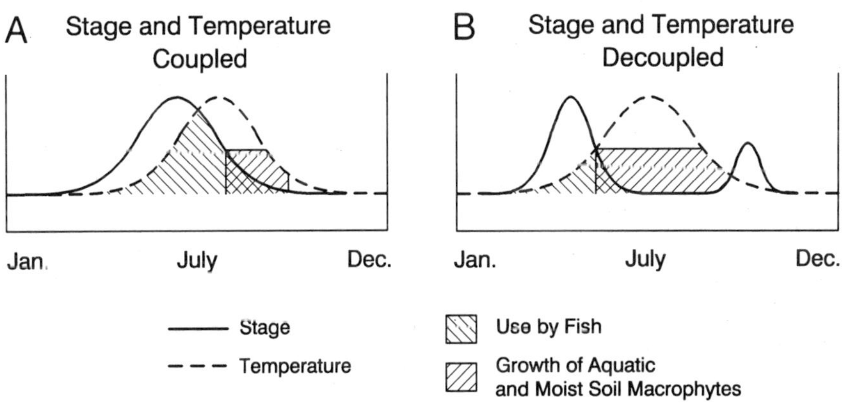

Figure 4. Hypothetical examples of river stage and temperature relationships for a large river floodplain in the north temperate zone. The temperature regime in both A and B is the same, but the flood pattern varies. A. A long duration spring flood that is coupled with warming temperatures provides an extended period for fish spauning, rearing, and feeding. Aquatic and moist soil plants grow during the summer when water levels decrease. However, there is no autumn flood to make the summer's net primary production readily available to dabbling ducks. B. An early, short duration spring flood is decoupled from temperature, restricting fish use but increasing the favorable growing period for plants. The increased primary production is accessible to dabbling ducks because of the autumn flood. The autumnal flood was typical of predam hydrographs in the Illinois and upper Mississippi rivers (adapted from Sparks et al. 1990).

chains into large predators. Water birds are behavioral opportunists; they can fly overland to the most favorable feeding grounds. Many river-floodplain inhabitants have long-lived adults, so that one year of reproductive failure because of unfavorable water levels does not cause local extinction of the species. Therefore, it is probably unnecessary and even undesirable to create an "ideal" flood regime every year, even if it were technically feasible to do so. The ideal regime is not necessary each year as long as a mosaic of habitat types exists so that spawning, feeding, and overwintering can occur somewhere in the river-floodplain ecosystem within accessible range of local populations, even if the same area is not used for the same purpose every year.

Some year-to-year variation in the flood pattern is probably necessary to maintain a full complement of species; e.g., some plants may require an unusually long summer low-flow period to set seed and replenish soil seed-banks. Extreme events may even have a rejuvenating function, in that they push geomorphic and biological succession back to an earlier stage. For example, record floods may rejuvenate some long abandoned side channels by scouring away accumulated sediment or causing more rapid senescence of floodplain pools and lakes by depositing sediment, depending upon the flow patterns (velocity, direction, turbulence), sediment load of the river water, and stability of the bottom material. The drought and low flows of 1976-1977 in the upper Mississippi basin apparently rejuvenated Stump Lake, a backwater lake near the junction of the Illinois and upper Mississippi rivers, which had become turbid and barren of aquatic vegetation because of accumulated fine sediments that were easily resuspended by wind-driven waves. The constantly roiled sediments reduced light penetration and did not provide a stable rooting medium (Sparks et al. 1990). The lake dried completely during the drought and was colonized by terrestrial herbaceous vegetation. The sediments became consolidated during drying and the stems and roots of the terrestrial vegetation anchored the bottom and dampened waves long enough for the aquatic vegetation to reestablish when the drought broke and the lake reflooded. The effects of extreme events depend greatly on local conditions, including geomorphology and vegetation type. The same drought that rejuvenated Stump Lake (pushed succession back to an earlier vegetational stage) advanced vegetational succession in an impounded reach of the upper Mississippi River near Keokuk, Iowa (Sparks et al. 1990). There the reduced runoff and soil erosion improved water transparency and allowed zones of submergent aquatic plants to advance and maintain themselves in water much deeper than would have been predicted under normal water and turbidity regimes (Sparks et al. 1990).

In summary, the naturally fluctuating water regime and the spatial heterogeneity of large river-floodplains are both important to the preservation of species occupying these ecosystems. The species are adapted to a relatively predictable annual cycle in terms of seasonal timing, duration, and rate of rise and fall. Spatial heterogeneity ensures that some portion of the floodplain will meet the requirements for a species during the flood and low flow periods, if not the same location every year. A depression that is ordinarily dry during moderate floods may become a spawning site during record floods, when traditional spawning sites are unusable because of excessive water velocities or sediment loads. If heterogeneity is reduced, by leveeing off the floodplain or by excessive filling of depressions and lakes with sediment, then there is so little habitat redundancy that small changes in the annual water regime become increasingly critical to species survival. When highly valued species become scarce, hunters and fishers may demand that the water regime be managed within increasingly narrow limits to favor the species of interest to them.

ALTERED WATER REGIMES

River-Climate Interactions

Risks are inherent in the common assumption of a steady state climate for project planning purposes. Weather patterns may cycle on greater-than-annual time scales and show long-term trends (e.g., global warming) that affect flow regimes and invalidate planning assumptions. Altered flow regimes in turn may have unanticipated effects on local and even regional climate.

Project lives are commonly assumed to be 30-50 years, although structures such as dams and allocations such as water rights usually last much longer than that, in many cases longer than the hydrological record on which the initial plans were based. The over allocation of water in the Colorado River system is at least partially attributable to flow measurements made during a period that was wetter than normal (Reisner 1986).

Sometimes the problem is not too little water, but too much. When water levels rise in the Great Lakes, shoreline erosion threatens property and the public may demand increased diversion of water from Lake Michigan into the Illinois River to lower the lake level, regardless of effects on the river (Chicago District Corps of Engineers 1981; Kilburn et al. 1981). The real problem is that water levels in the Great Lakes fluctuate on approximately an 11-year cycle, and when the water levels are low, structures are built too close to shore. Fortunately, public pressure to divert water has been resisted because the maximum possible diversion would affect lake levels by only a few centimeters but would have marked downstream effects, primarily in the Illinois River, but also in the St. Lawrence drainage (Chicago District Corps of Engineers 1981; Kilburn et al. 1981).

Superimposed on the 11-year rainfall cycle that drives the lake levels is a 20-year trend toward a wetter and cooler climate in the northern half of the Illinois River basin. Illinois River flows and flood peaks have increased significantly, thereby increasing bank erosion and flood damage and raising groundwater levels in levee protected farmlands on the floodplain (Singh and Ramamurthy 1990). In comparison to the previous 60 years, the precipitation increase during the four-month flood season (March-June) is about 25% (Singh and Ramamurthy 1990).

Although the meteorological mechanism for the observed trend in the Illinois River has yet to be determined, such mechanisms have been elucidated in other regions. Molles and Dahm (1990) report a strong correspondence between El Nino-Southern Oscillation (ENSO) phenomena and flow in the Gila and Pecos rivers in New Mexico, and Hastenrath (1990) reported a similar relationship for the Essequibo River in Guyana (in the northeastern portion of South America). Molles and Dahm (1990) also discuss the potential influence of other large scale phenomena (Eurasian snow cover and volcanism) on local precipitation and streamflow.

Extensive deforestation in the Amazon basin is expected to reduce greatly the transpired water that cycles back as rainfall, potentially converting much of the remaining forest to near desert (Gentry and Lopez-Parodi 1980)—an example of ecosystem collapse that would have drastic effects on the water regime of the Amazon River. This collapse could occur because the volume of water recycled in the basin is far greater than the volume entering the basin in the trade winds or leaving at the mouth of the Amazon (Gentry and Lopez-Parodi 1980).

Not only do climate changes affect flow regimes, but flow regimes may affect climate. At a global scale, the increased transport of nitrogen (Vorosmarty et al. 1986) and other nutrients in rivers (as a result of intensified agriculture, deforestation, and other land use changes) may increase primary productivity in coastal zones, thereby removing carbon dioxide from the atmosphere and reducing the rate of global warming attributable to accumulation of "greenhouse" gases. However, these effects are complicated by man-made changes in deltas: in the Mississippi delta, for example, the nutrients flow between levees directly into deeper water instead of accumulating in the floodplain and coastal marshes. The nutrients may stimulate phytoplankton that senesce and sink, using up oxygen and releasing carbon dioxide in the decay process. This may contribute to the spreading zones of oxygen depletion on the bottom, which adversely affect Gulf of Mexico fisheries (Turner and Rabalais 1991).

Alteration of flows in northward flowing rivers in Canada and Siberia may have local and regional effects on climate. The spring breakup of ice in the northern, downstream reaches of these rivers is governed by the pressure built up by the pulse of warmer water coming from the south (Baxter 1977). Alteration of the magnitude or timing of the spring flood could hasten or delay the beginning of spring in these rivers and their valleys and perhaps in the coastal zones as well. The effects could extend to the rest of the northern hemisphere depending on the extent to which sea temperatures and ice cover are altered.

In addition to these general risks of assuming that climate river interactions are in a steady state and will persist in a steady state when river flow regimes are altered are risks involved in specific types of alterations: diversions, consumptive uses (e.g., irrigation), changes in land use in drainage basins, and dams.

Withdrawals and Diversions

The most extreme alteration of flow, common in the arid West, is complete appropriation of water flowing in the channel, either by direct withdrawal or by pumping from the riparian zone. The issue of in-stream flow is being joined in other, wetter regions of the country as well. During the 1988-1989 drought in the upper Mississippi Basin, many municipalities asked for permits to withdraw virtually the entire flow of some rivers, such as the Mackinaw River in Illinois. Dessication of formerly perennial streams causes local extinction of most native aquatic life that does not have the life history adaptations to intermittency that characterize aquatic organisms in desert streams and rivers. Recovery of the community when flow returns depends on recolonization or survivors in pools that did not dry out or become anoxic.

Increasing flow in rivers involves risks, just as reducing flow involves risks. The annual flow pattern may be changed, with the increased flow occurring at the wrong time for the aquatic organisms. Irrigation reservoirs typically store the spring flood, reducing or eliminating the flood pulse downstream of the dam. Later, during the summer growing season, water returns to the river downstream of the irrigated fields. The normal pattern, thus, is reversed downstream of the dam: the river is low in the spring and higher in the summer. Some navigation dams have the same effect (see below).

Land Use

In urban and suburban areas, rain falls on impervious surfaces and is often directed into the nearest watercourse via storm sewers. In agricultural areas, drain tiles, ditches, and channelized streams have the same effect. Water drains much more rapidly from logged areas than from the original forest (Swank and Douglas 1974). The result of these land uses is that flood peaks are higher and low stages are lower and longer lasting than in the past because there is less retention of water in the basin itself (Bormann et al. 1969; Herricks and Osborne 1985; Karr and Dudley 1981). Changes in the flow pattern often trigger unwanted changes in deposition and erosion. Sediments may accumulate in formerly productive channels and backwaters downstream, or a process of headward erosion can begin. During droughts, base flow may be too small to support aquatic life and other beneficial downstream uses.

Modifications in floodplains, as well as on the upland drainage, have altered flow patterns. Flood protection levees permit the former floodplain to be used for agriculture, industry, or housing, but it is no longer available for fish and wildlife production, production of hardwood timber, recreation, or for the storage and conveyance of floods. It is ironic that these levees actually contribute to increased flood heights on the upper Mississippi River (Belt 1975; Grubaugh and Anderson 1989; Stevens et al. 1975). Sedimentation rates increase on the remaining unleveed floodplains to the point that the native vegetation, including valuable hardwoods, may be smothered (Turner 1929). Vegetation also changes due to the altered water regimes because individual species of plants vary in their seasonal tolerance for inundation and water saturated soil (Havera et al. 1980). In the Mississippi floodplain, 78% of the forested wetlands has been lost (MacDonald et al. 1979), greater than the 50% loss of all wetlands in the coterminous United States since the time of European settlement (Dahl 1990).

Vegetation is not only affected by changes in water regimes, but it may itself regulate water regimes. Vegetation affects water regime directly by transpiring water that then is unavailable in the ground or in stream channels. Vegetation affects water regime indirectly by shading stream channels, lowering water temperatures and thereby reducing the evaporation rate. Leaves and stems of plants mediate runoff by intercepting rain or fog and by accumulating as water absorbing organic matter on the ground and in the soil. Changing the type of vegetation can affect streamflow: replacement of hardwoods with pines reduced streamflow by 20% in the Appalachian Mountains (Swank and Douglas 1974). Leaves and stems offer resistance to flow in channels, riparian zones and floodplains; the result can be increased flood height and duration upstream and retardation and smoothing of the flood peak downstream.

Vegetation stabilizes banks and thereby controls channel morphometry. Westphal and Clemence (1976) indicate that removal of vegetation from the banks of the Mississippi in the 19th century contributed measurably to bank caving, river widening, and changes in the stage/discharge relationships between 1829 and 1881. In 1848, a single steamboat required approximately 150,000 trees per year for fuel; most of this timber was obtained as close to the river as possible. The amount of land cleared for fuel for a single boat could have amounted to 5000 to 8000 hectares (20 to 30 square miles) during the 40 years between 1830 and 1870 (Westphal and Clemence 1976)! During the same period, the Corps of Engineers also removed timber for a distance of 46 m (50 yards) from the river's edge as a way of preventing trees from falling in the river and becoming hazards to navigation (Westphal and Clemence 1976).

Dams

Dams have been constructed on virtually every type and size of flowing water from intermittent headwater streams to the Mississippi River. Croome et al. (1976) suggest that, by the year 2000, approximately 66% of the total streamflow of the world will be controlled by dams. Mermel (1976) used information from the *World Register of Dams* (International Commission on Large Dams 1973) to conclude that large dams are being built on the world's rivers at an average rate of two per day. Benke (1990) could find only 42 rivers in the United States that exceeded 200 km (160 miles) in length that did not have dams or other major structural alterations.

Dams change flowing water to standing water, alter downstream flow patterns of both water and sediment, block migrations of aquatic organisms, alter water quality, and initiate long-term changes in downstream channels, riparian zones, and floodplains. Secondary impacts include shifts in the patterns of waterborne diseases and displacement of human populations.

Health Risks for Livestock and Humans. Altered water regimes may create public health problems. Schistosomiasis, caused by blood flukes of the genus *Schistosoma*, usually increases near impoundments in the tropics because the slackwater habitat favors the snails that are the intermediate hosts (Egbuniwe 1976). The Tennessee Valley Authority in the United States fluctuates water levels in its reservoirs to prevent breeding of malaria breeding mosquitoes (Baxter 1977). Onchocerciasis, or river blindness, is transmitted by certain black flies of the genus *Simulium* that require running water. Impoundments may shift the location of this disease by flooding rapids, but create new habitat in the tailwaters below the dam (Baxter 1977). In a few cases, species shifts caused by changes in hydraulic regime turn out to be beneficial: hydroelectric dams in Canada caused *Simulium arcticum*, a black fly that severely injures cattle, to be replaced by *Simulium luggeri*, a more benign species (Fredeen 1977).

Social and Economic Risks. Massive hydroelectric and irrigation projects often are undertaken for the greater good of an entire region or nation. However, these create severe impacts on local human populations (displaced by dams or reservoirs) or on long-range effects of changes in water, sediment, and nutrient flows on fisheries and floodplain agriculture (see discussion of long-range effects below). Little or no information is available on subsistence and local market riverine fisheries because they are often widely dispersed in remote areas; therefore, these fisheries are overlooked in the cost/benefit analyses of water resource developments. However, Bayley and Petrere (1989) estimate that local fisheries account for 60% of the total fisheries yield in the Amazon. Disruption of these fisheries could disrupt local economies, cause migration of people from the rural areas into the already over burdened cities, and lead to the loss of a major protein source for human populations. Large impoundments in Africa have disrupted tribal communities and economies dependent on hunting, fishing, and trapping (Waddy 1975, as cited in Baxter 1977).

The human costs of disrupting riverine ecosystems have been ignored in the past in developed, as well as developing, countries. The commercial fishery along a 200-mile reach of the Illinois River produced 24 million pounds of fish in 1908, 10% of the total catch of the United States—more than any other river in North America (excluding rivers with anadromous fishes, such as the Columbia) (U.S. Department of Commerce and Labor 1911). Over 2000 commercial fishers were employed, and the yield was 200 kg/ha (178 pounds/acre) (Lubinski et al. 1981; U.S. Department of Commerce and Labor 1911). The fisheries declined due to a combination of factors, including: (1) drainage of the floodplains and their associated lakes, (2) the rapid (50 years) filling of the remaining lakes with sediments, (3) changes in water regime associated with navigation dams and diversions from Lake Michigan, and (4) pollution (Bayley, in press; Bellrose et al. 1983; National Research Council 1992; Sparks 1984; Sparks et al. 1990). Since the 1970s, the commercial fish yield has been a low 4.1 kg/ha (4 pounds/acre), totaling only 0.32% of the total freshwater harvest of the United States (Sparks 1984). In the last 5 years (1987-1991), the number of full time commercial fishers on the Illinois River has ranged from 4 to 13 (William Fritz, personal communication, October 17, 1991, Illinois Department of Conservation, Springfield, Illinois). In addition to commercial fishing, outdoor recreation associated with the river, including fishing and hunting, was important to the economies of many of the small towns along the Illinois River. These losses were not considered when modifications to the floodplains and river were planned.

Today, the Illinois River provides 2 million angling days per year, valued at $40 million annually (Conlin 1991). Hunters spend an additional $44 million per year (Conlin 1991). These values could be increased substantially by restoration of the floodplains and associated lakes. Moreover, the necessary relocation of flood prone structures, public acquisition of flood prone land, and conversion to parks and refuges would substantially reduce the risk of flood damage and the $25 million average annual cost of Illinois River flooding (National Research Council 1992; Wetmore 1987).

Ecosystem Risks. Changes in flow regime, water temperature, oxygen, turbidity, and quality and quantity of food particles in the river downstream of reservoirs shift the upstream-downstream patterns of biotic structure and function predicted by the river continuum concept of Vannote et al. (1980). A dam may make conditions more like those of the headwaters (an upstream shift), more like those downstream (a downstream shift), or it may have a negligible effect, according to the serial discontinuity concept of Ward and Stanford (1983).

If the new conditions persist many years without interruption, some adaptation may occur; e.g., some aquatic insects have adapted their life cycles to altered temperature regimes (Ward and Stanford 1982) and a local population of the red shiner, *Notropis lutrensis*, broadened its temperature preference and tolerance after living 40 years below a dam in Texas (Calhoun et al. 1982). The adaptation was the result of Darwinian selection for heterozygotic fish with six isozymes that had broader temperature tolerance than homozygotes with three isozymes. Species that formerly spawned at different times or places because they preferred different velocity and thermal regimes may spawn together in the disturbed environment, leading to hybridization and potential loss of one or both species (Zimmerman 1984). These effects may lead to the alteration or loss of an entire species, rather than a local population, if species endemic to a particular river concentrate together below dams that block their upstream spawning migrations. Also, adaptation of a few genetically or physiologically plastic species does not make a complete ecosystem and may pose new risks: if the operating characteristics of the dam change too quickly, because of breakdowns, scheduled maintenance, or new objectives, the populations may be adversely affected.

Hydroelectric dams that are used for peaking power are a special case because they release water during periods of peak electric demand, usually in the morning and evening, producing daily "tidal" fluctuations in ecosystems that are commonly adjusted to seasonal fluctuations. In contrast to a true marine or estuarine tidal system, however, there has been no long period of evolutionary adaptation to these daily tides, although the reach immediately below the dam may be colonized by a few headwater species that can resist or seek temporary refuge from spates, providing other conditions (temperature, oxygen, food supply) are suitable.

Long-Range Effects. Dams may have long-range effects in addition to the local effects discussed above. Flow regulation by dams reduced the area of floodplain wetlands downstream by 67% in a 145-km (90-mile) sample reach of the Missouri River (Whitley and Campbell 1974). Further reductions occurred because the dams slowed water flow and allowed sediments to settle out in the upstream reservoirs, increasing the erosive power and down-cutting of the middle Missouri River. When the channel down-cut, side channels, pools, and wetlands that once supported fish and wildlife were left high and dry (Hesse et al. 1982).

This reduction of sediment loading has had effects even farther downstream in the Mississippi delta. Closure of the major dams on the upper Missouri River was followed by a step decrease in the sediment load measured on the lower Mississippi at Tarbert's Landing, Mississippi, and Simmesport, Louisiana (Keown et al. 1981). Hence, less sediment is now available for maintaining the Mississippi delta against coastal erosion and subsidence (Penland and Boyd 1985). Instead of being allowed to spread over and replenish the delta during spring floods, the remaining sediments are carried in leveed channels across the delta and deposited off the edge of the continental shelf. The resulting coastal erosion threatens both man-made structures and the wetland nurseries of many freshwater and estuarine animals (Penland and Boyd 1985).

The trapping of nutrients and sediment behind the Aswan high dam on the Nile led to similar downstream problems, including reduction of sardine catches in the Mediterranean near the delta (George 1972) and dependence of agriculture in the lower Nile Valley on commercial fertilizer to replace nutrients once provided by the annual flood (Baxter 1977). The nutrients withheld in Lake Nasser contributed to some increase in the fishery there, but this was probably of little comfort to the sardine fishers on the coast (Ryder 1978; Ryder and Henderson 1975).

Navigation Dams

Much less has been written about the effects of low navigation dams on water regimes than on the more spectacular high dams used for water supply, flood control, and hydropower. Despite their modest size, navigation dams also alter water regimes and affect biological communities. The dams are of three types: (1) mainstem dams that maintain water depths at some minimum necessary for navigation, (2) rock dikes (wing dams) that extend perpendicularly part way from shore to confine the current to the main channel, and (3) other rock dams (closing structures) placed across side channels, also to confine flow to the main channel. The rock dams may be high enough to emerge from the water during low stages or they may be permanently submerged. In addition to stabilizing the position of the channel, the dams also promote scouring, reducing the need for dredging.

Although water depths could be maintained by a fixed weir and a navigation lock, a weir would increase flood heights above natural levels and increase the potential for flood damage to property. A fixed weir also would trap sediments that eventually would reduce water depths, thus negating the purpose of the structure. Instead, most navigation dams on the major rivers of the United States employ gates that can be raised from the bottom of the river completely out of the water (roller gates or tainter gates) or that fold into the bottom of the river (Chanoine wickets). The gates thus pass sediments and flood flows with little obstruction.

One major effect of navigation dams on the water regime is to increase the period of inundation of the floodplain, at least in the reach of the river closest to the dam. During normal low flow periods, dams keep the water levels from getting low enough to expose the floodplain (Fig. 5C, F). Reaches farther upstream of the influence of the dam may experience the full range of water level variation, from flood (Fig. 5A, D) to low flow (Fig. 5C, F). In the Illinois and upper Mississippi rivers, the reaches between navigation dams typically have a downstream, impounded portion and an upstream portion of approximately equal length that retains the floodplains and islands typical of the undisturbed river (Sparks et al. 1990).

Although navigation dams have little or no effect on water levels during major floods (because the gates are out of the water and the dams are overtopped, see Fig. 5A, D), the gates can be partially opened to control water levels at intermediate discharges (Fig. 5B, E). The operating goal is to maintain water depths for navigation while keeping the extent and duration of flooding to a minimum. This procedure minimizes the amount of land that the government has to acquire because of potential additional flooding due to operation of the dam; the government is not liable for

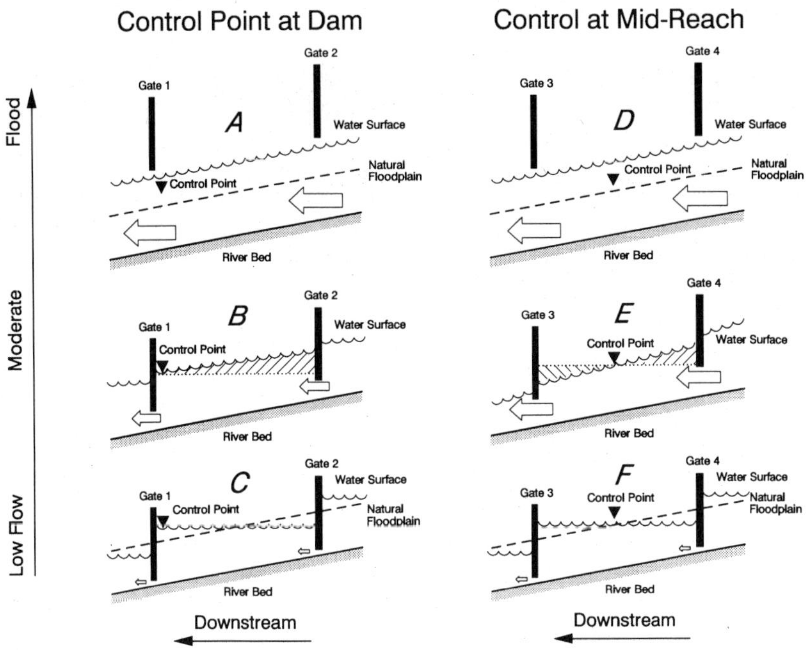

Figure 5. Effects of navigation dams on water regimes. Gates at the dams lift from the bottom and the size of the arrows indicates the magnitude of water flow. At high river flows, the gates are completely out of the water, the dam has no control over water levels, and the floodplain is inundated (A and D). At moderate river flows, the gates are operated to maintain stable water levels either at the dam (B) or at the midpoint of the reach (E). With the control point at the dam, the water surface tilts upstream, as though the control point were a hinge. With the control point at mid-reach (E), the water surface seesaws, as though the control point were a pivot. Downstream of the mid-reach control point, the water level *drops* as flow increases—just the opposite of the normal stage discharge relationship! At low river flows, the water surface is practically flat, regardless of the location of the control point (C and D); and the portion of the floodplain closest to the dam is inundated. The net difference between the two operating procedures is that with the control point at the dam the floodplain closest to the dam remains inundated at all river stages (A-C), whereas mid-reach control exposes portions of the floodplain during moderate river flows (D-F).

natural fluctuations of the river (Rock Island District Corps of Engineers 1991).

The decision whether to raise or lower the gates is made on the basis of a water level control point that may be located at the dam (Fig. 5A-C) or at a point midway to the next dam upstream (Fig. 5D-F). If the water is rising at the control point, the gates are opened to increase flow and maintain the control elevation; if the water is falling, gates are lowered. The water surface profile appears to be "hinged" at the dam if the control point is at the dam (Fig. 5B). If the control point is mid-pool, the water surface appears to seesaw about the control; the downstream end tilts down and the upstream end tilts up as discharge increases (Fig. 5E). Therefore, a control point at the dam is sometimes called a hinge point, and the mid-pool control a pivot point.

If the control point is at the dam, the floodplain near the dam is permanently inundated at all river stages (Fig. 5A-C). Also, water levels do not increase at the dam as discharge increases until all the gates are out of the water (compare Fig. 5B to 5A). Water levels upstream do increase in proportion to their distance from the dam (Fig. 5B). If the control point is at midreach, water levels near the dam actually *decrease* as discharge increases—just the opposite of the natural pattern (compare Fig. 5F to 5E)!

Reversal of the natural pattern is detrimental to plants and animals that need access to the floodplain. Waterfowl managers try to restore the natural water regime by erecting low levees on the floodplain and using pumps to raise water levels in spring and fall. Pumps or gravity drains remove water in summer to maintain low, stable water levels or mud flats during the growing season. These techniques are employed even in the mid- to upstream portions of the navigation pools on the Illinois and upper Mississippi rivers where the floodplain is still subject to natural fluctuations. In these areas, the purpose is not to overcome the effect of the navigation dams, but to make the water regime more predictable than it actually is for the benefit of a single group of organisms. Unfortunately, low levees cut off access by fish to the impounded areas during low river stages. In addition to fishes that use floodplains for spawning, nursery, and feeding areas in spring and early summer, freshwater drum and bluegill are known to winter in off-channel areas; consequently, biologists are likely to find more species as they do more winter sampling (Bodensteiner and Sheehan 1988; Bodensteiner et al. 1990). During floods, the low levees are overtopped and the fish have free access, but the fish may become entrapped when water levels drop and may die if the impoundments are completely dewatered. Water impoundments created by wildlife managers on the Mississippi delta curtailed the access and production of fish and crustaceans (Herke et al. 1987).

Closing structures and wing dams create their own problems and opportunities for fish and wildlife populations. In some side channels, closing structures reduce scouring flows and thereby cause them to fill with sediment; in others, the structures slow the rate of sedimentation by excluding the sand bed load that moves in the main channel. Wing dams accomplish their purpose of stabilizing the position of the main channel and promoting self-scouring; they also temporarily benefit some aquatic organisms. Rocks provide solid substrate for many kinds of invertebrates that serve as food for fish. Hard substrate is at a premium in alluvial rivers characterized by shifting sand and silt. The structures also create persistent eddies where fish can maintain position at low energy cost by avoiding the higher velocities of the main channel. Some of these eddies probably serve as hydraulic retention and distribution devices for detritus and dissolved organic matter, thereby creating local "hot spots" of enhanced secondary productivity (Adams 1986). The eddies distribute organic matter produced in inshore marshes and aquatic plant beds to offshore consumers, but also retain the material so that it is not lost rapidly to the main channel.

Unfortunately, these beneficial biological effects may be short lived because reduced current velocity increases sedimentation, which leads to the buildup of dry land outward from the shoreline between the wing dams. This not only reduces the aquatic habitat but also the total area accessible to the public in those states where accrued land belongs to the adjacent landowner.

PREDICTING AND REDUCING RISKS OF ALTERED WATER REGIMES

Prediction of risk depends upon understanding and information; reduction of risk requires the collective or political will to act upon that knowledge. The first step in predicting and avoiding risk is to ensure that all the physical, biological, and human dimensions of the river-floodplain ecosystem are considered and evaluated in the planning process. Risks are more likely to be considered and reduced if projects are broadly conceived in the context of *managing* a complex, dynamic system that fulfills many natural functions (e.g., transport of water, sediment, and nutrients and maintenance of deltas and estuaries) and human services, many of which have not been included in traditional analyses of costs, benefits, and risks in narrowly conceived economic development projects. Disciplines that should be represented on planning teams, in addition to economists and engineers, include fluvial geomorphologists, hydrologists, rural sociologists, fisheries ecologists, and systems ecologists with experience on large rivers. Epidemiologists should be included when waterborne diseases are likely to be a problem. These specialists can identify potential risks to the physical, biological, and societal components of the river system, including long-term and far-ranging effects, and determine whether the current data bases are adequate to quantify those risks. Projects with marginal economic benefits and high risks of throwing river-floodplain ecosystems into disequilibrium might be abandoned in the planning stage. The risks of losing attributes of the river-floodplain system (the floodpulse, floodplain habitats) that sustain fish and wildlife populations and local economies (e.g., recreational, commercial, and subsistence fisheries) should be evaluated, perhaps by considering what it would cost to replace (if replacement is even feasible) or maintain those services. Risk analysis should extend to fisheries and economies that may be distant but directly affected by the local project, as in the Nile and Mississippi deltas where fisheries were adversely affected by upstream dams. In view of the accumulated worldwide experience with water resource developments and their attendant risks, there is little excuse now for failing to evaluate risks during the planning stages of projects.

The same expertise used in planning new projects can assist in the modification of existing projects; e.g., the notching of wing dams to prevent excessive sedimentation and loss of aquatic habitats in the Missouri and upper Mississippi rivers. The need for appropriate conceptual understanding of fluvial systems may be even greater for planning restoration projects than development projects because restoration or mitigation activities are responses to relatively recent societal concerns about environmental degradation, and there is less accumulated experience on which to base planning and risk assessment (National Research Council 1992). Funds for planning, implementing, monitoring, and evaluating restoration and mitigation projects could be transferred from the maintenance costs of development projects that have proven to be uneconomical.

An essential element of risk assessment is to consider the status of knowledge about the ecosystem. The conceptual understanding of the physical system may be excellent, but, if few hydrological data are available to quantify the historical water regime, it may be impossible to predict what changes in the channel are likely to ensue, e.g., from withdrawal of water. A good understanding of the system sometimes extends the usefulness of an inadequate data base. For example, the developing understanding of the general relationship between El Nino-Southern Oscillation events in the Pacific and streamflow in the American Southwest will assist in reconstructing historical streamflows ("hindcasting") from relatively brief local records (e.g., see Molles and Dahm 1990). Such an analysis will prevent overallocating water based on streamflow measurements made during an exceptionally wet period, as in the case of the Colorado River (Reisner 1986). As information about mechanisms that drive weather patterns improves, so too will knowledge of long-term patterns in stream flow and ability to forecast extremes.

A good conceptual understanding of physical and biological processes in fluvial systems can be used to reason backwards from results to causes, thereby making it possible to reconstruct the approximate water and sediment regimes that created and maintained the present system. The present channel and floodplain morphology tells a fluvial geomorphologist much about the pattern of water and sediment loading of the river. Attributes of the water regime that are essential for the maintenance of biota can be intimated from life histories and physiological requirements. Such analysis is fairly straightforward for floodplain vegetation, which is often zoned according to land elevation and average water regime but has recently been applied also to young-of-the-year fish assemblages that reflect the reproductive potential of floodplain ecosystems for fishes (Copp 1989). Maintenance of biota depends on maintenance of the physical system that supports it. Observation of the present system, coupled with a good conceptual understanding of similar systems, can forestall risky manipulations and unpleasant surprises.

Almost any management scheme involves some departure from the unfettered water regime. What is particularly important is to avoid crossing thresholds that cause the physical or biological system to change from a desirable equilibrium state (usually, the predisturbance condition) to a less desirable state, usually with rapid and undesirable channel and species adjustments in between. Some thresholds are known, such as the water and sediment regimes that change a meandering channel to a braided one or turbidity levels that shift fish communities from sight-feeders to scent-feeders (Buck 1956).

There may be two opportunities to determine critical thresholds using planned or unplanned field experiments: once when an ecosystem degrades (analysis is often retrospective in this case, see Sparks et al. 1990) and again if it recovers. Planned restorations, thus, are important not only because of the value of the particular ecosystem that is restored but also because thresholds for maintenance of similar ecosystems might be determined and the risks of crossing those thresholds averted (National Research Council 1992).

Prediction and Avoidance of Problems

Attempts to mitigate the degradation of river-floodplain ecosystems are sometimes ineffective and expensive to maintain because they treat symptoms, rather than causes. Mitigation is often controversial because different species have different requirements and human advocates. For example, lateral water impoundments are proposed for some of the navigation pools on the upper Mississippi River where the normal floodpulse is inverted because of the method of dam operation (William Fritz, personal communication, October 31, 1991, Batchtown Habitat Rehabilitation and Enhancement Project, Upper Missippi River, Planning Division, St. Louis District Corps of Engineers). Pumps and gravity drains would be used to hold water for migratory waterfowl in spring and autumn and to lower water levels for the production of waterfowl food plants during summer. Fishery biologists oppose the plan because the impoundments would cut fish off from feeding, spawning, and wintering areas or entrap them during falling river stages and would destroy beds of mussels that prefer swift water. A similar argument was avoided in planning the Kissimmee River restoration by focusing on recreating the natural water regime that supported both fish and waterfowl (Loftin et al. 1990). Restoration of a more natural flood regime in the upper Mississippi River would require acquisition of flood prone land and modification of the operating procedures at the dams (Rock Island District Corps of Engineers 1991)—a solution that is systemic, rather than piecemeal, and probably cheaper in terms of extended operating costs (Bayley, in press).

Although the natural water regime might be suboptimal in a given year for fish, most warm water fishes are adapted to a variable system by means of a high reproductive potential that enables them to make up quickly for lost year classes (Bayley, in press; Junk et al. 1989). Migrant waterbirds have a very plastic, opportunistic behavior. If they fail to find food or suitable resting places, they move on. In contrast to fish, birds are able to move over levees and overland to other drainages and to inland cooling lakes and reservoirs created by man. When the food supply for diving ducks failed in the Illinois River in the late 1950s, the ducks shifted to the upper Mississippi River (Mills et al. 1966). The problem is that state and federal fish and wildlife agencies want to satisfy constituents who may view reduced wildlife populations on a favorite area in a given year as a failure of the agency. Public education undertaken by these agencies should include concepts such as compensatory recruitment, opportunistic behavior, and a longer term and larger scale view of biological productivity in river-floodplain ecosystems.

DISCUSSION

As the examples in this discussion indicate, the environmental effects of manipulating water regimes too often have come as unwelcome surprises. Past experience with environmental degradation and better application of principles and information from rapidly developing environmental sciences should lead to better management of rivers to meet demands for economic development as well as new societal demands for preservation of biodiversity, management of the "greenhouse" gases that regulate global warming, and outdoor recreation.

In past water resource developments, rivers were treated simply as hydraulic systems with little conception of their geological or ecological structure and function. Consequently, little concern was raised about the risks of altering water regimes that strongly influence both physical structure and biological communities. Integrative concepts from the disciplines of fluvial geomorphology and river ecology are important in understanding systems as complex as large rivers and in predicting their responses to manipulations. If the planners and designers of the large reservoirs on the upper Missouri River had understood rivers, both as physical systems that are in a dynamic equilibrium with their water and sediment loads and as life support systems, the consequences of trapping sediments in the reservoirs might have been anticipated and considered in the engineering design and economic evaluation of the project. However, fisheries and wildlife biologists were surprised later when the downstream channel downcut, leaving biologically productive lateral wetlands and channels high and dry (Hesse et al. 1982).

Until recently, ecological knowledge about river ecosystems was inversely proportional to their size and economic importance, and the ability to predict environmental effects lagged far behind the ability to engineer large rivers for economic development (Hynes 1989). For a time, one concept was dominant in river ecology—the River Continuum Concept (RCC) (Vannote et al. 1980)—and perhaps the warning of the French philosopher Emile Auguste Chartier should have tempered the enthusiasm of ecologists: nothing is more dangerous than an idea, when it's the only one we have (Bartlett 1980). The RCC is based on data from streams and small rivers draining forested watersheds and describes how biological processes and community structure in the channel change in a predictable way along a continuum from headwater streams to larger rivers; floods disturb or reset the continuum by shifting the substrate and flushing away accumulated organic matter, debris dams, and even organisms that do not have evasive behaviors.

The RCC stimulated development of concepts more applicable to altered rivers (the discontinuum concepts of Ward and Stanford 1983) and to large floodplain rivers (the floodpulse concept of Junk et al. 1989). According to Ward and Stanford (1983), dams shift the expected continuum, e.g., the reach immediately downstream of the dam becomes more like a headwater stream if cold, oxygenated water is released from the hypolimnion of the reservoir, or more like a lowland river if warm water containing plankton is released from the epilimnion.

In contrast to the view of floods as disturbances in the RCC, floods in large river-floodplain ecosystems are predictable annual events that facilitate the exchange of water, sediment, nutrients, and organisms between the channel and the floodplains that are the centers of biotic production. In these ecosystems, the *absence* of a flood constitutes a disturbance. The floodpulse is somewhat analogous to the tidal pulse of the coasts where biological communities have evolved to capitalize on the regular flushing and exchange of nutrients and organic matter, although the flooding frequency is much different. The water retention capacity of the floodplain is important in determining the duration of flooding and the rate of rise and fall of water: the greater the capacity in relation to the discharge capacity of the channel, the more long lasting the flood, the slower the advance and recession, and the greater the opportunity for organisms to take advantage of the floodpulse for spawning and feeding on the floodplain itself or in the expanded floodplain lakes and backwaters. In temperate zones, the timing of the floodpulse in relation to the annual temperature cycle is critical: a midwinter flood does not help floodplain-spawning fish.

Because biological productivity is highly dependent on the floodpulse, complete reversal of the flood pattern by some irrigation and navigation dams is a major disturbance. Irrigation dams typically store the spring flood for gradual release during the summer growing season, thereby reducing or eliminating the floodpulse downstream, and river levels sometimes rise in summer because of return flows from the irrigated fields. The floodplains immediately above some of the navigation dams on the upper Mississippi River are partially inundated during low flows to maintain adequate depths for navigation and are exposed during moderate-to-high flows because water levels are drawn down to minimize flooding. The floodpulse can be eliminated from floodplains by levees, as in the Illinois River, where half the floodplain has been drained primarily for agriculture (Mills et al. 1966). Disruption or loss of the floodpulse not only decreases the productivity of flood adapted fish and wildlife populations, but may require substitution of commercial fertilizer for the nutrient pulse once delivered to human agricultural systems, e.g., on the Nile floodplain in the Nile delta.

Optimizing water regimes for one or a few species in complex river ecosystems such as the Mississippi (which has thousands of species, including 260 kinds of fish) risks creating suboptimal conditions for others. A better approach is to approximate the natural flow regime that maintained the floodplain and channel morphology and the entire panoply of species. This, however, raises the question of what the natural flow regime was. In a few places, such as the upper Mississippi River, discharge and water elevation data are sufficiently long term to describe the predam water regime. Fluvial geomorphologists and river ecologists sometimes can deduce the past water regime from the present river morphology or life history adaptations of organisms. A retrospective modeling approach also might be used to reconstruct predisturbance flow regimes. Hydrological simulation models could be applied to watersheds that are "revegetated" according to soil types (soils are heritages of presettlement vegetation patterns).

Although complete absence or reversal of the timing of the floodpulse is clearly a major disturbance, much work remains to be done to quantify relationships between the floodpulse and productivity and to specify ranges within which the water regime can be altered without disequilibrating the physical or biological components of large river-floodplain ecosystems. It is possible to generalize about attributes of the floodpulse (seasonal timing, flood extent, duration, rate of rise, and rate of fall) that are important to biotic productivity, based on life history characteristics of organisms.

The scale of the system under analysis and the adequacy of knowledge about the system (conceptual understanding and data) should be carefully examined so that potential long-range and long-term risks are not overlooked. Water from the Colorado River was over-allocated partly because allocations were based on a relatively brief hydrological record collected during a period that was exceptionally wet (Reisner 1986). Knowledge of the mechanisms that govern global and regional weather patterns is rapidly expanding and will assist both in predicting and reconstructing flow regimes, e.g., a link between the El Nino-Southern Oscillation phenomenon in the Pacific Ocean and stream flow in parts of the Southwestern United States is now well established (Molles and Dahm 1990). The long-range delivery of sediments and nutrients from river basins to deltas stabilizes the deltas against subsidence and coastal erosion and maintains their productivity. On a global scale, the increased flux of nutrients from the land to the sea, resulting from anthropogenic disturbance, may increase marine photosynthesis and the uptake of carbon dioxide from the air, thereby reducing the rate of global warming, although the exact nature of these linkages and the direction of the trends remain highly speculative.

ACKNOWLEDGMENTS

Some of the research described in this chapter was supported by the National Science Foundation, Division of Biotic Systems and Resources, under the Long-Term Ecological Research (LTER) Program (grants BSR-8114563 and BSR-8612107); the Long-Term Resource Monitoring Program (LTRM) administered by the Environmental Management Technical Center of the U.S. Fish and Wildlife Service and the Illinois Department of Conservation; the Illinois Natural History Survey; and by the loan of equipment from the Upper Mississippi River Basin Association. I gratefully acknowledge the able assistance of Ruth M. Sparks, K. Douglas Blodgett, and Camilla S. Smith in preparing this article.

REFERENCES

ADAMS, J.R. (1986). Mechanics of a large eddy in the Mississippi River. In: *Advancements in Aerodynamics, Fluid Mechanics, and Hydraulics* (R.E.A. Arndt, H.G. Stefan, C. Farell, and S.M. Peterson, eds), pp. 645-652. Proceedings of the Specialty Conference, American Society of Civil Engineers, Minneapolis, MN.

BARTLETT, J. (1980). Familiar Quotations. Little, Brown and Company, Boston. 1540 pp.

BAXTER, R.M. (1977). Environmental effects of dams and impoundments. Ann. Rev. Ecol. Syst. 8:255-283.

BAYLEY, P.B. and PETRERE, M., JR. (1989). Amazon fisheries: Assessment methods, current status, and management options. In: *Proceedings of the International Large River Symposium (LARS)*. Can. Special Publ. Fish. Aquat. Sci. 106:385-398.

BAYLEY, P.B. (In press). The flood pulse advantage and the restoration of river-floodplain systems. Reg. Rivers, 25 pp.

BELLROSE, F.C., HAVERA, S.P., PAVEGLIO, F.L., JR., and STEFFECK, D.W. (1983). The fate of lakes in the Illinois River valley. Illinois Natural History Survey Biological Notes 119.

BELT, C. B., JR. (1975). The 1973 flood and man's constriction of the Mississippi River. Science 189:681-684.

BENKE, A.C. (1990). A perspective on America's vanishing streams. J. N. Am. Benthol. Soc. 9:77-88.

BODENSTEINER, L. and SHEEHAN, R. (1988). Implications of backwater habitat management strategies to fish populations. In: *Proceedings of the Forty-Fourth Annual Meeting of the Upper Mississippi River Conservation Committee*, pp. 60-65. Peoria, Illinois.

BODENSTEINER, L.R., LEWIS, W.M., and SHEEHAN, R.J. (1990). Differences in the physical environment of the Upper Mississippi River as a factor in overwinter survival of fish. In: *The Restoration of Midwestern Stream Habitat*, pp. 109-117. 52nd Midwest Fish and Wildlife Conference, Minneapolis, MN.

BORMANN, F.H., LIKENS, G.E., and EATON, J.S. (1969). Biotic regulation of particulate and solution losses from a forest ecosystem. BioScience 19:600-710.

BUCK, D.H. (1956). Effects of turbidity on fish and fishing. In: *Transactions of the Twenty-First North American Wildlife Conference* (J.B. Trifethan, ed), pp. 249-261. Wildlife Management Institute, Washington, D.C.

CALHOUN, S.W., ZIMMERMAN, E.G., and BETTINGER, T.L. (1982). Stream regulation alters acute temperature perferenda of red shiner, *Notropis lutrensis*. Can. J. Fish. Aquat. Sci. 39:360-363.

CHICAGO CORPS OF ENGINEERS. (1981). Increased Lake Michigan diversion at Chicago demonstration and study program information report to the Congress. Department of the Army, Chicago District, Corps of Engineers, Chicago, IL. 237 pp.

COLWELL, R.K. (1974). Predictability, constancy, and contingency of periodic phenomena. Ecology 55:1148-1153.

CONLIN, M. (1991). Illinois River fisheries and wildlife resources. In: *1991 Governor's Conference on the Management of the Illinois River System* (Korab, H., ed). Third Biennial Conference, Peoria, IL.

COPP, G.H. (1989). The habitat diversity and fish reproductive function of floodplain ecosystems. Environ. Biol. Fish. 26:1-27.

CROOME, R.L., TYLER, P.A., WALKER, K.F., and WILLIAMS, W.D. (1976). A limnological survey of the River Murray in the Albury - Wodonga area. Search 7(1):14-17.

DAHL, T.E. (1990). Wetland Losses in the United States: 1780's to 1980's. U.S. Department of Interior, Fish and Wildlife Service, Washington, D.C. 21 pp.

EGBUNIWE, H. (1976). Public health aspect of tropical water resources development. Water Resour. Bull. 12:393-398.

FREDEEN, F.J.H. (1977). A review of the economic importance of blackflies (Simuliidae) in Canada. Quaest. Ent. 13:219-229.

FREMLING, C.R., RASMUSSEN, J.L., SPARKS, R.E., COBB, S.P., BRYAN, C.F., and CLAFLIN, T.O. (1989). Mississippi River fisheries: A case history. In: *Proceedings of the International Large River Symposium*. Can. Special Publ. Fish. Aquat. Sci. 106:309-351.

GENTRY, A.H. and LOPEZ-PARODI, J. (1980). Deforestation and increased flooding of the upper Amazon. Science 210:1354-1356.

GEORGE, C.J. (1972). The role of the Aswan High Dam in changing the fisheries of the Southeastern Mediterranean. In: *The Careless Technology* (M.T. Farvar and J.P. Milton, eds), pp. 159-178. Natural History Press, Garden City, NY.

GOULDING, M. (1980). The Fishes of the Forest. University of California Press, LTD, London. 280 pp.

GRUBAUGH, J.W. and ANDERSON, R.V. (1988). Spatial and temporal availability of floodplain habitat: long-term changes at pool 19, Mississippi River. Am. Midl. Nat. 119:402-411.

GRUBAUGH, J.W. and ANDERSON, R.V. (1989). Long-term effects of navigation dams on a segment of the Upper Mississippi River. Reg. Rivers Res. Manage. 4:97-104.

HASTENRATH, S. (1990). Predictability of anomalous river discharge in Guyana. Nature 345:53-54.

HAVERA, S.P., BELLROSE, F.C., ARCHER, H.K., PAVIGLIO, JR., F.L., STEFFECK, D.W., LUBINSKI, K.S., SPARKS, R.E., BRIGHAM, W.U., COUTANT, L., WAITE, S., and MCCORMICK, D. (1980). Projected effects of increased diversion of Lake Michigan water on the environment of the Illinois River valley. U.S. Army Corps of Engineers, Chicago District. XXV + 550 pp. + appendices.

HAWKINSON, B. and GRUNWALD, G. (1979). Observation of a wintertime concentration of catfish in the Mississippi River. Minnesota Department of Natural Resources, Division of Fish and Wildlife Section of Fisheries, Investigational Report No. 365. 13 pp.

HERKE, W.H., KNUDSEN, E.E., KNUDSEN, P.A., and ROGERS, B.D. (1987). Effects of semi-impoundment on fish and crustacean nursery use: Evaluation of a "solution." Coastal Zone '87, May:2562-2576.

HERRICKS, E. E. and OSBORNE, L.L. (1985). Water quality restoration and protection in streams and rivers. In: *The Restoration of Rivers and Streams: Theories and Experience* (J. A. Gore, ed), pp. 1-20. Butterworth Publishers, Stoneham, MA.

HESSE, L. W., HERGENRADER, G.L., LEWIS, S.L., REETZ, S.D., and SCHLESINGER, A.D. (1982). The middle Missouri River. A collection of papers on the biology with special reference to power station effects. The Missouri River Study Group, Norfolk, NE. 301 pp.

HOLLAND, L.E. (1986). Distribution of early life history stages of fishes in selected pools of the Upper Mississippi River. Hydrobiologia 136:121-130.

HUGHES, R.M., WHITTIER, T.R., ROHM, C.M., and LARSEN, D.P. (1990). A regional framework for establishing recovery criteria. Environ. Manage. 14(5):673-683.

HYNES, H.B.N. (1989). Keynote address. In: *Proceedings of the International Large River Symposium (LARS)*. Can. Special Publ. Fish. Aquat. Sci. 106:5-10.

INTERNATIONAL COMMISSION OF LARGE DAMS. (1973). World register of dams.

JUNK, W.J. (1985). Temporary fat storage, an adaptation of some fish species to the water level fluctuations and related environmental changes of the Amazon river. Amazoniana IX:315-351.

JUNK, W., BAYLEY, P.B., and SPARKS, R.E. (1989). The flood pulse concept in river-floodplain systems. In: *Proceedings of the International Large River Symposium (LARS)*. Can. Special Publ. Fish. Aquat. Sci. 106:110-127.

KARR, J.R., and DUDLEY, D.R. (1981). Ecological perspective on water quality goals. Environ. Manage. 5:55-68.

KEOWN, M.P., DARDEAU, E.A., JR., and CAUSEY, E.M. (1981). Characterization of the suspended-sediment regime and bed-material gradation of the Mississippi River Basin. Potamology Program (P-I). Report 1, Volume II. U.S. Army Engineer District, New Orleans, LA. 375 pp.

KILBURN, P.D., TATE, D.J., and WHITE, S. (1981). An assessment for an increased diversion at Chicago. American Society of civil Engineers, St. Louis, Missouri. Preprint 81-524.

KWAK, T.J. (1988). Lateral movement and use of floodplain habitat by fishes of the Kankakee River, Illinois. Am. Midl. Nat. 120:241-249.

LAMBERT, T.R. and HANDLEY, J.M. (1984). Setting an instream flow requirement for smallmouth bass (*Micropterus dolomieui*). In: *Regulated Rivers* (A. Lilehammer and S.J. Saltveit, eds), pp. 387-406. Columbia University Press, NY.

LOFTIN, K.A., TOTH, L.A., and OBEYSEKERA, J.T.B. (1990). Kissimmee River restoration: Alternative plan evaluation and preliminary design report. South Florida Water Management District, West Palm Beach, FL.

LUBINSKI, K.S., WALLENDORF, M.J., and REESE, M.C. (1981). Analysis of Upper Mississippi River system correlations between physical, biological and navigation variables. Technical Report in partial fulfillment of Contract No. 895-305, Upper Mississippi River Basin Commission, St. Paul, MN.

LUBINSKI, K.S. (1984). Winter diving surveys of main channel microhabitats and fish populations in Mississippi River reaches subjected to thalweg disposal. Illinois Natural History Survey, Aquatic Biology Technical Report 1984(13). 86 pp.

MACDONALD, P.O., FRAYER, W.E., and CLAUSER, J.K. (1979). Documentation, chronology, and future projections of bottomland hardwood habitat losses in the Lower Mississippi alluvial plain. U.S. Department of the Interior, Fish and Wildlife Service, Washington, D.C. 2 Volumes.

MERMEL, T.W. (1976). International activity in dam construction. Water Power Dam Constr. 28(4):66-69.

MILLS, H.B., STARRETT, W.C., and BELLROSE, F.C. (1966). Man's effect on the fish and wildlife of the Illinois River. Illinois Natural History Survey Biological Notes No. 57, Urbana, IL. 24 pp.

MOLLES, M.C., Jr., and DAHM, C.N. (1990). A perspective on El Nino and La Nina: Global implications for stream ecology. J. N. Am. Benthol. Soc. 9:68-76.

NATIONAL RESEARCH COUNCIL. (1992). Restoration of Aquatic Ecosystems: Science, Technology, and Public Policy. National Research Council, Water Science and Technology Board, Washington, D.C.

OMERNIK, J.M. (1987). Ecoregions of the conterminous United States. Ann. Assoc. Am. Geogr. 77:118-125.

OSBORNE, L.L. and HERRICKS, E.E. (1987). Microhabitat characteristics of *hydropsyche* (Trichoptera: Hydropsychidae) and the importance of body size. J. N. Am. Benthol. Soc. 6:115-124.

PENLAND, S. and BOYD, R. (eds). (1985). Transgressive Depositional Environments of the Mississippi River Delta Plain: A Guide to the Barrier Islands, Beaches, and Shoals in Louisiana. Guidebook Series No. 3. Louisiana Geological Survey, Baton Rouge, LA. 233 pp.

PIANKA, E.R. (1970). On r- and K-selection. Am. Nat. 104:592-597.

POFF, N.L. and WARD, J.V. (1989). Implications of streamflow variability and predictability for lotic community structure: A regional analysis of streamflow patterns. Can. J. Fish. Aquat. Sci. 46:1805-1818.

REISNER, M. (1986). Cadillac Desert: The American West and Its Disappearing Water. Penguin Books, NY. 582 pp.

ROCK ISLAND DISTRICT CORPS OF ENGINEERS. (1991). Identification of constraints on regulation of Upper Mississippi River System Lock and Dam 18. U.S. Army Corps of Engineers, Rock Island, Illinois, for U.S. Fish and Wildlife Service, Environmental Management Technical Center, Onalaska, WI. 54 pp.

RYDER, R.A. and HENDERSON, H.F. (1975). Estimates of potential fish yield for the Nasser Reservoir, Arab Republic of Egypt. J. Fish. Res. Bd. Can. 32:2137-2151.

RYDER, R.A. (1978). Fish yield assessment of large lakes and reservoirs-a prelude to management. In: *Ecology of Freshwater Fish Production* (S.A. Gerking, ed), pp. 403-423. Blackwell Scientific Publications, London.

SHEAFFER, W.A. and NICKUM, J.G. (1986). Backwater areas as nursery habitats for fishes in Pool 13 of the Upper Mississippi River. Hydrobiologia 136:131-140.

SINGH, K.P. and RAMAMURTHY, S.R. (1990). Climate change and resulting hydrologic response: Illinois River Basin. In: *Watershed Planning and Analysis in Action Symposium Proceedings of Illinois River Conference Watershed Management*, pp. 28-37. Illinois River Division, Durango, CO.

SPARKS, R.E. (1984). The role of contaminants in the decline of the Illinois River: Implications for the Upper Mississippi. In: *Contaminants in the Upper Mississippi River* (J.G. Wiener, R.V. Anderson, and D.R. McConville, eds), pp. 25-66. Proceedings of the 15th Annual Meeting of the Mississippi River Research Consortium. Butterworth Publishers, Stoneham, MA.

SPARKS, R.E., BAYLEY, P.B., KOHLER, S.L., and OSBORNE, L.L. (1990). Disturbance and recovery of large floodplain rivers. Environ. Manage. 14:699-709.

STEVENS, M.A., SIMONS, D.B., and SCHUMM, S.A. (1975). Man-induced changes of middle Mississippi River. J. Waterways Harbors Coastal Eng. Div. May:119-133.

STRAHLER, A.N. (1952). Hypsometric analysis of erosional topography. Bull. Geol. Soc. Am. 63(6):1117-1142.

STRAHLER, A.N. (1957). Quantitative analysis of watershed geomorphology. Am. Geophysic. Union Trans. 38(6):913-920.

SWANK, W.T. and DOUGLAS, J.E. (1974). Streamflow greatly reduced by converting deciduous hardwood stands to pine. Science 185:857-859.

SWANSON, F.J. and R.E. SPARKS. (1990). Long-term ecological research and the invisible place. BioScience 40(7):502-508.

TALBOT, M.J. (1982). Catfish wintering habitat evaluation. In: *Mississippi River Work Unit Annual Report 19, 1981-1982.* Wisconsin Department of Natural Resources, LaCrosse, WI.

TALBOT, M.J. (1984). Evaluation of flathead catfish habitat selection. In: *Mississippi River Work Unit Annual Report 1982-1983.* Wisconsin Department of Natural Resources, LaCrosse, WI.

THURSTON, H. (1991). Power in a land of remembrance. Audubon 93(6):52-59.

TODD, B.L., DILLON, F.S., and SPARKS, R.E. (1989). Barge effects on channel catfish. Illinois Natural History Survey Technical Report 89/5, Champaign, IL.

TURNER, L.M. (1929). The 1926-27 floods and the Illinois river valley vegetation. Trans. Ill. State Acad. Sci. 22:95-97.

TURNER, R.E. and RABALAIS, N.N. (1991). Changes in Mississippi River water quality this century. Implications for coastal food webs. Bioscience 41:140-147.

U.S. DEPARTMENT OF COMMERCE AND LABOR. (1911). Special reports. Fisheries of the United States 1908. Government Printing Office, Washington, D.C. 324 pp.

VANNOTE, R.L., MINSHALL, G.W., CUMMINS, K.W., SEDELL, J.R., and CUSHING, C.E.C. (1980). The river continuum concept. Can. J. Fish. Aquat. Sci. 37:130-137.

VOROSMARTY, C.J., GILDEA, M.P., and MOORE, B. (1986). A global model of nutrient cycling: II. Aquatic processing, retention and distribution of nutrients in large drainage basins. In: *Watershed Research Perspectives* (D.L. Correll, ed), pp. 32-56. Smithsonian Institution Press, Washington, D.C.

WADDY, B.B. (1975). Research into the health problems of man-made lakes, with special reference to Africa. Trans. R. Soc. Trop. Med. Hyg. 69:39-50.

WARD, J.V. and STANFORD, J.A. (1982). Thermal responses in the evolutionary ecology of aquatic insects. Ann. Rev. Ent. 27:97-117.

WARD, J.V. and STANFORD, J.A. (1983). The serial discontinuity concept of lotic ecosystems. In: *Dynamics of lotic ecosystems* (T.D. Fontaine, III and S. M. Bartell, eds), pp. 29-42. Ann Arbor Science Publishers, Ann Arbor, MI.

WELCOMME, R.L. (1985). River fisheries. FAO Fisheries Technical Paper 262. Food and Agriculture Organization of the United Nations, Rome, Italy. 330 pp.

WENDLAND, W.M. (1987). Illinois rivers, Illinois. In: *The Climates of the Long-Term Ecological Research Sites* (D. Greenland, ed), pp. 34-39. Institute of Arctic and Alpine Research, University of Colorado, Occasional Paper No. 44.

WESTPHAL, J.A. and CLEMENCE, S.P. (1976). SLD potamology study (S-7), pp. 1-59. The Department of the Army, St. Louis District, Corps of Engineers, University of Missouri, Rolla, MO.

WETMORE, F. (1987). Flood damage protection programs. In: *Management of the Illinois River System: The 1990's and Beyond*, pp. 89-102. Illinois River Resource Management. Peoria, IL.
WHITLEY, J.R. and CAMPBELL, R.S. (1974). Some aspects of water quality and biology of the Missouri River. Trans. Mo. Acad. Sci. 8:60-72.
ZIMMERMAN, E.G. (1984). Genetic and physiological correlates in fish adapted to regulated streams. In: *Regulated Rivers* (A. Lilehammer and S.J. Saltveit, eds), pp. 273-292. Columbia University Press, NY.

CHAPTER 10

ECOLOGICAL IMPACTS AND RISKS ASSOCIATED WITH FOREST MANAGEMENT

L. A. Morris,[1] P. B. Bush,[2] and J. S. Clark[3]
[1]Warnell School of Forest Resources
[2]Cooperative Extension Service
[3]Department of Botany
The University of Georgia
Athens, Georgia

INTRODUCTION

Forest management revolves around two basic paradigms: (1) benefits accrued from management activities should exceed the cost of production and (2) sustained continuity of forest production (or production capabilities) must be assured (Chapman 1950). Activities that are consistent with these paradigms constitute forest management. Thus, forest management can cover a range of activities including periodic selection of individual trees for harvest, establishment of plantations of genetically selected trees bred in seed orchards or propagated by tissue culture, and intensive cultural practices such as pesticide application, fertilization, selective thinning (removing some trees to allow other trees room to grow), and pruning (removing branches to promote certain tree growth forms). All of the activities constitute forest management, and all work within the two fundamental paradigms discussed by Chapman.

Included within the boundaries of forest management is the lowest intensity management possible—taking no active role in forest development. What distinguishes such management from no management is that decisions are made on the basis of the benefits sought and the capability of such actions (no active action) to continue providing these benefits. It is in the establishment of desired benefits, the cost/benefit accounting algorithm, and our understanding of continuing production that many issues in management of forests arise.

Human activities that initiate change and pose risks to forest ecosystems can be grouped into these general categories (Franklin 1978):

(a) alteration of basic climatic conditions within which forests exist.
(b) introduction of exotic organisms into the ecosystem.

1. Corresponding Author: Larry A. Morris, School of Forest Resources, The University of Georgia, Athens, GA 30602.
2. Key Words: biodiversity, endangered species, harvest, long-term productivity, nutrient balance, pesticides, water quality, wildlife habitat.

(c) elimination of natural biotic components from the ecosystem.
(d) alteration of natural patterns of disturbance and succession through harvesting plants or control of composition.
(e) removal of material, especially organic matter and nutrients, from the forest.
(f) physical alteration of the site, such as disturbances associated with soil compaction and drainage.
(g) addition of materials to the forest, such as pesticides or fertilizers.

Of these activities, the last four (d-g) are directly related to management decisions for specific forest stands or landscapes. The first three (a-c) may be associated with specific forest management decisions, but are more typically associated with human activities on a global scale. For instance, conversion of forests to other uses, such as pasturelands or commercial croplands, is not included in forest management; neither is clearcutting of tropical forests followed by slash burning and planting of agricultural food crops, even if the land reverts to forest after being abandoned. In contrast, clearcutting of tropical forests followed by slash burning and regeneration of trees is considered forest management when these activities are judged to regenerate forests to meet desired goals.

Equally as significant as these activities are the impacts of exotic species introduced into forests through activities unrelated to forest management. For instance, the American chestnut *Castanea dentata*, which was the most important hardwood tree in North America at the turn of this century, was eliminated from eastern forests by a blight disease (*Endothia parasitica*) introduced into New York City in 1904 on a shipment of horticultural plants (Beattie and Miller 1954). Similarly, the gypsy moth (*Lymantria dispar*), one of the most destructive defoliators of eastern hardwood forests, was introduced into this country when it escaped from captivity near Medford, MA, where attempts were being made to cross it with silkworms (Baker 1972). This insect is still shaping forests of the eastern United States.

Agriculture has also had major effects on forests. For example, conversion of hardwood forests, which once dominated the Piedmont region of the southern United States, to stands dominated by loblolly and shortleaf pine had little to do with forest management. Rather, it was a consequence of clearing these lands for agriculture and then abandoning them after years of farming. Pines were more suited for colonizing the eroded and nutrient depleted land that was left than were the hardwood species that thrived prior to cropping.

Ecological risks associated with forest management are reviewed here in the context of these large "accidental" changes in forest ecosystems and with an understanding that any management activity will cause ecosystem changes. A brief introduction covers various approaches to forest management. The remainder of this chapter is organized into sections that evaluate risks associated with: altered forest structure and diversity, long-term soil productivity, water quality and yield, pesticide use, and atmospheric changes. Relative risks within these categories are then briefly summarized.

FOREST MANAGEMENT SYSTEMS

Numerous factors are involved in determining the type of forest management and specific activities that occur within a particular forest area. Paramount in these factors are landowner or government objectives. Lands owned or leased by the forest industry are often managed to maximize fiber production with minimized per unit cost of lumber or fiber on a continuing basis. Secondary objectives may include maximizing net returns from hunting leases, protection against catastrophic loss, and promotion of public image. Similar objectives may be true for forests managed by government agencies; however, more typically, these agencies consider the promotion of recreational uses, provision of range, watershed protection, habitat preservation, and promotion of biodiversity as important as wood or fiber production. Management objectives for non-industrial, privately held forests cover a spectrum ranging from preservation of unique undisturbed forest communities (such as typified on lands owned by The Nature Conservancy) to optimization of short-term revenues from timber harvest. Whether timber harvest can be considered forest management or resource mining depends on the consideration and planning for continuation of a forest; a clearcut harvest followed by sale of the land for real estate development is not considered forest management since regeneration of the forest and its future use are not planned.

Forest management that includes wood utilization typically involves four classes of on-the-ground activities: inventory, development of access, harvesting in the current forest, and regeneration of future forests. It is convenient to consider the last three of these within broad groupings based on the level of physical and technological input as illustrated in Figure 1.

Managed Forests

Protected Wild Forests. These forests are managed with the lowest level of physical and technological input and, in the United States, are best represented by wilderness areas. They are protected either from resource utilization and its associated road construction and harvesting or from devastating natural processes such as wildfires. Both protection activities are consistent with the management objectives of providing a forest in which plant associations develop and succession proceeds uninterrupted. More recently, goals for management of these forests have been expanded to include management for biodiversity. Achieving such goals may not always be compatible with complete protection (Sharitz et al., in press). For instance, long-leaf pine/wire grass communities of the U.S. Southeast are not easily maintained in the absence of fire.

Regulated Forests. Management in regulated forests encompasses the forest tradition brought from Europe in the late 1800s. Tree species composition and structure of the forest are controlled by the number, species, and age

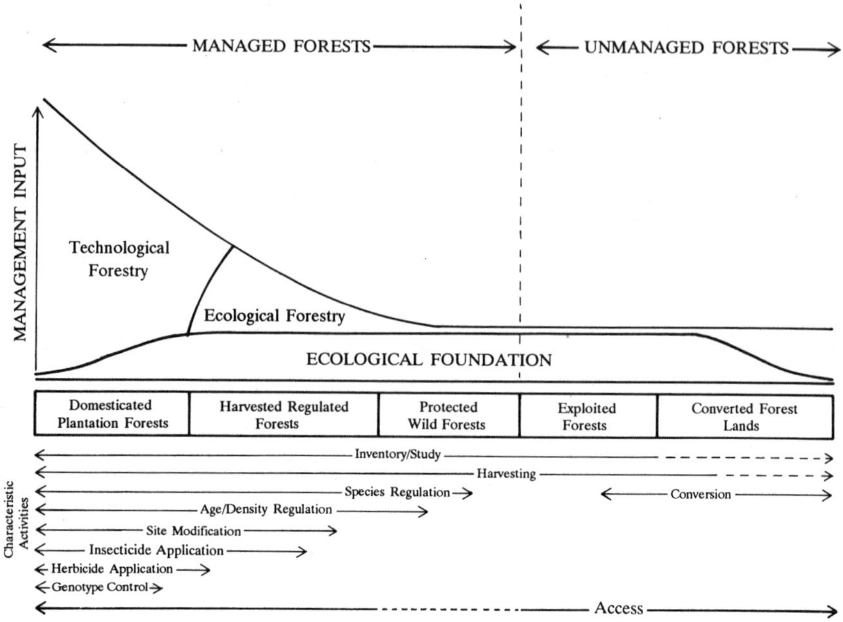

Figure 1. Classification of forests along a managerial purpose and physical input gradient with characteristic activities indicated (adapted from Stone 1975).

of the trees harvested; forest regeneration is accomplished by encouraging recruitment and growth of desired species. Depending on the species that are to be favored, the forest is either gradually harvested through removal of individual trees or small groups of trees over an extended period of time or harvested by removal of virtually all trees in a relatively large area. Harvesting systems that create only small openings in the forest canopy encourage growth and reproduction of shade tolerant tree species (e.g., red oak, *Quercus rubra*; eastern hemlock, *Tsuga canadensis*; sugar maple, *Acer saccharum*). Young trees that exist in the understory or seeds from mature trees that remain following each partial harvest provide for regeneration. The resulting forest will include trees of various ages and will not be dominated by any single age class. (An age class is defined as a group of trees established in response to a single disturbance event. Because seedling germination and establishment may proceed for a number of years, all trees within the same age class are not the same age. Generally, trees are considered to be within the same age class if the age variation is less than 20% of the age of the oldest trees in the stand [sensu Smith 1986].) Harvesting systems that create large, cleared openings encourage reproduction and growth of species that are most competitive in full sunlight. Fast growing pioneer species, including many species of pine (*Pinus* spp.) as well as most poplars (*Populus* spp.), are included within this group. In these systems, the forest is regenerated by seeds left in-place from harvested trees, stored in the forest floor, blown in from adjacent stands, or from a few widely scattered mature trees left on the site following harvest. Forests developed after such harvests are dominated by trees of a single age class and are referred to as even-aged forests.

Many hardwood tree species sprout readily from stumps or roots after harvest. Because they begin with reserves of stored carbohydrates and a developed root system, sprouts arising in this manner grow much faster than seedlings of equal age. The use of such sprouts as the primary means of regeneration following partial or clearcut harvest is known as coppicing. Management of forests for fuelwood by coppice methods has been used in Europe since 1086 (Rackham 1986) and continues today in both undeveloped and developed countries (Savil and Evans 1986).

Several silvicultural (silviculture is the practice of controlling forest establishment, composition, and growth through application of forest ecological principles) systems exist within each of the following general harvesting and regeneration regimes: single tree selection, group selection, shelterwood, strip shelterwood, group shelterwood, seed tree, and clearcutting (Smith 1986). Essential differences in the structure of forest stands resulting from these methods are illustrated in Figure 2. Heterogeneity is greatest in uneven-aged stands produced by single-tree or group selection methods. Seed-tree methods produce the most uniform overstory conditions.

Domesticated Plantation Forests. These forests are managed by physical and technological activities that are, in many ways, analogous to traditional agriculture. They are typified by assemblages of relatively large 50- to 500-hectare blocks of planted trees of a single species. Within each block, all trees are of a single age; however, age differences occur among blocks. Unlike regulated forests, in which species composition, stand structure, and growth are controlled by harvesting, these forests are controlled by planting considerations and cultural treatments in established stands. As with any type of timber utilization, road construction (development of access) signals the first step in the harvest. On poorly drained sites, this may be preceded or associated with artificial drainage. Clearcut harvesting is the norm. Following harvest, pre-planting site preparation is usually required. This may include reduction of logging slash and dead wood material by chopping, burning, or pushing them into localized piles. The soil may be tilled and/or restrictive subsoil conditions ameliorated. On wet sites, soil may be mounded to create planting spots that are elevated above the water table (Fig. 3). Fertilization, particularly with phosphorus fertilizers, is often part of preparing the site for planting. Following preparation, the site is planted with nursery grown seedlings from parent trees selected on the basis of growth rate, resistance to selected insects or disease, shape, or other characteristics. Following site preparation or planting, growth of undesirable plant species is reduced by mechanical control or herbicide application. Additional fertilization, often associated with thinning, is common as planted trees begin to mature.

Clearly, the potential for ecological change is greatest within domesticated plantation forests. Not only is the stand and species structure dramatically

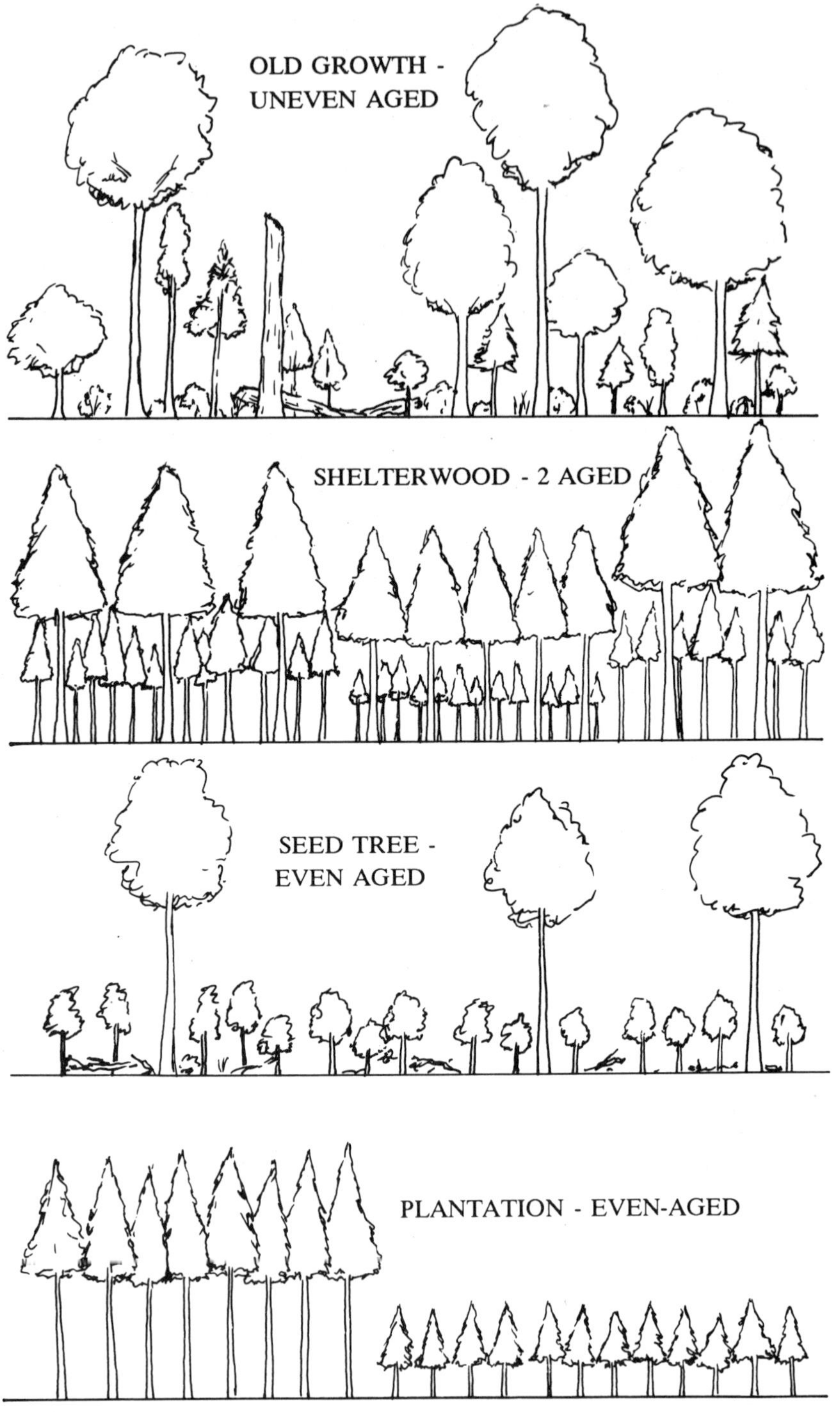

Figure 2. Representative stand structure of forests managed under different silvicultural systems from protected old-growth to plantation forests.

Figure 3. Raised beds formed during site preparation for plantation establishment on a wet Coastal Plain site in the southeastern United States. Note seedlings on planted beds.

changed, but drainage, soil disturbance, fertilizer addition, and use of pesticides increase the potential for secondary impacts.

Unmanaged Forests

Exploited Forests. Exploited forests are used by man for wood, forage, or other raw materials with little regard to the consequences of these actions. Unlike protected forests, access is uncontrolled. Tropical forests subject to slash-and-burn agriculture are included within this type of forest. Any positive results from such activities are entirely due to serendipity. For this reason, they are categorized as unmanaged forests.

Converted Forest Lands. Forests have been managed, exploited, and, in some cases, eradicated, either purposefully or unintentionally, throughout human history. Lebanon was the source of cedar for Solomon's temple, and Pliny the Elder discussed trade in ash cut from Mount Ida, Greece, in the first century (Makkonen 1967). Neither country is now known for its forests. Britain was once largely forested, and open areas were confined to natural openings along river valleys, wetlands, and other places where soil conditions were severe or by openings maintained by large herbivores or cleared by beaver (Warren and Key 1991). By the Iron Age (500 BC), 50% of these woodlands had been cleared; 15% remained 1500 years later (1000 AD) (Rackham 1986), and 2% remain today (Warren and Key 1991). Forest loss has accelerated during the last 150-200 years, particularly in the tropics. For example, within the last 35 years, 75% of Madagascar's rainforests have been converted to other uses (Green and Sussman 1990). The ecological risks associated with such deforestation are global in nature.

FOREST STRUCTURE AND DIVERSITY

When used as a qualitative term, the meaning of *diversity* is apparent: diversity relates to numbers and arrangement of items. Biodiversity is diversity restricted to biological items such as plants, mammals, insects, or any other subset of living things. Intuitively, diverse communities are communities with many species occurring in combination.

Although biodiversity is most often considered with respect to the number and arrangement of species (Franklin et al. 1981), species represent only one level of organization. Biodiversity includes both lower levels of organization, such as genetic diversity, as well as higher levels of organization, such as structural or functional diversity. In managed forests, concern is with (1) diversity of gene pools within a species (e.g., genetic manipulation in domesticated plantation forests), (2) diversity of species, and (3) functional diversity such as associated with differences in organic matter decomposition and accumulation among plantation forests, regulated forests, and old-growth.

Diversity is important for several reasons. First, diverse systems may be less susceptible to devastation or destruction by pests or other disturbances. (Although it was once believed that diverse systems were always more stable, it is now generally recognized that high species diversity does not guarantee stability [Wilson 1989], rather that this must be evaluated in the context of the specific community and type of disturbance.) Catastrophic disturbances can affect important forest ecosystem functions such as water storage and supply, erosion protection, and soil development. Second, loss of diversity is associated with an increased potential for species extinction or, at a minimum, reduction of the genetic base of extant species, many of which have significant unforseen value. The concern here is not with local changes in diversity (such as within a single forest stand) but with regional and global populations. For instance, the pacific yew (*Taxus brevifolia*) has been considered an undesirable tree in forests managed for timber production. Recently, it was discovered that this plant produces a chemical called taxol that is useful in the treatment of ovarian cancer. At present, no synthetic substitute for natural taxol exists (Murray 1991), and availability of this treatment depends on harvest of this tree. Third, diverse communities are considered more aesthetically appealing, which apparently relates to the perception of randomness (Gleick 1987).

Plant Species Diversity

In an introduction to his popular text on silviculture, Smith (1986) lists six major tools of forest management. Listed first among these tools is control of species composition. Such positioning is no accident for, even in the least intrusive types of forest management, changes in species composition are an inevitable and desirable impact of forest management. Even a decision to manage forests as wilderness will result in species change through succession or natural disturbance.

Responses of plant species diversity to forest management are highly variable because of the ways plants adapt (MacArthur 1972; Tilman 1982). Tradeoffs in the allocation of carbon and nitrogen to resource acquisition activities, defense, growth, and reproduction occur among all species (e.g., Chapin et al. 1990). Excelling in any one limits the extent to which the species can deal with other aspects of growth and reproduction. The maintenance of diversity, then, is dependent on the diversity of opportunities available for specialization. Structural, edaphic, and biotic factors all contribute to habitat variability, which in turn permits coexistence of large numbers of different species and genotypes within species.

Diversity During Stand Development. After a major disturbance, sites are initially colonized by rapid growing, short-lived herbaceous species that are able to exploit high light availability. The composition of this initial assemblage depends on the availability of seed and site factors such as soil fertility, moisture, temperature, organic content of soil, and soil physical conditions. Harvesting and forest regeneration activities affect all of these soil conditions (Morris and Lowery 1988; Vitousek and Matson 1985) and affect plant diversity.

Even minor management activities can affect understory populations. For example, thinning has been shown to reduce populations of *Senecia sylvaticus*, a summer annual forb, allowing a winter annual variety to develop. Soil tillage during plantation establishment has been shown to favor populations of figwart (*Scrophulaira rodosa*) that can acidify soil in the root zone and, thus, increase uptake of iron. Apparently, plants lacking this characteristic suffer from iron deficiency when the acidifying effects of the forest floor are eliminated by tillage (Ernst 1984).

Diversity of canopy species that succeed early successional forbs and grasses largely depends on initial establishment success. Trees that establish following disturbance continue to dominate the forest canopy for a full generation. Open areas are few in such stands, and the closed canopy restricts further seedling recruitment (Oliver 1981; Sprugel 1985). This closed canopy is maintained as long as growth rates of canopy individuals is sufficiently high so that mortality is compensated by continual lateral growth of individual trees. During this phase, herb and shrub layer development is restricted.

As trees mature and growth rates slow, canopy gaps begin to appear (Oliver 1981; Peet 1981; Sprugel 1985). Gaps in the forest canopy created by removal of even a few large trees change the climate at ground surface, resulting in changes in available sunlight, temperature, moisture, and nutrients. The plant communities that establish themselves in these gaps differ from the surrounding forest. With the opening of the canopy, recruitment to the understory increases, horizontal and vertical structure increases, and diversity may increase. Over time, the resulting forest is a mosaic of age and density classes, providing a spectrum of different environments and species. In managed forests, subtle and gradual changes in plant communities are replaced by more dramatic changes. In the most intensively managed forest plantations, harvesting can be likened to a natural catastrophe that kills many of the overstory trees. Windthrow associated with infrequent major storm events, fire, and insect or disease outbreaks are examples of such natural catastrophes. Because site conditions created by such events provide a more uniform habitat than that created by maturation of undisturbed forests, the variety of tree species that are competitive for the site resources is more limited. Domesticated forest plantations and even-aged stands that naturally occur following disturbance are characterized by low tree species diversity; however, the diversity of understory communities may, initially, be greater. For instance, Swindel et al. (1987) and Conde et al. (1983a,b) examined the effect of intensive site preparation treatments (burning, bedding, windrowing, and disking) on plant diversity in Coastal Plain flatwoods. They determined that plant species diversity 5 years after site preparation and planting of slash pine (*Pinus elliotti*) was greater than in stands prior to harvest. A recent analysis of burning and fertilization in pine/wire grass systems of southern Alabama showed this same trend (C.E. Lewis, personal communication).

Reduction of vegetation regrowth that competes with planted trees is one of the most important steps in establishing plantation forests. Herbicides are used extensively for this purpose in developed nations where labor is expensive and concerns about accelerated soil erosion from mechanical treatments, which can also be used for this purpose, are great. The commonly accepted wisdom has been that such herbicide applications delay plant succession in forests but do not alter the plant diversity. However, recently published reviews (Miller et al. 1990a,b) have concluded that there is little hard evidence to support this claim. Herbicide application may merely retard succession and diversity development or it may propel diversity on a completely different trajectory, leading to either increased or decreased diversity (Fig. 4).

Landscape Level Diversity. Diversity can also be evaluated on a scale larger than a typical forest stand of 100 hectares or less. At a landscape level, diversity includes the mixture of species, age, and density classes among as well as within individual forest stands. Biological processes important in plant community establishment, such as mean dispersal distances of seed and critical patch size, determine conditions under which populations become isolated (Okubo 1980) and are of particular significance to landscape level diversity.

Forest practice involves harvest schedules that are incompatible with those related to the life history of some plant species. Many weedy and early

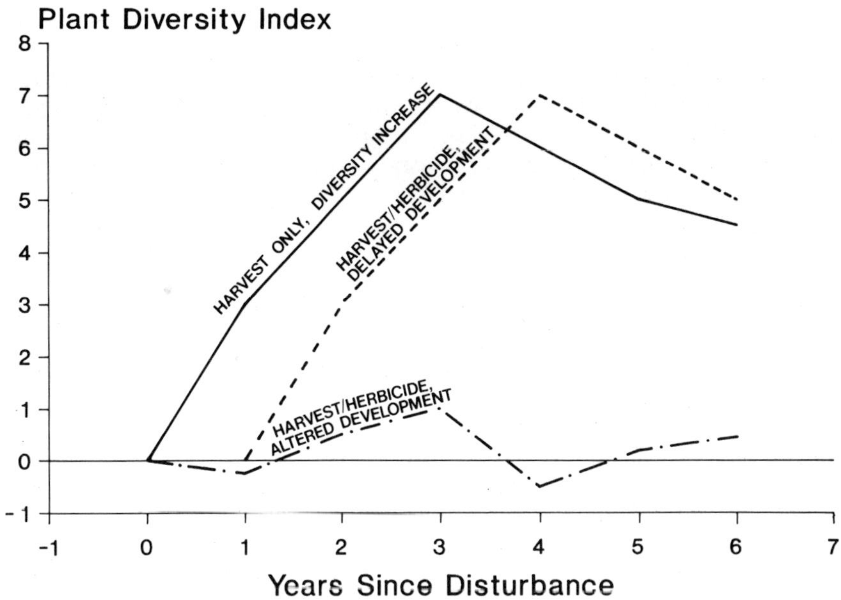

Figure 4. Potential changes in plant species diversity following commercial clearcut harvesting with and without herbicide application (after Neary et al. 1990).

successional shrubs and trees rely on the availability of disturbed patches at frequent intervals to maintain populations. Species that have been abundant in the United States since the time of European settlement such as birch (*Betula* spp), aspen (*Populus tremuloides*), and ragweed (*Ambrosia* spp.) would be relatively rare in the absence of continued human disturbance. Likewise, long lived perennials, such as many late successional trees, are uncommon in landscapes that experience frequent disturbances. On a landscape level, diversity is maximized by a combination of recently disturbed lands, lands in intermediate stages of disturbance, and undisturbed old-growth. Within a stand, diversity would be maximized by single tree and group selection methods that mimic the openings created by natural gaps. On a landscape level, combinations of old-growth, selection methods, and even-aged plantations would create a mosaic that maximizes biodiversity.

Plant Species Extinction. The most alarming threat to biodiversity is associated with extinction of individual species. Species extinction has occurred at varying rates since life on this planet began, but five episodes of mass extinctions have been identified (Pielou 1991). The most recent mass extinction occurred at the end of the Pleistocene Epoch 12,000 to 30,000 BC. Loss of species appears to have accelerated since the Industrial Revolution. Since 1600, *known* extinctions of 384 species of vascular plants have occurred (Reid and Miller 1989). These losses represent about 2% of known plant species.

The contribution of forest management to plant species extinction is not obvious. As already discussed, forest management is not reasonably equated with deforestation and land conversion, slash and burn agriculture, or exploitation. Lowest intensities of regulated forest management, which utilize single tree or group selection (Fig. 2), are unlikely to contribute significantly to species extinction except in cases where a rare species is selectively or unwittingly harvested.

While the potential for loss of plant species certainly exists in intensively manged forests, increasing site management is often required for perpetuation of populations of endangered species (Falk 1990). For instance, the known population of Virginia roundleaf birch (*Betula uber*) consists of 20 plants located on protected land in a single stream valley in Virginia (Little 1981). Appropriate forest harvest of areas adjacent to these trees could provide open areas necessary for regeneration of the species and might aid in stabilizing the population.

Insect and Other Animal Habitat

Role of Coarse Woody Debris. Undisturbed forests are dominated by mature and declining trees and are characterized by an abundance of standing dead trees together with fallen trees, branches, and foliage on the forest floor in various stages of decomposition. In cool climates, this debris accounts for 20-30% of the plant biomass in the forest (Harmon and Hua 1991). These organic materials serve as an energy source for saprogenic organisms and provide habitat. Large woody material (either standing or dead) is particularly important because of its longevity. In addition to serving as an energy source, this material provides unique habitats for plants (Harmon et al. 1986), microorganisms, arthropods (Maser and Trappe 1984), and other animals (Forsman et al. 1984); may be important sites for nutrient flux, particularly for nitrogen fixation; and contributes to soil structure.

Regardless of the forest region, the amount of coarse woody debris present in the forest decreases as management increases from protected forests to intensively managed, domesticated forests. For instance, in old-growth, Douglas fir forests of the U.S. Northwest, coarse debris in all stages of decomposition may exceed 90 tons/acre. In managed forests and plantations, debris is less than 10 tons/acre (Table 1).

Habitat provided by dead trees and woody debris can be particularly important to insect populations. Because of the low occurrence of suitable habitat, the richest assemblages of insects requiring mature and senescent trees will generally occur within larger areas of undisturbed timber (Warren and Key 1991). Reductions and fragmentation of areas of older timber will reduce the abundance of species dependent upon these niches.

Some species of larger animals, particularly birds, also depend on habitat associated with standing dead or fallen woody debris. In their review, Maser et al. (1988) list 18 habitat functions associated with standing dead trees in forests of the northwestern United States: nesting cavities, nesting platforms, feeding substrate, plucking posts, singing or drumming functions, food caches, courtship rituals, overwintering sites, roosting, lookout posts, hunting perches, fledgling sites, dens, loafing sites, underbark nesting sites, nursery colonies, anvil sites, and thermally regulated nests. Even allowing for significant overlap among these uses, the various roles this material may play in ecosystems are apparent.

One of the most notable larger animals affected by forest management is the northern spotted owl (*Strix occidentalis*) that nests in old-growth timber (Solis 1983; Forsman et al. 1984). Because this is a threatened species, the harvest of old-growth forest in the northwestern United States has been severely limited. The old-growth forests provide multistoried canopies and large quantities of standing and fallen coarse woody debris. Single tree selection systems of management provide stand structures that most closely match these conditions. A combination of selection harvest, shelterwood, and undisturbed old-growth habitat may allow timber utilization with minimal risk to spotted owls or species with similar old-growth requirements. In this case, as in other cases, lack of complete understanding of how species interact with old-growth habitat is the major obstacle to developing appropriate management strategies.

Disturbance and Edge Effects. Considerable evidence indicates that local extinctions of forest and woodland species have occurred as a result of forest clearing and conversion in temperate climates during the last 4 millennia. For instance, in Britain, eight species of Coleoptera associated with mature forest became extinct during the last 400 years (Hammond 1974). Shirt (1987) listed 14 species of saprogenic insects extinct in Britain. The number of insect extinctions associated with land conversion in the tropics is not known. Only a small portion of the estimated 10-30 million insect species (Sutton and Collins 1989), most of which occur in the tropics, has been described, and insufficient evidence generally exists to document extinction for those that have been described. Estimates of species extinction of 15%, however, are typical.

For populations of species that rely on the combination of habitat provided only by old-growth stands, harvest and more intensive management will decrease numbers. However, major shifts in insect and animal species will occur apart from reduction of old-growth. Populations of butterflies that depend on open areas in British forests were common 50-150 years ago; however, they have declined in recent years (Warren and Key 1991). A change in forest management from hardwood coppice systems, which create numerous clearings, to extensive plantings of conifer species managed on longer rotations appears responsible for the decline. The young conifer plantations provided acceptable habitat for many of these species; however, populations that could not survive the extended period between harvests died as the plantations matured (Warren 1987). In contrast, species asso-

TABLE 1
Biomass distribution in undisturbed old growth and managed forest ecosystems

				Living Trees		
Forest type	Management	Location	Age	Foliage	Branches	Bole
Noble-Fir (1)[a]	Old-growth	Washington (subalpine)	175	15.7	17.7	303.7
Douglas-fir/hemlock (2)	Old-growth	Oregon (Temperate)	450	6.2	26.3	323
Douglas-fir	Plantation	Oregon (Temperate)	60	6.0	7.0	145
Douglas-fir	Natural	Oregon (Temperate)	32	5.3	4.6	23.3
Mixed hardwood (3)	Natural	New Hampshire (Temperate)	53	3.1	32.9	84.3
Mixed hardwood (4)	Single-tree-Selection	Belgium (Temperate)	75	3.5	35.9	73.3
Loblolly pine (5)	Plantation	North Carolina	16	5.4	10.4	82.3
(*Cavanillesia-Ceiba*) (6)	Old-growth	Panama (Tropical Moist)	unk[c]	11.3	— 354.7 —	
Deciduous (*Celtis-triplochiton*) (7)	Natural	Ghana (Tropical Moist)	50	ND	261.6	ND
Evergreen (*Musanga cero-piodes*)	Natural	Zarre (Tropical Moist)	18	6.5	ND	116.5
Gmelina (8)	Plantation	Nigeria	10	7.5	46.8	202

[a]Number in parentheses refers to forests for which nutrient content is presented in Table 3.
[b]Not determined.
[c]Converted to organic matter assuming an organic carbon content of 49%.
[d]Unknown.

ciated with a continuous forest canopy appeared to have been unaffected or expanded their numbers during the same period (Warren and Key 1991).

Shifts in animal populations similar to shifts in insect populations also occur in managed forests. A recent study of a chronosequence of forests in the western Cascades of Oregon (Corn et al. 1988) documented shifts in populations of small mammals (Table 2). While species abundance and diversity varied, only one species of small mammal, the red tree vole (*Arborimus longicadis*), was so strongly associated with old-growth that it was seldom trapped in younger aged forests.

Even where old-growth stands are not required to support animal populations, change is associated with management. The landscape mosaic of

TABLE 1 continued
Biomass distribution in undisturbed old growth and managed forest ecosystems

Other Vegetation	Roots	Dead: Standing	Dead: Coarse	Dead: Fine	Soil	Total in Vegetation and Debris	Reference(s)
10⁶g/ha							
17.7	ND[b]	127.7	10.0	43.5	244.0[c]	552	Turner and Singer (1976)
ND	76.6	— 97 —		26	56	555	Harmon et al. (1990)
ND	34.6	— 3.8 —		7.1	56	204	Harmon et al. (1990)
3.2	20.7	— 2.6 —		35.0	ND	95	Heilman (1961)
0.5	26.2	— 2.9 —		19.4	24.6	83	Whittaker et al. (1979)
4.7	35.0	— 1.6 —		3.2	120	157	Duvigneaud and Denaeyer-De-Smet (1970)
ND	22.8	6.1	ND	18.4	ND	161	Kinerson et al. (1977)
1.7	9.9	— 14.6 —		2.9	ND	395	Golley et al. (1975)
14.4	24.7	71.8	25.4	2.3	ND	400	Greenland and Kowal (1960)
ND	31.4	17.4	ND	5.6	ND	177	Greenland and Kowal (1960)
10.4	ND	ND	ND	6.7	ND	273	Nwoboshi (1983)

relatively small stands (<100 ha) of different age and composition created in most areas of commercially managed forests results in production of large areas of "edge" habitat. Creation of such habitat has traditionally been considered beneficial to wildlife and is a basic paradigm of wildlife management. Interior forest species can, however, be adversely affected by edge because of a lack of specific vegetation requirements and by increased predation and parasitism (Reese and Ratti 1988).

The assumption that harvest or other disturbances in managed forests will increase risks to insect and animal diversity needs to be qualified. When such disturbances threaten populations, such as associated with the destruction of nesting habitat of red-cockaded woodpeckers in the southeastern United States by storm winds of Hurricane Hugo in 1989, these disturbances increase risk. When such disturbances maintain habitat of endangered species, some types of risk may be reduced. An excellent example of such reduction is the role that fire plays in maintaining stands

TABLE 2
Average[a] abundance of small mammals captured in forests of different age in Douglas-fir forests in the Cascade Mountains of Oregon (source: Corn et al. 1988)

Species	Old growth >195 yr	Mature 80-195 yr	Young 30-80 yr	Recent plantation <10 yr
Trowbridge's shrew	56	62	91	25
Montane shrew	16	15	20	8
Vagrant shrew	3	4	8	25
Pacific marsh shrew	4	5	5	<1
Northern watershrew	0	0	1	0
Pacific shrew	<1	0	<1	0
Shrew mole	2	2	5	<1
Coast mole	3	2	2	2
Western red-backed vole	6	13	19	<1
Creeping vole	1	<1	1	11
Red tree vole	2	1	<1	<1
Water vole	<1	0	1	<1
Heather vole	<1	0	0	1
Townsend's vole	<1	1	0	<1
Deer mouse	2	2	3	10
Pacific jumping mouse	<1	<1	5	1
Western pocket gopher	0	1	<1	5
Others	1	0	1	1

[a]Average for stands and moisture classes.

of young jack pine (*Pinus rigida*) for nesting sites of Kirtland's warbler (Radtke et al. 1983). Thus, it is not disturbance *per se* that increases risks resulting from changes in species composition and stand structure, only disturbance that destroys critical habitat provided by unique combinations of species and stand conditions.

Recent studies have demonstrated that certain components of redeveloping early successional communities can differ drastically among various site preparation and herbicide treatments (Hurst 1988; Hurst and Blake 1987; May et al. 1982; McComb and Hurst 1987; Miller and Witt 1991). Habitat values related to these differences vary with the habitat requirements of the particular wildlife species studied. In the southeastern United States, forb and woody vegetation production following mechanical site preparation normally provides abundant forage for white-tailed deer (*Odocoileus virginianus*). Although total production of forbs and woody vegetation is often less on chemically treated areas, several preferred forb species may be more abundant. In a Piedmont study (Bush, P.B., unpublished), hexazinone treatment increased the amount of legumes favored by bobwhite quail (*Colinus virginianus*). The heavy grass cover on the chemically prepared sites likely favored many rodent species and provided summer foraging sites for wild turkey (*Meleagris gallopavo*). Finally, the presence of snags on these areas also provided habitat for several cavity nesting or bark foraging birds and raptors.

In another study (Miller and Witt 1991), the response of wildlife food plants to various mechanical and chemical site preparation treatments at 2-4 years post-treatment was evaluated. Chemical treatments examined included imazapyr, triclopyr, picloram + triclopyr, and hexazinone. Vegetation was categorized as quail foods, soft mast, or winter browse. Although vegetation category means were variable among all sites, treatments, and age classes, trends in vegetation responses were apparent. Imazapyr, as well as mechanical site preparation, and hexazinone treatments tended to enhance the production of quail foods. In general, soft mast and winter browse were more abundant on the hexazinone and triclopyr treatments than the control.

Generally apparent for the species studied is that wildlife habitat is not adversely affected by site preparation or herbicide treatments used for plantation establishment. Both more intensive site preparation and herbicide applications produce a shift in plant species composition from one dominated by woody plants to one dominated by forbs. In most cases, this shift results in much higher habitat values for wildlife species. In addition, certain herbicides may enhance the production of valuable wildlife food plants. For example, sites prepared using hexazinone often produce an abundance of legumes that are important wildlife food plants and also fix nitrogen. Likewise, imazapyr promotes legumes but it does not kill blackberries (*Rubus* spp.), another extremely important wildlife food plant. However, few studies of habitat requirements for rare nongame species have been completed. Again, lack of information on effects limits any ability to quantify risk.

CATASTROPHIC DISTURBANCE

Managed forests can be completely or nearly completely destroyed by natural events that have only limited effects in unmanaged natural stands. For instance, even-aged plantations that have been recently thinned are more susceptible to windthrow than unthinned plantations or natural stands. In 1974, a windstorm, with sustained winds of 50 km/hr occurring at a time when soils were wet, blew down over 22% of the trees in a recently thinned radiata pine (*Pinus radiata*) plantation near Canberra, Australia. Less than 0.2% of the trees in adjacent stands undisturbed by recent felling were affected (Cremer et al. 1977).

In undisturbed, mixed-species, and/or uneven-aged forests, an abundance of nonhost trees, occurrence of edge, and spatially varied environment all interfere with the spread of insects and disease. Management of forests for timber production tends to replace variety with similarity, and, consequently, the susceptibility to outbreaks of insect and disease tend to increase with increasing intensities of forest management. Increased damage from the spruce budworm (*Choristoneura fumiferana*) in the United States and Canada appears related to increasing proportions of balsam-fir in the East and Douglas-fir in the West (Savil and Evans 1986). Infection of trees by root rotting fungi such as *Heterobasidium annosum* is increased in managed forests, particularly in thinned stands where cut stumps provide a site for infection and root grafts allow spread of the disease. The potential for epidemic disease or insect outbreak is even greater in exotic plantations where natural controls on pest populations may be absent. Adaptation of native insect populations to the introduced species or introduction of an exotic pest can destroy large areas of plantation. An example of the former is widespread loss of *Pinus caribaea* planted in southeast Asia following attack by native pests (Speight and Speechly 1982); an example of the latter is damage to loblolly pine planted in South Africa following introduction of two exotic pine aphids (*Cinara cronartii*, *Eulachnus rileyi*) (Donald 1984).

Although risk of insect and disease outbreak in some managed forests appears increased, it is incorrect to assume that managed forests, or even monocultures, are always more insect and disease prone. When alternate hosts are removed from the stand, such as for rust fungi, tree mortality can be reduced. Replacement of mature trees established after a natural disturbance event by young trees of the same species can reduce many types of insect damage including that of spruce budworm and southern pine beetle (*Dendroctonus frontalis*), major insect pests in the United States. Introduction of specific species to managed stands can also reduce risk. For instance, Shea et al. (1978) have suggested that control of fungus-caused (*Phytophtora cinnamomi)* dieback in jarrah (*Eucalyptus margnata*) can be reduced by planting legumes in the understory.

The risks of immediate insect or disease outbreak from native pests are greatest where introduced species have close native relatives; however, it seems unlikely that exotic forests can be indefinitely protected from introduced pests. While widespread catastrophic loss of managed and plantation forests has been the exception, lack of knowledge and ability to predict epidemics are limited and some risk remains. These risks can be minimized for plantation forestry by maintaining a landscape mosaic of several species and ages with provision for maintenance of within-species diversity.

LONG-TERM PRODUCTIVITY

The risk of reducing the potential of a site to grow trees as a result of forest management will vary depending on the initial conditions and the management methods employed. Furthermore, the degree of risk will also depend on the definition of forest productivity. Two contrasting definitions of productivity must be considered. The first, referred to as gross primary productivity, refers to the assimilation and flow of carbon and energy within a forest ecosystem. It encompasses the interaction of the total assemblage of microbes, insects, plants, and larger animals within the forest. This definition of productivity has theoretical advantages as it provides a basis for evaluating system response to stress and the risk of catastrophic change; however, practical application of it to management-induced changes is limited by the extraordinary information requirements. The second definition, referred to as site productivity, is the capacity of the land to grow selected "indicator" tree species to a merchantable size. Such a definition is analogous to the understanding of crop land productivity and is closely related to net primary production. Productivity so defined is closely associated with edaphic site and soil characteristics affecting tree establishment, rooting, water storage, and the availability and supply of essential plant nutrients. This utilitarian approach to productivity has been used in most long-term studies of plantation forestry, harvesting, slash reduction, soil tillage, fire, and fertilization.

Use of Monocultures

Modern concerns regarding the influence of forest management on long-term site productivity originated in central Europe at the turn of the 20th century in response to reports of growth declines in spruce (*Picea abies*) plantations in the Black Forest of Germany (Weidemann 1923). As a consequence of this and other early research, a concept that conifers are soil degraders emerged. Later reports of productivity declines in plantations of exotic species in Australia (Bednall 1968; Keeves 1966), New Zealand (Whyte 1973), and Africa (Evans 1975) appeared to lend credence to this viewpoint. More recent evidence, which includes more detailed analyses of earlier reports, has not generally indicated a reduction in site productivity simply as a result of the establishment of plantations. Rather, productivity declines are associated with specific management techniques (Cellier et al. 1985; Evans 1982; Squire et al. 1985).

Nutrient Depletion and Addition

Although little evidence exists that site productivity is decreased solely through the planting of a single tree species, removal of trees or portions of the trees in harvest has potentially negative effects on site productivity through continued removal of nutrients in biomass and a decline in soil fertility.

Storage and Supply of Nutrients in Forests. In forest ecosystems undisturbed by harvesting or large wildfires, a gradual increase occurs in the quantity of essential plant nutrients stored within the system due to atmospheric inputs, rock weathering, and the contribution of nitrogen fixing bacteria and plants. The distributions of nutrients in representative mature forest stands within the boreal, temperate, and tropical forest biomes are presented in Table 3. Variability between stand types occurs within each climatic region, but comparatively larger quantities of N, P, K, Ca, and Mg are stored in the vegetation and less in the forest floor and soil of tropical forests than in forests of temperate and cool climates. Forests growing on young soils tend to have lower than average contents of nitrogen.

TABLE 3
Nutrient content and distribution in above ground biomass and in soil of forest ecosystems

Forest type	Vegetation					Organic Debris				
	N	P	K	Ca	Mg	N	P	K	Ca	Mg
	10^3g/ha									
Noble-fir (1)[b]	347	54	843	1025	133	675	57	312	568	73
Douglas-fir/hemlock (2)	566	86	189	687	105	445	62	80	619	160
Spruce-fir	387	52	159	413	36	1465	100	1052	253	154
Jack pine	346	29	146	294	52	544	40	37	254	33
Mixed hardwood (3)	375	36	156	405	156	1104	67	110	206	38
Mixed oak	415	29	185	1250	47	150	12	20	160	20
Mixed oak (4)	406	32	245	868	81	33	1	15	74	5
Beech	285	39	187	152	43	180	11	20	51	14
Loblolly pine (5)	320	48	225	ND	ND	306	30	28	ND	ND
Mixed slash-Longleaf pine	140	12	43	150	34	271	10	9	96	20
Douglas-fir	326	67	227	342	ND	175	26	32	137	ND
Celtis-triplochiton (7)	1530	103	702	2140	318	514	34	204	530	72
Cavanillesa-ceiba (6)	ND	241	4598	4702	437	ND	ND	ND	ND	ND
Gmelina (8)	1014	372	ND	2466	630	30	1	ND	48	11

[a] To approximate rooting depth using a variety of extraction procedures. Refer to original citation for information on soil depth and method utilized.
[b] Number in parentheses indicates comparable stand in Table 1.
[c] Not determined.

The potential for reductions in productivity associated with nutrient removal in biomass harvest will partially depend on the storage of nutrients; however, the flow of nutrients across ecosystem boundaries is also of significance for evaluating change over long periods of continuous forest management. Essential plant nutrients enter a forest ecosystem as ions dissolved in rainwater, particulate fallout, trapped aerosols, or directly absorbed and biologically fixed gasses. Precipitation inputs, which include both dissolved ions and particulate fallout, have been the most thoroughly studied of these inputs, and wide ranges of values have been reported for all elements. In general, annual inputs of most nutrients are greatest near oceans and decrease inland; however, anthropogenic sources may increase concentrations over large regions. For example, Swank and Henderson (1976) reported greater inputs of both cations (Ca, Mg, K, and NH_4) and anions ($N0_3$ and $P0_4$) in watersheds that were located near coal burning electric generating stations (Walker Branch, TN) than in watersheds located in more isolated areas (Coweeta, NC) within the Appalachian Highlands of the United States.

TABLE 3 continued
Nutrient content and distribution in above ground biomass and in soil of forest ecosystems

Mineral soil[a] N	P	K	Ca	Mg	Source
		10^3g/ha			
15,855	3212	85,780	180,960	136,837	Turner and Singer (1976)
4560	660	2040	34	560	Cole and Rapp (1981)
559	114	9383	766	1113	Weetman and Webber (1972)
5554	495	500	1727	289	Green and Grigal (1980)
ND[c]	ND	ND	ND	ND	Whittaker et al. (1979)
3725	1041	23,205	6270	7417	Johnson and Todd (1987)
4480	920	26,800	132,900	6460	Duvigneaud and Denaeyer-DeSmet (1970)
6640	42	254	365	249	Ovington (1962)
1752	370	403	ND	ND	Wells and Jorgensen (1975)
2959	24	82	396	208	Morris and Pritchett (1982)
2809	3878	234	741	ND	Cole et al. (1967)
4592	13	650	2576	3969	Greenland and Kowal (1960)
ND	33	508	18,582	2830	Golley et al. (1975)
1725	54	ND	3481	1508	Nwoboshi (1983)

Nutrients are trapped by forest vegetation in excess of those measured in precipitation. In most forests, the magnitudes of these inputs are similar to those associated with precipitation (Mayer and Ulrich 1974; Miller et al. 1976). Nitrogen is also added to the forest through the activities of microorganisms that incorporate atmospheric nitrogen into their bodies; when they die, this nitrogen is released within the forest ecosystem. Two types of such fixation occur: (a) fixation by free living microorganisms (non-symbiotic fixation) and (b) fixation by microorganisms that grow in association with the roots of higher plants (symbiotic fixation). Rates of non-symbiotic fixation in soil are generally low, ranging from 1-3 kg/ha/year in soil of temperate forests and 2-20 kg/ha/year in tropical forests (Boring et al. 1988). Inputs of N through symbiotic fixation can be extremely high in forests containing tree species that form these associations. Newton et al. (1968) estimated that the N-fixation in red alder (*Alnus rubra*) could exceed 300 kg/ha/year in forest stands.

Additions of nutrients to forest ecosystems can be offset by losses associated with leaching, surface runoff, erosion, or fire. Surface runoff is the most thoroughly documented of these losses, and, in undisturbed forests, it is considered the largest loss. The balance between precipitation input and runoff export of macronutrients for a number of forest ecosystems is presented in Table 4. When parent materials are young and reserves of base cations of K, Ca, and Mg are high, a net export occurs. Where reserves of base cations are low, such as reported for watersheds in Florida, these cations accumulate. Virtually all undisturbed forested watersheds accumulate N and P.

It is generally acknowledged that erosion rates and the loss of nutrients associated with sediment are low in forested watersheds (Fredriksen et al.

TABLE 4
Annual elemental input in precipitation and export in runoff from undisturbed forested watersheds

Location		N	P	K	Ca	Mg	Reference(s)
		10^6g/ha					
British Columbia	Input	4.0	<0.1	1	7	1	Feller and Kimmins (1984)
	Export	0.8	<0.1	1.5	27.2	4.1	
	Net Gain	3.2	—	−0.5	−20.2	−3.1	
New Brunswick (Narrows Mtn)	Input	2.7	0.11	1.0	2.2	0.4	Krause et al. (1979)
	Export	2.8	0.10	2.9	36.9	6.4	
	Net Gain	−0.1	0.01	−1.9	−34.6	−6.0	
North Carolina (Cowetta)	Input	3.4	0.19	1.6	4.9	1.1	Swank and Henderson (1976) Cole and Rapp (1981)
	Export	0.2	ND[a]	4.5	5.9	ND	
	Net Gain	3.2		−2.9	−1.0		
Tennessee (Walker Branch)	Input	7.0	0.55	3.0	15.7	2.9	Swank and Henderson (1976) Cole and Rapp (1981)
	Export	1.8	0.02	6.8	147.5	77.1	
	Net Gain	5.2	0.53	−3.8	−131.8	−74.2	
New Hampshire (Hubbard Brook)	Input	5.8	0.10	1.1	2.6	0.7	Whittaker et al. (1979)
	Export	2.3	0.02	1.7	11.7	2.8	
	Net Gain	3.5	0.08	−0.6	−8.5	−2.1	
Oregon (H.J. Andrews)	Input	2.0	0.30	0.9	3.6	ND	Sollins et al. (1980) Cole and Rapp (1981)
	Export	1.7	0.80	9.7	121.8	10.7	
	Net Gain	0.3	−0.50	−8.8	−118.2		
Florida (IMPAC)	Input	7.8	0.9	6.0	15.1	ND	Riekerk (1981)
	Export	6.0	0.4	2.3	5.8		
	Net Gain	1.8	0.5	2.7	9.3		
Panama	Input	ND	1.0	9.5	29.3	4.9	Golley et al. (1975)
	Export		0.7	9.3	163.2	43.6	
	Net Gain		+0.3	+0.2	−133.9	−38.7	

[a]Not determined.

1975). McColl and Grigal (1979) used data from several undisturbed forests in the eastern and western United States to calculate sediment associated nutrient losses. They estimated losses of N, K, Ca, and Mg to be less than 0.1 kg/ha/year. These losses are much smaller than those reported for soluble losses of these nutrients (Table 4). Phosphorus is strongly adsorbed by soil and, unlike other nutrients, is primarily lost as sediment (Sharpley and Syers 1979). Research in forested watersheds indicates that P losses associated with sediment can be twice as great as dissolved losses, although these losses still average less than 0.2 kg/ha/year (Duffy et al. 1978; Singer and Rust 1975).

Gaseous losses of N result from biological denitrification and from volatilization during wildfire. Volatile losses of N during fires can be great. Estimates of such losses range from 855 kg/ha during wildfires in boreal forests (Grier 1975) to losses of only 26 kg/ha during prescribed burns in young pine plantations in the southeastern United States (Kodoma and Van Lear 1980). Hot fires can also result in significant losses of non-volatile elements in fly ash. For instance, in the aforementioned study, Grier estimated that K, Ca, and Mg losses were 282, 75, and 33 kg/ha, respectively.

Harvest Removal. Nutrient imports and exports shown in Table 4 provide a framework within which the significance of differences in nutrient input and export from managed forests can be evaluated. Accelerated losses of nutrients in runoff and sediment are generally low in relation to removal during harvest (Feller and Kimmins 1984; Mann et al. 1988; Neary et al. 1984). Nutrients lost through these processes may pose concerns related to water quality (see chapter 3), but they are not of critical concern from a nutrient budget standpoint. Losses are typically less than 10% of those associated with biomass removals. Nutrient removal associated with periodic harvest of woody portions of stems is relatively low and generally is compensated for by precipitation and other inputs during periods between harvests. Even removal of all tree stems during clearcut harvesting of plantations can be balanced by nutrient inputs on many sites. However, as the period between harvesting is shortened (rotation lengths decreased) and more complete utilization of the harvested trees (including foliage and small branches with high nutrient concentration) is instituted, a net loss of nutrients can occur (Table 5). Over numerous rotations, such losses may lead to declines in site productivity unless nutrients are replaced.

Most studies that have compared the influence of different levels of harvest utilization on the growth of the subsequent stand have found little immediate impact (Allen et al. 1991; Williams et al. 1989). The reasons for this are not entirely clear but are, at least in part, due to short-term changes in nutrient availability that do not reflect expected long-term trends (Morris and Lowery 1988; Vitousek and Matson 1985) and the limited period for which growth data are available. Some results from older experiments do indicate growth reductions associated with increased biomass harvest (Compton and Cole 1991).

TABLE 5
Comparison of nutrient removals with different levels of biomass utilization and rotation length

Forest Type	Harvest Utilization	Rotation Length (Age)	Biomass Removed	Nutrient Removal N Rotation	Nutrient Removal N Annual	Nutrient Removal P Rotation
			—10^6g/ha—			
Loblolly pine plantation	Merchantable stem	20	ND[a]	293	14.6	17
		40	ND	464	11.6	28
	Whole-tree	20	ND	385	19.2	27
		40	ND	570	14.2	40
Douglas-fir	Merchantable stem	53	281	478	9.0	56
	Whole-tree clearcut	53	318	728	13.7	96
Mixed oak	Sawlog stem selection	38	64	110	1.7	7
	Whole-tree clearcut	38	166	315	4.4	22
Cottonwood	Whole-tree	4	34	128	32.0	22
		12	81	17.8	17.8	29
		20	136	12.8	12.8	47
Spruce	Merchantable-stem	18	9	20	1.1	3
		50	218	217	4.3	20
		85	328	328	3.03	23
	Whole-tree	18	25	163	9.1	25
		50	284	842	16.8	80
		85	382	722	8.5	78
Gmelina	Thin 25% of stems	10	14	66	6.6	9.0
	Thin 50% of stems	10	22	134	13.4	32.7
	Stem coppice with standards	10	135	384	38.4	184.9
	Whole-tree harvest	10	271	959	95.9	371.5

[a]Not determined.

Slash Disposal and Site Preparation. Although the risks to site productivity associated with biomass removal during harvest are equivocal, evidence indicates that activities associated with reduction of logging slash and other structural organic matter prior to planting may affect long-term site productivity. In particular, the practice of mechanically piling this material into windrows has detrimental effects on site productivity (Dyck et al. 1989; Fox et al. 1988). This effect is evidenced as decline in the growth rate 6-10 years following planting and appears to be a consequence of topsoil and nutrient displacement during the piling operation. Losses due to piling operations can exceed removals in crop harvests by five-fold or more (Morris et al. 1983; Tew et al. 1986; Webber 1978).

In many forests, fire is also routinely used to reduce slash and woody material prior to planting and for fire prevention and brush control in

TABLE 5 continued
Comparison of nutrient removals with different levels of biomass utilization and rotation length

Nutrient Removal							
P	K		Ca		Mg		Reference(s)
Annual	Rotation	Annual	Rotation	Annual	Rotation	Annual	
			10^3g/ha				
0.8	169	8.4	178	8.9	ND	ND	Switzer and Nelson (1973)
0.7	293	7.3	304	7.6	ND	ND	
1.4	212	10.6	205	10.2	ND	ND	
1.0	346	8.6	348	8.7	ND	ND	
1.1	225	4.2	ND	ND	ND	ND	Bigger and Cole (1983)
1.8	326	6.2	ND	ND	ND	ND	
0.2	36	0.9	410	10.8	ND	ND	Johnson et al. (1982)
0.6	120	3.4	1090	28.7	ND	ND	
5.5	118	29.5	262	65.5	30	7.5	Switzer et al. (1976)
2.4	173	14.4	384	32.0	45	3.8	
3.9	277	13.9	588	29.4	66	3.3	
0.2	18	1.0	16	0.9	ND	ND	Tamm (1969)
0.4	248	5.0	278	5.6	ND	ND	
0.3	161	1.9	390	4.6	ND	ND	
1.4	106	5.9	157	8.7	ND	ND	
1.6	442	8.8	463	9.2	ND	ND	
0.9	330	3.9	521	6.2	ND	ND	
0.9	ND	ND	124	12.4	15.1	1.5	Nwoboshi (1983)
3.2	ND	ND	205	20.5	26.3	2.6	
18.5	ND	ND	1135	113.5	280.7	28.1	
37.2	ND	ND	2424	242.4	614.7	61.4	

established stands. The risk to site productivity associated with either practice appears minimal. Studies in the southern and northwestern United States have generally failed to show a sustained decline in soil nutrient concentrations or tree growth (Curran and Ballard 1990; Miller and Bigley 1990). In many cases, positive effects of fire on soil fertility and tree growth were observed. In addition, fire is essential to maintenance of specific plants and plant communities within forests with an extensive fire history (Sharitz et al., in press).

One concern about herbicide usage in forest management is the potential for long-term effects on soil productivity. Early pine growth responds positively to herbicide applications. Elimination of competing plants early in the rotation of a stand can have significant impact on short-term and long-term productivity (Glover 1985; Michael 1980, 1985). Allocation of plant nutrients and water to target species is increased by removing competition.

It is unlikely that this change in allocation will affect the potential for sustained forest production.

Fertilization. Reviews of long-term changes in site productivity tend to focus on management practices that have potentially negative impacts. In forests managed by harvest regulation, such a bias is appropriate. However, just as the potential risks associated with intensive plantation management are greater, so are the potential gains. A number of practices have the potential to increase site productivity throughout an entire rotation and, perhaps, for several rotations. For example, fertilization of phosphorus deficient sites at planting can improve tree growth throughout an entire rotation and into subsequent rotations. Growth of radiata pine (*P. radiata*) fertilized with P at the start of one 20-year rotation continued to exceed growth of non-fertilized trees through the sixteenth year of the second rotation following fertilization (Gentle et al. 1986). Such persistence in response can be explained by the difference in the high rates of P normally applied in forest management (50-100 kg/ha) and the low removals in forest harvest (5-70 kg/ha) (Ballard 1984).

Physical Soil Damage

Apart from effects on nutrient budgets, harvesting and other forest management activities can have direct and indirect effects on soil factors affecting site productivity. In particular, the use of heavy equipment to remove trees from the site can lead to soil compaction (particularly within the 10-25% of the site in skid trails) and reduced tree growth (Froelich et al. 1986; Hatchell et al. 1970; Helms and Hipkin 1986). Recovery from such damage can occur quickly, as reported by Thorud and Frissell (1976) following harvesting in Minnesota, or can require more than 50 years as reported by Perry (1964) in North Carolina and Greacen and Sands (1980) in Australia. The risk to long-term site productivity depends on the speed of recovery. Risks are low when compaction damage is limited to the surface soil and/or the impacts of freezing and thawing, wetting and drying, insect activity, and root growth are great. Deep compaction poses a significant long-term risk, particularly when associated with a degradation of soil structure (puddling) in warm climates with low shrink-swell soils.

WATER YIELD AND QUALITY

Water Yield

Any management activity that reduces the surface area of foliage and the capacity of vegetation to transpire water will normally increase discharge from the watershed. Timber harvest, physical removal of vegetation during mechanical site preparation, and herbicide application increase water yield. Increases are generally greatest immediately after disturbance, before vegetation regrowth occurs, and are roughly proportional to the amount of leaf surface area removed. For forest harvest, leaf area removal is closely related to basal area reduction (Fig. 5). Removal of a small portion of the basal area (<15%) is normally not detected by increased streamflow; how-

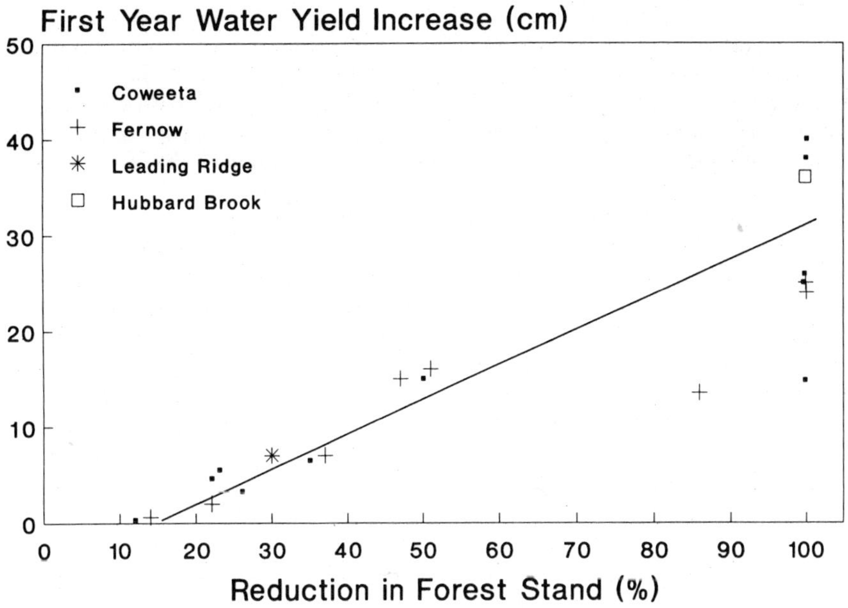

Figure 5. Increase in runoff from forested watersheds as affected by the proportion of trees removed (after Douglas and Swank 1972).

ever, removal of all trees can result in streamflow increases of as much as 30%.

The long-term influence of streamflow increases is probably not large. The major increase in streamflows diminishes as regrowing vegetation develops foliage area. By 3-5 years following disturbance, streamflows approach predisturbance conditions. While flows may remain elevated for some time, they will be within the range associated with normal variation. Exceptions to this pattern most likely occur when streamflow increases are associated with both reduced transpirational capacity and reduced infiltration of water into soil as a result of forest floor disturbance and mineral soil compaction. Since road construction and tree removal occur as part of harvesting, reported increases in water yield probably include an element of both; however, when soil conditions have been adversely affected by management, both water yields and sediment delivery to streams may be increased.

Water Quality

Landslides. Landslides are a rare but natural part of geological weathering in sloping terrain. They occur only when rare combinations of high rainfall and wet soil conditions occur in areas that are preconditioned to structural failure. Although areas where landslides occur certainly suffer from productivity loss, the area impacted by landslides is very small. Rather, concern associated with landslide occurrence focuses on the impacts on water quality and stream habitat.

An increase in frequency of landslide events appears to be associated with road construction and harvesting. In a survey of mass movement following unusually heavy rainfall in Oregon, Dryness (1967) found that 34 of 47 mass wasting events were directly associated with roads and 8 of 47 events occurred in harvested areas. Only 5 of 47 events occurred in undisturbed areas even though these represented the greatest area (Table 6).

When mass failure occurs in association with roads, it generally occurs either at the toe of the slope above the roadcut because support has been removed from the upslope soil mass or through failure of poorly consolidated fill material. Deterioration and decay of tree roots is the major cause of mass failure in harvested areas. The greatest potential for mass movement to occur is between 3 to 10 years following harvest when strength of roots from harvested trees is low and rooting systems of vegetation regrowth are not fully developed (Fig. 6). Control of competing vegetation regrowth through herbicide application is likely to extend the period of increased landslide risk since it is unlikely the additional growth of trees will compensate for the loss of the competing vegetation.

Sediment Production. Modest increases in soil erosion occur following harvest and site preparation. These increases are generally of short duration and are small when evaluated in the context of rotations in regulated forests or in most plantation forests (Patric 1976). Large losses are generally limited to mountainous areas such as the western United States (Fredriksen 1970). Felling trees does not, in itself, increase erosion. Most harvesting related erosion results from road construction either in association with increased surface erosion or through increased probability of mass movement. Fredriksen (1970) reported that sediment yield from a small watershed in western Oregon, which was harvested using a skyline system that did not require road construction, was only three times greater than from an undisturbed control watershed during the first post-harvest year.

TABLE 6
Frequency of mass failures caused by a 33-cm precipitation event in the H.J. Andrews Forest located in the Cascades Mountains of Oregon[a]

Location	No. of Events	Mass Movement
Roads		-m^3-
Fill failure	12	31,955
Cut slope failure	5	11,250
Fill and cut slope failure	6	29,944
Caused by road drainage	8	67,542
Stream crossings	3	3,398
Harvested areas	8	17,305
Undisturbed areas	5	110,318

[a]Source: Dryness (1967).

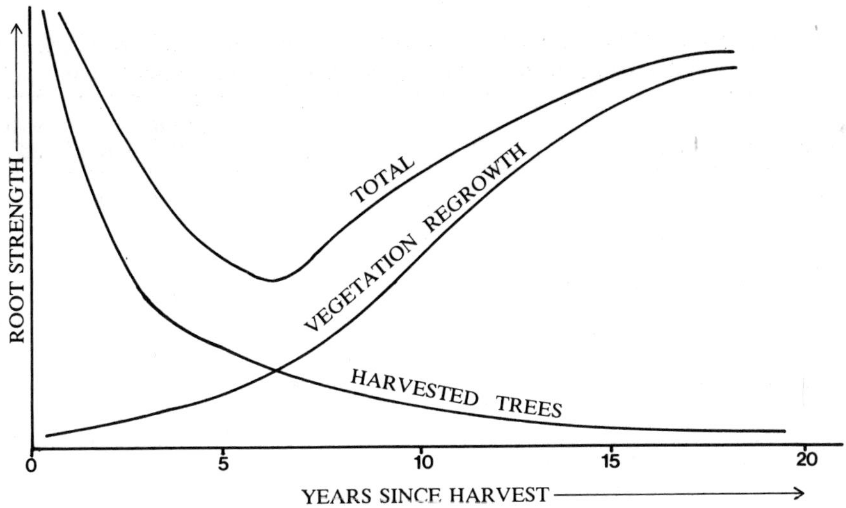

Figure 6. Contribution of decaying roots from harvested vegetation and regrowing vegetation to slope stability (after Sidle 1991).

The sediment yield from a conventionally harvested watershed, which required road construction, was 100 times greater than that from the undisturbed control.

The impacts of forest management activities on sediment production in two diverse forest ecosystems are summarized in Table 7. The sediment yields from undisturbed forests and plantations are low, ranging from trace quantities to a maximum of 0.7 metric tons ha/yr. In thinned forests or forests harvested by selection harvest, no detectable increase in sediment production may occur (Beasley and Ganillo 1988). Sediment production from burned or carefully clearcut forests are generally within this range; however, large increases in sediment production were associated with careless clearcuts and site preparation. In all cases, sediment production diminishes rapidly following disturbance.

The influence of prescribed fire and herbicide application has also been investigated in several forest ecosystems. Because these treatments do not normally disturb the soil, sediment production increases are negligible (Cushwa et al. 1971; Douglas and Van Lear 1983; Lewis 1974; Richter et al. 1982; Ursic 1970). Exceptions occur when excessively hot slash disposal fires destroy the organic layers protecting the mineral soil. In these cases, sediment yields can increase dramatically (Brown and Krygier 1971).

Stream Temperature. Removal of vegetative cover increases incident radiation and stream temperatures. Increases in stream temperature can directly influence its suitability as habitat (Levno and Rothacher 1969; Brown and Krygier 1970; Swift and Messer 1971) or indirectly influence

TABLE 7
Sediment losses from managed watersheds in the the Southern United States and a tropical forest during the first post-treatment year

Management Activity	Range of Sediment Yields Reported
	10^6g/ha/year
Southern Pine (U.S. Southeast)[a]	
Undisturbed natural forests	trace - 0.72
Pine plantations	trace - 0.18
Forest thinning	trace - 0.13
Periodic burning	0.02 - 0.52
Annual burning	0.74 - 17.74
Careful clearcut	0.13 - 0.38
Careless clearcut	3.02
Mechanical site preparation	12.5 - 14.2
Cassaus (wet tropics)[b]	
Undisturbed forest	trace
Manual harvest only	trace
Manual harvest-tilled	3.78
Shear harvest only	0.97
Bulldozer clearing with root rake	0.28
Bulldozer cleary with root rake-tilled	4.19

[a]Source: Yoho (1980).
[b]Source: Lal (1986).

its suitability through its influence on plant growth. Whether these changes are beneficial or detrimental will vary. Such changes may increase overall productivity (Meehan et al. 1969). They may also result in critical tolerances being exceeded for cool water species as, for example, salmon in headwater streams of the northwestern United States.

Dissolved Nutrients. Investigations of water quality changes following forest harvest leave little doubt that water quality can be adversely affected by forest management. As previously discussed, the accelerated export of soluble nutrients that results from forest management activities is relatively small in comparison to other nutrient removals and losses. While such changes pose little risk to long-term nutrient budgets, they can have detrimental impacts on streams and lakes. Of particular concern are increased concentrations of soluble phosphorus and nitrogen because of their association with stream and lake eutrophication. Increased nitrogen concentration is an additional concern because of the potential human health risks associated with high nitrate-nitrogen concentrations.

Research at Hubbard Brook watershed reported in the early 1970s has become the benchmark against which research on water quality changes is measured. This is unfortunate since the results from the study do not represent changes associated with normal forest management. In the original Hubbard Brook experiment, all trees within a small watershed were cut and left on site (i.e., no biomass was removed). Following cutting, herbicides were applied to the watershed to prevent vegetation regrowth. Accelerated decomposition of the felled material and forest floor litter coupled with a suppression of vegetation uptake resulted in large increases in streamwater nutrient concentrations, particularly nitrate-nitrogen, that exceeded standards for drinking water. Subsequent investigations of water quality changes following commercial harvesting and regeneration operations in the same vicinity and elsewhere have not found changes of this magnitude (Table 8). Generally, streamwater N concentrations increase for a few years after harvest, site preparation, thinning, or herbicide application, but these increases tend to be small, with total inorganic nitrogen concentrations and nitrate concentrations remaining below 3 mg/l and 2 mg/l, respectively.

Lotic Habitat. Changes in temperature, runoff, sediment loads, and water chemistry resulting from road construction, harvest, and regeneration activities alter lotic habitats. These changes can temporarily alter the system or can push the system into a new condition from which recovery to previous conditions is not possible (Holling 1973). The risk that forest management poses is difficult to judge, but depends on both the nature of the disturbance and characteristics of the community itself. Harvesting activities that change the temporal pattern of streamflow and dissolved nutrients do not represent novel physiographic stresses, and ecological theory suggests that communities could completely recover from such stresses, particularly when the stresses do not greatly exceed natural variation (sensu Poff and Ward 1990). Thus, increases in streamflow and the small increases in dissolved nutrient concentrations associated with forest management would appear to pose little risk to lotic communities. Since nutrient availability can limit productivity in streams draining undisturbed forested watersheds, short-term increases in productivity will often occur (DeAngelis et al. 1990). In some instances, increases will be too small to affect lotic populations (Everest and Harr 1982). In contrast to these short-term natural stresses, changes in stream temperature resulting from harvesting an entire watershed without a streamside management zone can result in long-term changes in systems.

Poff and Ward (1990) evaluated the potential effects of various disturbance types on lotic communities and concluded that slow ecosystem recovery was most likely to be associated with chronic changes occurring over large areas. Exceptions were associated with lethal doses of short-lived pesticides or chronic doses of persistant toxicants. Using their concept of recovery time as an index of ecosystem risk, we would conclude that the greatest potential for non-pesticide related risk to stream ecosystems is associated with widespread erosion and chronic sublethal temperature increases resulting from removal of streamside vegetation. Both of these impacts can be minimized by state-of-the-art road construction and harvesting practices. Low vegetation can protect streams from solar radiation as effectively as

TABLE 8
Influence of forest management activities on nitrogen concentrations in watershed runoff. Data are volume-weighted mean concentrations during year of maximum unless otherwise noted

Forest Cover	Disturbance	Nitrate-N	Ammonium-N	Reference
		10^{-3}g/*l*		
Mixed conifers (Oregon)	Undisturbed	0.002-0.07[a]	0.005[b]	Fredriksen et al. (1975)
	25% watershed stem harvested and slash burned	0.079	ND[c]	
	100% watershed stem harvested	0.600	ND	
	Area Fertilization	0.177	0.048	
Mixed hardwoods (New Hampshire)	Undisturbed	0.9 (3.5[c])	ND	Likens et al. (1970)
	Cut/regrowth eliminated for 2 years	53 (82[c])	ND	
	Commercial stem harvest	6.4	ND	Pierce et al. (1972)
Pine-hardwoods (West Virginia)	Undisturbed	0.02	0.01	Van Lear et al. (1985)
	Urea fertilization	0.03	0.01	
Mixed hardwoods (West Virginia)	Undisturbed	0.211[c]	0.13	Aubertin et al. (1973)
	Urea fertilization	4.86[c]	0.20	
Radiata pine (New Zealand)	Undisturbed	0.01	ND	Dyck et al. (1983)
	Commercial stem harvest	1.0[b]	ND	
Douglas-fir/western hemlock (British Columbia)	Undisturbed	0.01	<0.01	Feller and Kimmins (1984)
	19% of watershed commercial stem harvest and slash burned	0.17	<0.01	
	61% of watershed commerical stem harvest	0.52	<0.01	
Mixed tropical (British Columbia)	Undisturbed	0.06	ND	Lamb (1990)
	11% of watershed clearcut	0.04	ND	
	61% of watershed commercial stem harvest	0.45	ND	

[a]Range of maximum concentrations reported.
[b]Highest recorded concentration.
[c]Not determined.

large trees, and narrow buffer strips along streams appear capable of accomplishing most of the functions of continuous forest cover (Swift and Messer 1971). Careful location and construction of roads, particularly when combined with group selection systems of forest harvest, will lead to minimal increase in sediment production (Heede and King 1990). Because the pattern of forest management activity is dispersed both spatially and temporally and refugia is high, the overall risk to stream ecosystems appears low in forests managed with appropriate stream protection.

PESTICIDE USE AND FATE

The movement, effect, and fate of pesticides in a forest ecosystem are governed by a complex interaction of physical, chemical, biological, and hydrologic processes. These include pesticide characteristics such as solubility, volatility, and photochemical degradation rate; climatic and hydrologic conditions such as rainfall, temperature, and evapotranspiration; soil and microbiological processes such as infiltration capacity, organic matter content, particle size distribution, and microbial activity. For herbicides, vegetation uptake and recycling can also be key processes for determining their fate and impact on the ecosystem.

Uses of Pesticides

The use of insecticides over extensive forest areas is generally restricted to cases where trees with high value are threatened or where epidemic outbreaks (e.g., gypsy moth, tussock moth, spruce budworm, or pine bark beetle), usually of introduced pests, occur. In such cases, entire watersheds may be treated and entire forests impacted.

Plantation forestry relies heavily on genetically improved seedling stock to increase forest regeneration and productivity. One of the largest uses of insecticides is for control of seed and cone insects in seed orchards (Ebel et al. 1980) where uncontrolled infestations can cause seed crop losses greater than 90% (Barry et al. 1984). Some insecticides currently in use include carbofuran, azinphosmethyl, malathion, fenvalerate, and permethrin.

A major use of pesticides for plantation forestry is in tree seedling nurseries. Each nursery uses a number of different pesticides to suppress weeds, insects, fungi, and diseases that destroy seedlings or impair their growth. Most nurseries operate on a 3-4 year plant growth cycle. During the initial year of the cycle, the nursery bed is kept in cover crops such as sorghum, rye, oats, winter wheat, winter rye, vetch, or sudan grass. Herbicides such as atrazine, glyphosate, 2,4-D, dicamba, and oxyfluorfen may be used to control weeds in the cover crop. At the end of a season's growth, these cover crops are worked into the soil, the soil is fumigated, and seeds produced in seed orchards from selected trees are planted. A number of herbicides, fungicides, and insecticides are applied as needed to control the growth of weeds, diseases, and insects. The pesticides used vary with the species of tree being grown and the target species of pest. Pesticides may be repeatedly applied during the 1-3 years needed to produce seedlings for outplanting. Pesticides are, therefore, applied intensively to relatively small acreage (<100 ha each). Potential for off site drift from these applications is generally minimal since application is via irrigation or ground sprayers. Since many of the pesticides are applied in irrigation water (chemigation), the potential for off site leaching to shallow groundwater exists. On the site, changes in soil microflora and soil productivity will occur.

In addition to pesticide use in nurseries and seed orchards, pesticides are commonly used in plantation management. Herbicides are used to assist in regeneration. In these cases, herbicides are applied once or twice in a 20-80 year rotation on relatively small blocks (25-150 ha) at one time. Entire watersheds are seldom treated at one time, and major impacts are restricted to those associated with plant species diversity and wildlife habitat previously discussed.

Forms and Modes of Action

Many of the pesticides (insecticides, fungicides, herbicides, and fumigants) used in forest pest management are listed in Table 9. The first group, the organophosphate insecticides (OP), are irreversible acetylcholinesterase inhibitors where acetylcholine functions as a neural transmitter substance. The carbamates (CAR) are reversible inhibitors of acetylcholinesterase mediated nerve impulse transmission. These compounds are readily metabolized, broken down in the environment, and generally do not bioaccumulate.

TABLE 9
Classification and characteristics of pesticides used in forest management[a]

Chemical	Usage[b]	Classification: Type of[c] Chemical	Classification: Toxicity Category[d]	Classification: LD_{50} (mg/kg)[e]	Runoff Potential: kd[f]	Runoff Potential: $t^{1/2}$[g]
Azinphosmethyl	SO	OP	I	13-16.4	10	40
Lindane	SO, HVT	CH	II	88-125	11	400
Chlorpyrifos	SO, HVT, N	OP	II	97-276	61	30
Dimethoate	SO, HVT, N	OP	II	250	0.2	7
Diazinon	SO, N	OP	II	300-400	5.7	40
Malathion	SO	OP	III	1375	18	1
Fenvalerate	SO, HVT, N	SyP	III	>3200	53	35
Permethrin	SO	SyP	III	>4000	10	32
Dimilin	HVT	IGR	III	>4640	100	10
Bacillus thuringiensis	HVT	M	IV	non-toxic to mammals		
Carbaryl	N	CAR	II	270	3.0	10

TABLE 9 continued
Classification and characteristics of pesticides used in forest management[a]

Chemical	Usage[b]	Classification: Type of[c] Chemical	Classification: Toxicity Category[d]	Classification: LD_{50} (mg/kg)[e]	Runoff Potential: kd[f]	Runoff Potential: $t^{1/2g}$
Triadimefon	N	TR	II	363	3.0	26
Thiram	N	DC	III	560	6.7	15
Metalaxyl	N	AC	III	669	0.5	70
Maneb	N	DC	III	4500	200	70
Captan	N	DO	IV	9000	2	2.5
Benomyl	N	BZ	IV	>10000	19	67
Chlorothalonil	N	SA	IV	>10000	13.8	30
DCNA	N	SA	IV	>10000	10	60
2,4-D	SPH	P	II-III	375-1200	0.2	10
Triclopyr	SPH	Misc	II-III	630-3000	0.3	46
Tebuthiuron	SPH	SU	III	644	-NA[h]-	
Dicamba	SPH	B	II-III	757-2900	<0.1	14
2,4-DP	SPH	P	III	800	-NA-	
Atrazine	SPH	T	III	1400-3000	1.0	60
Hexazinone	SPH	Misc	III	1690-2012	0.5	90
Glyphosate	SPH	Misc	III	4320-5600	240	47
Simazine	SPH	T	IV	>5000	1.3	60
Dalapon	SPH	Misc	IV	7570-9330	-NA-	
Picloram	SPH	Misc	IV	8200	0.2	90
Fosamine Ammonium	SPH	PGR	IV	24000	-NA-	
Amitrole	SPH	Misc	III-IV	1100-25000	-NA-	

[a]Compiled from Delaplane 1991, USDA 1984, USDA 1986a,b, USDA 1987a,b, and Wauchope et al. 1991.
[b]SO = Seed orchards; HVT = high value trees in parks, recreation and urban areas, N = seedling nurseries, SPH = site preparation and release herbicides.
[c]Acylalamines (**AC**), benzoic acids (**B**), benzimidazoles (**BZ**), carbamates (**CAR**), chlorinated hydrocarbons (**CH**), dithiocarbamates (**DC**), dicarboximides (**DO**), insect growth hormones (**IGR**), microbials (**M**), miscellaneous (**Misc**), organophosphates (**OP**), phenoxies (**P**), plant growth regulators (**PGR**), substituted aromatics (**SA**), synthetic pyrethrins or pyrethroids (**SyP**), substituted ureas (**SU**), triazines (**T**), trizoles (**TR**)
[d]Based on the following table (from Delaplane 1991):

Toxicity Category	Oral LD_{50} (mg/kg)	Dermal LD_{50} (mg/kg)	Equivalent Oral Dose for Adult Humans
I	0-50	0-200	A few drops to 1 tsp.
II	50-500	200-2000	1 tsp. to 2 Tbsp.
III	500-5000	2000-20000	1 oz. to 1 pt (1 lb)
IV	>5000	>20000	1 pt. or more

[e]Based on acute oral doses in rats.
[f]Distribution coefficient between adsorbed and (mobile solution) phase. High coefficients correspond to strongly sorbed compounds with low mobility.
[g]Half life of compound in soil (days).
[h]Not available.

The first generation chlorinated insecticides (CH), which were heavily used prior to 1972, have been phased out. These insecticides inhibited electrical impulse transmission along nerve cell membranes and included such compounds as DDT, dieldrin, and endrin. They were persistent in the environment and tended to bioaccumulate in the food chain. As a result, numerous secondary impacts of their use were observed. For example, egg shell thinning in fish eating and insectivorous birds was attributed to the use of these chemicals. Of the chlorinated hydrocarbons, only lindane is still extensively used under controlled conditions to control pine bark beetle outbreaks in high valued trees.

The latest generation of insecticides are the synthetic pyrethrins (SyP). These compounds are chemically synthesized analogues of the naturally occurring pyrethrin. They inhibit neural electrical impulse transmission via interruption of Na-K ATPase activity. They are complex organic molecules that are active at very low rates, readily metabolized or broken down in the environment, and do not tend to bioaccumulate.

The phenoxy herbicides, benzozoic acid, and picolinic acid herbicides (PGR), which are normally used during site preparation for planting, interfere with nucleic acid (DNA and RNA) and protein synthesis resulting in rapid, uncontrolled plant cell division and elongation. Plant vascular tissues become plugged, resulting in diminished food and water transport. They are broken down in the environment by photolysis, hydrolysis, and microbial activity. Their half-life is 0.5-4 weeks, depending on environmental conditions. These materials are generally water soluble and tend to leach.

The sulfonylurea herbicides are also absorbed by both the roots and shoots of susceptible plant species. These herbicides inhibit cell division by interfering with synthesis of the amino acids valine, leucine, and isoleucine. They are active at low application rates of 10-30 g/ha.

The triazine herbicides (T) inhibit electron transport in the light dependent phase of photosynthesis. Membrane disruption due to formation of toxic photosynthetic byproducts occurs. Like the triazines, the substituted urea (SU) tebuthiuron inhibits electron transport in the light dependent phase of photosynthesis.

Glyphosate is a broad spectrum herbicide that is absorbed through the leaf and stem tissue and is translocated in the photosynthetic stream to underground storage organs. Its herbicidal activity is attributed to inhibition of amino acid synthesis.

Impacts to Non-target Organisms

Insecticide applications on a watershed scale (i.e., gypsy moth control) produce direct and indirect effects on lotic communities and ecosystems. Catastrophic drift, reduced benthic abundances, and reduced emergence of adult insects are among the most obvious direct effects (Courtemanch and Gibbs 1980; Cuffney et al. 1984; Ide 1967; Wallace and Hynes 1975; Wallace et al. 1986). Insecticides may also produce a number of indirect effects on populations, functional structure of communities, and ecosystem level processes. For example, elimination of some predaceous taxa of Plecoptera and Tricoptera following treatment with DDT appears to have resulted in a resurgence of black fly larvae in some cases (see Wallace and Hynes 1981). Elimination or reduction of prey may also result in major dietary shifts of invertebrate predators (Wallace and Hynes 1981), salamanders (Lugthart 1987), and fish (Wallace and Hynes 1981).

Additionally, insecticides may influence patterns of energy flow and nutrient cycling within lotic ecosystems. Large increases in algae have been observed following elimination of herbivores (Yasuno et al. 1982) and, in heavily shaded headwater streams, massive detritivore reduction also reduced leaf litter processing rates and seston concentration (Cuffney et al. 1984; Wallace et al. 1982). These impacts are highly variable. Factors such as turbidity, temperature, and hydrologic factors such as flow are among the many variables that influence the impact of pesticides on stream fauna (Wallace and Hynes 1981).

The toxic effects of site preparation herbicides have been reviewed by Morrison and Meslow (1983), Newton and Norris (1976), USDA (1986a, 1988), and the Weed Science Society of America (1979). In general, much of the public concern over the use of herbicides, particularly aerial application, stems from their classification as pesticides, along with insecticides and fungicides. The vast majority of the studies on commonly used forestry herbicides indicate that they have relatively low mutagenicity and are either weakly or non-ontogenetic (USDA 1988). Mammalian elimination rates are rapid, ranging from about 74% in 4 days to 100% in 48 hr, depending on the chemical. In addition, since these chemicals are not deposited in adipose tissue as are some insecticides, they do not bioaccumulate. Forestry herbicides also have a short half-life in the environment ranging from <10 (sulfometuron methyl) to 63 (picloram) days (USDA 1988).

Micro- and macroorganisms found in the forest floor and soil horizons are critical to the functioning of forest ecosystems. Processes in which they are involved include organic matter decomposition, nutrient mineralization, nitrogen transformations, respiration, stabilization of soil structure, and porosity formation. All can be affected by herbicide applications. Overviews of herbicide effects on soil organisms are provided by Eijasackers and van de Bund (1980), Greaves and Malkoney (1980), Greaves et al. (1976), and Martin (1963). Stimulatory as well as inhibitory responses to herbicide application have been observed in microorganisms. Effects are greatly dependent on the herbicide, application rate, and soil environment factors. Herbicide usage at normal forestry rates does not reduce the overall activity of microorganisms but clearly seems to shift organism populations.

Surface and Groundwater Contamination
Contamination of surface and groundwater is both an ecological and health risk factor in pest control. Simulation models are the most widely used tool for assessing these risks (Dowd et al. 1990). Such models are simplifications of the complex real processes that allow many factors to be analyzed in short time frames that would be impossible through field experimentation. These models provide the equivalent to economic forecast models and, thus, can introduce environmental considerations into the land management process.

In a given situation, pesticides with the highest water solubilities; most resistance to physical, chemical, and biological degradation; lowest affinities for adsorption onto organic matter; and the highest application rates have the greatest potential for movement in the environment. This potential is increased dramatically by its placement in the landscape. Water moves through a forested watershed via various pathways, and the particular route it takes affects the fate of any applied pesticide. A significant portion of streamflow from forested watersheds can be generated from a small proportion of the watershed area. These areas, known as variable source areas (Hewlett and Hibbert 1963) because their contribution will change through storms and seasons, are located adjacent to stream networks. These areas are at or near saturation, and during a storm they are supplied by direct precipitation and with water from up slope. When the subsurface flow exceeds the capacity of the soil profile to transmit it, the water comes to the surface and the channel length grows. Under these conditions, water is transmitted rapidly, usually as saturation overland flow, to the channel. Any pesticides sprayed in such areas will be rapidly transported to the stream.

During pesticide application in forest watersheds, movement of residues into surface or subsurface water occurs on a broad landscape scale. The main movement mechanisms are drift, leaching, and surface stormflow. Drift can be controlled by selecting appropriate application equipment, use of granular formulations, utilizing adequate buffer strips along streams, adding anti-drift adjuvants into spray mixes, and avoiding windy conditions or inversions. Leaching can be minimized by selecting the appropriate pesticide or herbicide for the hydrologic conditions of individual soils. Off site movement in stormflow can likewise be minimized by careful application techniques.

Aerial application results in roughly 25.1% loss of the organophosphate azinphosmethyl (Table 10). As a result, only 74.9% of the applied active ingredient reaches the foliage, litter, or soil. With the more volatile organophosphate, Naled, less than 5% of the applied material can be accounted for on foliage, litter, or soil. For the tree spacing common in seed orchard operations, one-half the material applied aerially is intercepted by foliage and the remainder reaches the forest floor. With more complete canopy closure, less active ingredients will reach the forest floor. Aerial or ground application of a pelleted formulation results in 90-100% of the

TABLE 10
Distribution of applied pesticides (source: Bush et al. 1989)

Pesticide	Canopy	Application Method	Application Rate	Recovery Total	Recovery Foliar Interception	Recovery Ground
			10^3g/ha	%	%	%
Azinphosmethyl	Seed orchard	Helicopter	3.4 ai[a]	74.9	33.1	41.8
Fenvalerate	Seed orchard	Helicopter	0.8 ai	71.0	49.9	21.1
Permethrin	Seed orchard	Helicopter	0.8 ai	24.2	18.2	6.0
Malathion	Closed canopy 15 yr-old pine stand	Hydraulic sprayer	3.4 ai	ND[b]	ND	3.8
Lindane	Closed canopy 15 yr-old pine stand	Hydraulic sprayer	9.0 ai	ND	ND	49.6
Naled	Closed canopy 15 yr-old pine stand	Fixed wing	1.1 ai	< 5%	ND	2.5

[a]Active ingredient.
[b]Not determined.

applied material reaching the forest floor. Ground application of liquid pesticides using mist blowers, boom nozzles, or hydraulic sprayers results in less evaporative losses and more residue on the foliage surfaces than aerial application. The partitioning of the pesticide between initial volatilization losses, foliar interception, and the forest floor has a strong impact on the potential for off site movement in surface runoff or migration to groundwater.

Insecticides. Because of the high toxicities and repeated use of insecticides for seed orchard and nursery management (Table 9), these pesticides are considered to present the greatest risk to surface and groundwater. CREAMS computer simulations have been used to evaluate off site movement risks associated with insecticide use in seed orchard management (Bush et al. 1990; Nutter et al. 1990). Multiple applications of fenvalerate, permethrin, or azinphosmethyl are needed to control the cone and seed insect complex. Repeated applications of fenvalerate or permethrin resulted in low lethal concentration probabilities for sunfish ($p < 0.02$

and 0.001, respectively); however, multiple azinphosmethyl applications produced a moderate to high risk of stormflow containing lethal concentrations. Simulations results indicate this risk can be reduced by alternating applications of fenvalerate with azinphosmethyl.

GLEAMS has been used to evaluate potential seed orchard insecticide impact on groundwater (Nutter et al. 1990). Simulations of carbofuran, azinphosmethyl, fenvalerate, and permethrin applications to four southern U.S. seed orchards show that the high water solubility and extended half life of carbofuran resulted in a potential movement to groundwater. Carbofuran loss in percolate decreased as the time of application extended into the growing season and soil water content decreased. Results from this study illustrate the influence of site characteristics on risk. While contamination risks were low at three of four sites studied, higher insecticide leaching losses occurred at a site in Arkansas as a result of shallower rooting depth and higher conductivity in soil under wet conditions.

Although much less frequently, insecticides are also used in managed forest stands. PRZM simulations have also been conducted for insecticide application in a loblolly pine plantation in Georgia for a range of pesticides. Results from this study indicated that pesticide half life ($t^{1/2}$) and soil absorption (Kd) have a significant impact on pesticide leaching below the root zone. A half life greater than 50 days was necessary to produce detectable residues in soil leachate below the root zone. PRZM predicted that compounds with a Kd > 10 (classical chlorinated hydrocarbon insecticides, pyrethroids, and organophosphates) will not leach below the root zone within 2 years of application. When pines are present to increase evapotranspiration and reduce leaching, only about 0.3 kg/ha of a pesticide with Kd of 0.1 is predicted to leach below the root zone. Compounds with Kd values of 0.1 to 10 may leach if wet hydrologic conditions follow application. None of the insecticides listed in Table 9 are both persistent ($t^{1/2} > 50$ days) and sufficiently desorbed (Kd < 1.0) for potential movement to groundwater.

Insecticides commonly used for larger scale insect control, such as for the gypsy moth (dimilin, malathion, acephate, carbaryl, phosmet, and trichlorfon) or southern pine beetle (lindane, chlorpyrifos) are either not sufficiently persistent or are adsorbed to such an extent (Kd > 10) that they do not pose a groundwater contamination threat when carefully applied to most sites. Direct application to surface waters during aerial application poses a threat to aquatic systems; thus, adequate nontreated buffers must be maintained during forest treatment.

Herbicides. Herbicides generally pose less contamination risk than insecticides due to their lower animal toxicity (Table 9). The fate and movement of herbicides in the southeastern United States was recently reviewed by Michael and Neary (1989). Studies generally showed that surface waters are more likely to be contaminated than groundwater and the risk of contamination was greatly affected by method of application, with aerial applications posing the greatest risk and stem injection posing the least. Soil spot application resulted in only slightly greater levels of contamination than stem injection. The greatest risk of surface water contamination occurred when aerial application was used without a suitable untreated buffer zone adjacent to drainage channels.

Residues from herbicide application may persist in stream baseflow or stormflow for up to 2 years after application. Monitoring of picloram application to a southern Appalachian watershed for site conversion from low quality mixed hardwood to more productive eastern white pine has shown low level residue levels in springs and down gradient discharges. Computer modeling of this application revealed that the predicted pesticide concentrations were most sensitive to differences in parameters that directly affect pesticide availability: rate of application, partition coefficient (Kd), and pesticide decay rate (Bush et al. 1986).

Monitoring a herbicide application for site conversion in the upper Piedmont of Georgia revealed that stormflow events with the highest hexazinone concentrations were those closest to the application date. The source area for stormflow was the ephemeral stream channel and the immediate surrounding area. As time progressed, contributions to observed low level residues were from upslope subsurface movement. These residue levels were low and did not approach toxic levels (Neary et al. 1983).

Unlike agricultural practices, herbicide use in forestry involves one or two applications over a span of the rotation (20-80 years). Assessment of observed concentrations in light of current toxicological data and the long periods between applications indicates they pose little toxicological risk. Caution is appropriate because toxicological testing methods are based on static exposures and do not reflect the pulsed movement of herbicides in streams from treated sites. Adequate data are not available on the impacts of low-level, ephemeral herbicide concentrations on aquatic ecosystems.

The greatest impact for stream community changes resulting from pesticide use is associated with the occurrence of a widespread lethal dose, such as might be associated with runoff of an insecticide applied over an entire watershed, or, repeated use of pesticides that produce chronic sublethal doses. From the previous discussion, the former would most likely occur either (1) when insecticide applications to an entire watershed were followed by a rare, heavy precipitation event or (2) as a result of direct upstream spills. Influences of chronic sublethal doses would be confined to streams draining areas where nursery or seed orchard management was practiced.

CLIMATE CHANGE

Rising levels of atmospheric CO_2 and other greenhouse gases associated with global warming reflect a carbon cycle in which CO_2 contributions to the atmosphere exceed losses to various sinks. As both a source of atmospheric carbon and the largest terrestrial sink for carbon, forest ecosystems are considered to play an important role in global carbon budgets. While it is generally agreed that fossil fuel combustion, which released about $5.3 \cdot 10^{15}$g/year of carbon to the atmosphere during the 1980s, is the major source of atmospheric carbon increase, the effects of deforestation and the release of CO_2 formerly sequestered in vegetation and soil are also significant. The global contribution of such clearing is debatable, but probably amounts to $2 \cdot 10^{15}$ gC/year (Dale et al. 1991). Most of this contribution is driven by tropical deforestation; non-tropical terrestrial ecosystems appear to be in approximate carbon balance due to reforestation of areas formerly in agricultural uses (Sedjo 1990).

While the current contribution of deforestation to atmospheric carbon is only 20-30% of the contribution from fossil fuel combustion, it is considered important because these contributions cannot only be controlled, but other sources of input can be at least partially mitigated through reforestation or by adapting forest management practices (Sampson 1988). Thus, both the potential risks and benefits associated with forest management should be considered in regard to global warming.

Data on organic matter accumulation presented in Table 1 illustrate differences in organic matter accumulation between old-growth and managed forests. The accumulation of organic carbon in old-growth forests represents a relatively stable carbon sink. Undisturbed, these forests make little contribution to global carbon budgets because assimilation of carbon into biomass during photosynthesis is approximately balanced by evolution of carbon during decomposition.

Replacement of old-growth forests by managed plantation forests leads to an overall decrease in on-site carbon storage. In an analysis of the conversion of old-growth Douglas-fir/hemlock forests to 60-year-old, naturally regenerated Douglas-fir forests (Table 1), Harmon et al. (1990) concluded that such change reduced carbon storage by $305 \cdot 10^6$ gC/ha for one 60-year rotation. Simulation of multiple rotations over a 300-year period indicated reduction in carbon storage between 100 and $300 \cdot 10^6$ gC/ha. Differences in estimates depended on assumptions regarding wood decay rates, utilization, and end use.

Conversely, forest management can also contribute to carbon storage. In contrast to old-growth forests, which have little potential to store additional carbon, managed forests can store significant quantities of carbon. This storage potential is the basis for proposals to offset atmospheric CO_2 through plantation establishment. The projected storage depends on estimates of tree growth, treatment of soil organic carbon (which is the largest source of terrestrial carbon), and the end use of the wood produced. Estimates for all of these are poor. If wood is used for energy, carbon stored in biomass is released during combustion and its impact on global carbon is small and limited to the carbon stored in biomass during a rotation.

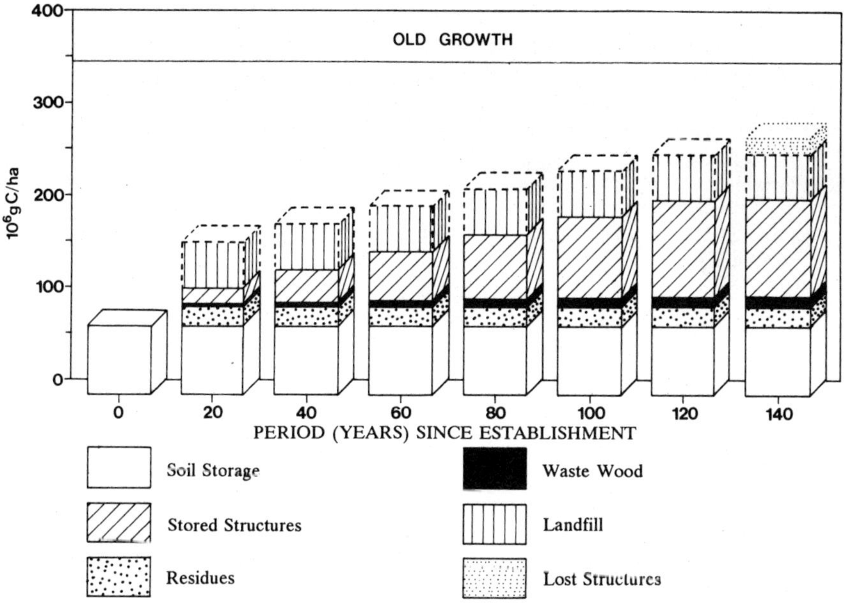

Figure 7. Disposition of carbon and potential accumulation in 20-year rotations of managed forests. Old-growth accumulation indicated as in Harmon et al. (1990). See text for description of assumptions.

TABLE 11
Merchantable wood yields of managed forests and plantations (sources: Evans 1986 from Wood 1975; Locke 1978; Dawkins 1967)

Region	Merchantable Growth[a]
	m^3/ha/year
Boreal and Temperate	
Canada average	1.0
British Columbia	1.5 - 5.3
U.S. Average	2.6
U.S. South [b]	5.3
U.K. average (conifers)	11.0
Spruce	9-15
Douglas-fir	10-14
Larch	6-11
U.K. average (board leaved)	5
New Zealand (pines)	18-30
Tropics	
Moist tropical (natural)	6-15
Moist tropical (hardwood plantations)	10-20
Moist tropical (conifers and eucalypts)	20-35

[a]Stem wood to 7 cm top diameter.
[b]25-year-old plantations on average sites (source: USDA Forest Service 1988).

TABLE 12
Summary of ecological impacts associated with forest management systems and specific activities. Ecological risks are ranked relative to protected wildland forests by risk groups

Risk Group	Management System	Activity	Direct Impact
Structure and species change	Regulated forests	Selection harvests	Removal of old trees and reduction of coarse woody debris
			Commercially important species favored
		Clearcut harvest	Old trees removed, coarse woody debris virtually eliminated
	Plantation forests	Use of even-aged tree monoculture	Forest with low structural and genetic diversity and high edge created
		Herbicide use	Delays understory succession and favor early successional species
Long-term productivity	Regulated forests	Selection harvests	Nutrient removal in biomass
			Accelerated erosion
			Soil physical conditions
		Clearcut harvest with whole tree utilization	Nutrient removal in biomass
			Accelerated erosion
			Damaged soil physical conditions
	Plantation forests	Use of monocultures	Soil chemical and biological change
		Slash disposal	Nutrient loss
			Accelerated erosion
			Soil physical conditions

TABLE 12 continued
Summary of ecological impacts associated with forest management systems and specific activities. Ecological risks are ranked relative to protected wildland forests by risk groups

Potential Consequences	Relative Risk	Options for Reducing Risk
Reduced number of species relying on habitat provided by dying and dead trees	Small increase	Snag management
Reduced insect and disease hazard	Small decrease	
Reduced diversity, species extinction	Negligible	
Species relying on habitat provided by dying and dead trees eliminated locally	Large increase	Snag management, establish reserves
Only early sucessional plant and animal species prosper	Large increase	Vary species and age classes on landscape combine other management systems
Species extinction	Large increase	Survey and protect critical areas
Specific host-habitat lost	Large increase	
Insect and disease epidemics	Medium increase	Monitoring and study of benefits resulting from interplanting
Increased abundance of early successional species including many game species	Negligible	
Non sustainable production	Negligible	
Non-sustainable production	Negligible	State-of-the-art road construction
Non-sustainable production	Small increase	Control harvesting operations
Non-sustainable production	Small to medium increase	Fertilization
Non-sustainable production	Negligible	
Non-sustainable production	Negligible to small increase	Control harvesting operation, plan skid trails, avoid wet weather harvest
Declining growth, soil deterioration	Negligible	
Non-sustainable production	Medium to large increase	Prepare site for planting using systems that leave fine organics on site, avoid hot fire
Non-sustainable production	Small increase	Avoid mechanical site preparation on sloping sites with erosive soils, avoid hot fire
Non-sustainable production	Medium increase to medium decrease	Avoid use of heavy equipment when soils are wet, match site preparation to site conditions

TABLE 12 continued
Summary of ecological impacts associated with forest management systems and specific activities. Ecological risks are ranked relative to protected wildland forests by risk groups

Risk Group	Management System	Activity	Direct Impact
		Fertilization	Increased nutrient availability
Watershed values	Regulated forests	Selection harvest	Water yield
			Sediment production
			Landslide frequency
			Stream temperatures
			Dissolved nutrients
	Plantation forests	Harvest and regeneration	Water yield
			Sediment production
			Landslide frequency
			Stream temperatures
			Dissolved nutrients
Contaminants	Regulated and plantation forests	Control of insect epidemics	On-site toxicity to non-target organisms Human health
	Plantation forests	Control of insects in seed orchards and nurseries	On-site toxicity to non-target organisms

TABLE 12 continued
Summary of ecological impacts associated with forest management systems and specific activities. Ecological risks are ranked relative to protected wildland forests by risk groups

Potential Consequences	Relative Risk	Options for Reducing Risk
Non-sustainable production	Negligible to medium decrease in risk of productivity decline	Long-term gains from phosphorus and micronutrient additions
Alteration of stream habitat due to increased flows	Negligible	Flow increase proportional to trees harvested limit basal area removed during any 5-year period to <15% of watershed total
Altered stream habitat due to increased turbidity and sediment	Small to medium increase	State-of-the-art road construction eliminates the major source of sediment
Altered stream habitat due to increased turbidity and sediment	Small increase	Road design can limit risk
Lethal or sublethal temperature can kill fish and reduce habitat suitability	Medium increase	Leave streamside vegetation
Alter stream productivity	Negligible	
Alteration of stream habitat due to increased sediment	Small increase	
Alteration of stream habitat due to increased sediment	Negligible to medium increase	Avoid mechanical operations that completely disturb surface soil, use streamside management zones
Alteration of stream habitat due to increased flow	Small increase	Road design can limit risk
Lethal or sublethal temperatures can kill fish and reduce habitat suitability	Large increase	Leave streamside vegetation
Alter stream productivity, alter habitat suitability	Medium increase	Vegetation regrowth with reduce nutrient concentrations, allow ample buffer for fertilizer applications
Catastrophic shifts in predator/prey relationships	Small to medium increase	Avoid repeated use
Contamination of water supplies	Small to medium increase	Use adequate buffers to avoid direct application to water
Catastrophic shifts in predator/prey relationships	Large increase	Minimize use, particularly of most persistent chemicals (these operations depend on constant monitoring and chemical control)

TABLE 12 continued
Summary of ecological impacts associated with forest management systems and specific activities. Ecological risks are ranked relative to protected wildland forests by risk groups

Risk Group	Management System	Activity	Direct Impact
			Human health
		Herbicide application	On-site toxicity
Global climate	Regulated forests	Harvest and utilization	Reduced carbon storage
	Plantation forests	Harvest and utilization	Reduced or increased carbon storage

Clearing mature forests to establish such plantations would lead to a decrease in total carbon storage. If wood from managed forests is used in structures that are themselves long lived or as paper that fills landfills where it decomposes slowly, an increase in long-term carbon storage is possible.

Recently, Vitousek (1991) analyzed the potential impacts of plantation establishment on global carbon cycles. He concluded that worldwide establishment of plantations on potentially available land could remove approximately $1 \cdot 10^{15}$ gC/year from the atmosphere. This amounts to about 15% of annual fossil fuel input. This ability is, however, limited to a relatively short period because there is no suitable method for placing assimilated carbon into long-term storage as illustrated in Figure 7. Merchantable yields of managed forests and plantations in temperate and tropical areas range from 1 to 35 m^3/ha/year (Table 11). Using an average yield of 15 m^3/ha, an average wood density of $0.5 \cdot 10^6$ g/m^3, and assuming organic matter is 45% carbon by weight, a reasonable estimate for carbon fixation is about $3.5 \cdot 10^6$ gC/ha/year. If current utilization patterns in the United States are applied, 10% of this would be bark, 50% would become solid wood products that can be used in long lived structures, 5% would become waste sawdust, 21% would be used in paper products, and 14% would be by-products and waste from paper production. If it is further assumed that 50% of all wood is used in structures with a life of 100 years (after which they are destroyed and the wood burned or allowed to decompose) and 20% of paper products end up in landfills where they decompose at a rate of 1% a year, an average hectare of managed forest can account for storage of $140 \cdot 10^6$ gC/ha in wood structures, $35 \cdot 10^6$ gC/ha in merchantable wood biomass, and $10 \cdot 10^6$ gC/ha in landfilled material. Additionally, some carbon

TABLE 12 continued
Summary of ecological impacts associated with forest management systems and specific activities. Ecological risks are ranked relative to protected wildland forests by risk groups

Potential Consequences	Relative Risk	Options for Reducing Risk
Contamination of water supplies	Medium increase	Limit use of soluble and persistent chemicals
Catastrophic shifts in microbial, insect, or animal populations	Small increase	
Contribution to greenhouse warming	Small increase	Conversion of old-growth to regulated forests will increase greenhouse effect
Contribution to alleviation of greenhouse warming	Small increase to small decrease	Conversion of old-growth to plantations will slightly increase greenhouse effect; conversion of non-forest lands to plantations will slightly decrease greenhouse gases

is stored in residue left on site and as carbon in soil. This storage will be lower in managed plantations than in old-growth and greater than in non-forested soils use. Inclusion of this material in estimates would change basic results only slightly. Overall, where managed forests replace non-forest uses, increases in carbon storage can be achieved. Where managed forests replace old-growth ones, reductions in storage will occur. Either of the above changes is small in relation to impacts of fossil fuel combustion or changes in other sinks when expanded to a global basis.

SUMMARY

Forests are dynamic systems undergoing constant change. Attempts to manage forests include redirecting the natural, often slow, changes to achieve specific objectives of the various private, cooperate, or public owners. Such changes will inevitably result in increases in some of the multiple values expected from forests and the diminishment of others. Whenever such changes lead to species extinction or destabilize unique ecosystems, they pose a major ecological risk. Risk is diminished when changes temporarily alter ecosystems without placing them in a condition from which recovery is impossible.

This chapter has considered ecological risks that include change from which recovery to predisturbance conditions is difficult or impossible. A summary of major risks and our subjective ranking of these risks for various intensities and forest management activities is presented in Table 12. In comparison with risks associated with other human activities, most of these risks are small or can be minimized through modifications in management practices.

REFERENCES

ALLEN, H.L., MORRIS, L.A., and WENTWORTH, T.R. (1991). Productivity comparisons between successive loblolly pine rotations in the North Carolina Piedmont. In: *Long-term Field Trials to Assess Environmental Impacts of Harvesting*, Proceedings of the IEA/BE A3 Workshop (W.J. Dyck and C.A. Mees eds), pp. 125-136. New Zealand For. Res. Inst. Bull. 161.

AUBERTIN, G.M., SMITH, D.W., and PATRIC, J.H. (1973). Quantity and quality of streamflow after urea fertilization on a forested watershed: First five year results. In: *Forest Fertilization Symposium*, Gen. Tech. Rep. NE-3. USDA Forest Service, Washington, DC.

BAKER, W.L. (1972). Eastern Forest Insects. USDA Forest Service Misc. Publ. 1175. U.S. Government Printing Office, Washington, DC.

BALLARD, R. (1984). Fertilization of plantations. In: *Nutrition of Plantation Forests* (G.D. Bowen and E.K.S. Nambiar, eds), pp. 327-360. Academic Press, London.

BARRY, J.W., BARBER, L.R., KENNY, P.A., and OVERGAARD, N. (1984). Feasibility of aerial spraying of southern pine seed orchards. South. J. Appl. For. 8:127-131.

BEASLEY, R.S. and GANILLO, A.B. (1988). Sediment and water yields from managed forests on flat Coastal Plain sites. Water Resour. Bull. 24:361-366.

BEATTIE, R.K. and MILLER, J.D. (1954). Fifty years of chestnut blight in America. J. For. 52:323-329.

BEDNALL, B.H. (1968). The problem of lower volumes associated with second rotations in *Pinus radiata* plantations in South Australia. Woods For. Dept. Bull. 17.

BIGGER, C.M. and COLE, D.W. (1983). Effects of harvest intensity on nutrient losses and future productivity in high and low productivity red alder and Douglas-fir stands. In: *Symposium on Forest Site and Continuous Productivity*, Gen Tech. Rep. PNW-163, pp. 167-187. USDA Forest Service, Washington, D.C.

BORING, L.R., SWANK, W.T., WARD, J.B., and HENDERSON, G.S. (1988). Sources, fates and impacts of nitrogen inputs to terrestrial ecosystems - review and synthesis. Biogeochemistry 6:119-159.

BROWN, G.W. and KRYGIER, J.T. (1970). Effects of clearcutting on stream temperature. Water Resour. Res. 6:1133-1140.

BROWN, G.W. and KRYGIER, J.T. (1971). Clearcut logging and sediment production in the Oregon Coast Range. Water Resour. Res. 7:1189-1198.

BUSH, P.B., NEARY, D.G., DOWD, J.F., ALLISON, D.C., and NUTTER, W.L. (1986). Role of models in environmental impact assessment. Weed Sci. Soc. 39:502-512.

BUSH, P.B., NUTTER, W.L., NEARY, D.G., and TAYLOR, J.W. (1989). Pesticide movement from southern pine seed orchards: Use of CREAMS model to facilitate evalaution of off-site pesticide movement. In: *CREAMS/GLEAMS Symposium* (Beasley, D., Knisel, W.G., and Rice, A.P., eds). University of Georgia Coastal Plains Experimental Station, Tifton, GA.

BUSH, P.B., NEARY, D.G., TAYLOR, J.W., and DOWD, J.F. (1990). Use of Computer Models to Evaluate Potential Herbicide Runoff from Silvicultural Operations in the Southeastern U.S. Sixth Biennial Southern Silvicultural Research Conference, Memphis, TN.

CELLIER, K.M., BOARDMAN, R., BOOMSMA, D.B., and ED, P.G. (1985). Response of *Pinus radiata* D. Don to various silvicultural practices on adjacent first and second - rotation sites near Tantanoola, South Australia. I: Establishment and growth up to age 7 years. Aust. For. Res. 15:431-447.

CHAPIN, F.S., SCHULE, E.D., and MOONEY, H.A. (1990). The ecology and economics of storage in plants. Ann. Rev. Ecol. Syst. 21:423-447.

CHAPMAN, H.H. (1950). Forest management. In: *Fifty Years of Forestry in the USA* (Winters, R.K., ed), pp. 78-91. Society of American Foresters, Washington, DC.

COLE, D.W. and RAPP, M. (1981). Elemental cycling in forest ecosystems. In: *Dynamic Properties of Forest Ecosystems* (D.E. Reichle, ed), pp. 341-409. Cambridge University Press, London.

COLE, D.W., GESSEL, S.P., and DICE, S.F. (1967). Distribution and cycling of nitrogen, phosphorus, potassium and calcium in a second growth Douglas-fir ecosystem. In: *Proceedings of the Symposium on Primary Productivity and Mineral Cycling in Natural Ecosystems* (H.E. Young, ed), pp. 197-232. University of Maine Press, Orono, ME.

COMPTON, J.E. and COLE, D.W. (1991). Impact of harvest intensity on growth and nutrition of successive rotations of Douglas-fir. In: *Long-term Field Trials to Assess Environmental Impacts of Harvesting*, FRI Bull 161 (Dyck, W.J. and Mees, C.A., eds). New Zealand Forest Research Institute, Rotorua, New Zealand.

CONDE, L.F., SWINDEL, B.F., and SMITH, J.E. (1983a). Plant species cover, frequency, and biomass: Early responses to clearcutting, burning, windrowing, discing, and bedding in *Pinus elliottii* flatwoods. For. Ecol. Manage. 6:319-331.

CONDE, L.F., SWINDEL, B.F., and SMITH, J.E. (1983b). Plant species cover, frequency, and biomass: Early responses to clearcutting, chopping, and bedding in *Pinus elliottii* flatwoods. For. Ecol. Manage. 6:307-317.

CORN, P.S., BURY, R.B., and SPIES, T.A. (1988). Douglas-fir forests in the Cascade Mountains of Oregon and Washington: Is the abundance of small mammals related to stand age and moisture? In: *Proceedings of the Symposium on Management of Amphibians, Reptiles and Small Mammals in North America*, Gen. Tech. Rep. RM-166. USDA Forest Service, Washington, DC.

COURTEMANCH, D.L. and GIBBS, K.E. (1980). Short- and long-term effects of forest spraying of Carbaryl (Sevin-4-oil) on stream invertebrates. Can. Entomol. 112:271-276.

CREMER, K.W., MYERS, B.J., VAN DER DUYS, F., and CRAIG, I.E. (1977). Silvicultural lessons from the 1974 windthrow in radiata pine plantations near Canberra. Aust. J. For. 40:274-292.

CUFFNEY, T.F., WALLACE, J.B., and WEBSTER, J.R. (1984). Pesticide manipulation of a headwater stream: Invertebrate responses and their significance for ecosystem processes. Freshw. Invertebr. Biol. 3:153-171.

CURRAN, M.P. and BALLARD, T.M. (1990). Some slashburning effects on soil and trees in British Columbia. In: *Sustained Productivity of Forest Soils*, Proceedings of the Seventh North American Forest Soils Conference (S.P. Gessel, D.S. Lactate, G.F. Weetman, and R.F.Powers, eds), pp. 355-356. University of British Columbia, Faculty of Forestry, Vancouver, BC.

CUSHWA, C.T., HOPKINS, M., and MCGINNES, B.S. (1971). Soil Movement in Established Gullies after a Single Prescribed Burn in South Carolina Piedmont, Res. Paper SE-153. USDA Forest Service, Washington, DC. 4 pp.

DALE, V.H., HOUGHTON, R.A., and HALL, C.A.S. (1991). Estimating the effects of global land-use change on global atmospheric CO_2 concentrations. Can. J. For. Res. 21:87-90.

DAWKINS, H.C. (1967). Productivity of tropical rain forests and their ultimate value to man. In: *The Ecology of Man in the Tropical Environment*, Publ. Ser. 4, pp. 178-182. International Union for Conservation of Nature and Natural Resources Morges, Switzerland.

DEANGELIS, D.L., MULHOLLAND, P.J., ELWOOD, J.W., PALUMBO, A.V., and STEINMAN, A.D. (1990). Biogeochemical cycling constraints on stream ecosystem recovery. Environ. Manage. 14:685-697.

DELAPLANE, K.S., ed. (1991). Georgia Pest Control Handbook, Sp. Bull. No. 28. USDA Cooperative Extension Service, University of Georgia, Athens, GA. 438 pp.

DONALD, D.G.M. (1984). Silviculture and yield. In: *IUFRO Symposium on Site and Productivity of Fast Growing Plantations*. South Africa Forest Research Institute, Pretoria.

DOUGLAS, J.E. and SWANK, W.T. (1972). Streamflow modification through management of eastern forests, Research Paper SE-94. USDA Forest Service, Washington, D. C.

DOUGLAS, J.E. and VAN LEAR, D.H. (1983). Prescribed burning and water quality of ephemeral streams in the piedmont of South Carolina. For. Sci. 29:181-189.

DOWD, J.F., BUSH, P.B., NEARY, D.G., TAYLOR, J.W., and BERISFORD, Y.C. (1990). Modeling pesticide movement in forested watersheds: Use of PRZM for evaluating pesticide options in loblolly pine stand management. Environmental Toxicology and Chemistry Society (SETAC) Annual Meeting, Arlington, VA.

DRYNESS, C.T. (1967). Mass soil movements in the H. J. Andrews Experimental Forest, Research Paper PNW-42. USDA Forest Service, Washington, DC.

DUFFY, P.D., SCHREIBER, J.D., MCCLURKIN, D.C., and MCDOWELL, L.L. (1978). Aqueous- and sediment phase phosphorus yields from five southern pine watersheds. J. Environ. Qual. 7:45-50.

DUVIGNEAUD, P. and DENAEYER-DESMET, S. (1970). Biological cycling of minerals in temperate deciduous forests. In: *Analysis of Temperate Forest Ecosystems* (D.E. Reichle, ed), pp. 199-225. Springer-Verlag, NY.

DYCK, W.J., GOSZ, J.R., and HODGKISS, P.D. (1983). Nitrate losses from disturbed ecosystems in New Zealand: A comparative analysis. N.Z.J. For. Sci. 13:14-24.

DYCK, W.J., MEES, C.A., and COMERFORD, N.B. (1989). Medium-term effects of mechanical site preparation on radiata pine productivity. In: *Research Strategies for Long-term Site Productivity*, Proceedings of the IEA/BE A3 Workshop (W.J. Dyck and C.A. Mees, eds), pp. 79-92. New Zealand For. Res. Inst. Bull. 152.

EBEL, G.B., FLAVELL, T.H., CRAKE, L.E., YATES, H.O. III, and DEBARR, G.L. (1980). Cone and Seed Insects of the Southern Pines, Res. Paper SE-185. USDA Forest Service, Washington, DC. 24 pp.

EIJASACKERS, H. and VAN DE BUND, C.F. (1980). Effects on soil fauna. In: *Interactions Between Herbicides and the Soil* (R.J. Hance, ed), pp. 255-305. Academic Press, NY.

ERNST, W.H.O. (1984). The effects of forest management on the genetic variability of the herb layer. In: *Population Genetics in Forestry*, Proceedings of the IUFRO Meeting (H.R. Gregorius, ed). Springer-Verlag, Berlin.

EVANS, J. (1975). Two rotations of *Pinus patula* in the Usutu Forest, Swaziland. Comm. For. Rev. 54:64-81.

EVANS, J. (1982). Plantation Forestry in the Tropics. Oxford University Press, NY.

EVEREST, F.H. and HARR, R.D. (1982). Influence of Forest and Rangeland Management on Anadromous Fish Habitat in Western North America: Silvicultural Treatments, Gen. Tech. Rep. PNW-134. USDA Forest Service, Washington, DC.

FALK, D.A. (1990). Endangered forest resources in the U.S.: Integrated strategies for conservation of rare species and genetic diversity. For. Ecol. Manage. 35:91-117.

FELLER, M.C. and KIMMINS, J.P. (1984). Effects of clearcutting and slash burning on streamwater chemistry and watershed nutrient budgets in southwest British Columbia. Water Resour. Res. 20:29-40.

FORSMAN, E.D., MESLOW, E., CHARLES, E., and WIGHT, H.M. (1984). Distribution and Biology of the Spotted Owl in Oregon, Wildlife Monogr. 87. The Wildlife Society, Washington, DC.

FOX, T.R., MORRIS, L.A., and MAIMONE, R.A. (1988). The impact of windrowing on the productivity of a rotation age loblolly pine plantation. In: *Proceedings of the Fifth Biennial Southern Silviculture Research Conference*, Gen. Tech. Rep. 5074, pp. 133-139. USDA Forest Service, Washington, DC.

FRANKLIN, J.F. (1978). Wilderness ecosystems. In: *Wilderness Management*, Misc. Publ. 1365 (J.C. Hendee, G.H. Stankey, and R.C. Lucas, eds), pp. 191-212. USDA Forest Service, Washington, DC.

FRANKLIN, J.F., CORMACK, K., DENISON, W., MCKEE, A., MASER, C., SEDELL, J., SWANSON, F., and JUDAY, G. (1981). Ecological Characteristics of Old-growth Douglas-fir Forests, Gen. Tech. Rep. PNW 118. USDA Forest Service, Washington, DC.

FREDRIKSEN, R.L. (1970). Comparative water quality - natural and disturbed streams. In: *Proceedings of the Symposium on Forest Land Uses and Stream Environments*, pp. 125-136. USEPA, Corvallis, OR.

FREDRIKSEN, R.L., MOORE, D.G., and MORRIS, L.A. (1975). The impact of timber harvest, fertilization, and herbicide treatment on streamwater quality in western Oregon and Washington. In: *Forest Soils and Forest Land Management* (B. Bernier and C.H. Winget, eds), pp. 283-313. Laval University Press, Quebec.

FROELICH, H.A., MILES, D.W.R., and ROBBINS, R.W. (1986). Growth of young *Pinus ponderosa and Pinus contorta* on compacted soils in central Washington. For. Ecol. Manage. 15:285-294.

GENTLE, S.W., HUMPHREYS, F.R., and LAMBERT, M.J. (1986). An examination of a *Pinus radiata* phosphate fertilizer trial fifteen years after treatment. For. Sci. 32:822-829.

GLEICK, J. (1987). Chaos-making of a new science. In: *A Geometry of Nature*, pp. 81-118. Penquin Books, NY.

GLOVER, G.R. (1985). Growth response of southern pines to control of competing vegetation. In: *Herbicides for Southern Forestry*, Proceedings of the 1985 Annual Spring Symposium (N.A. Pywell, D.G. Neary, and B. Law, eds), pp. 212-222. School of Forest Resources and Conservation, University of Florida, Gainesville, FL.

GOLLEY, F.B., MCGINNIS, J.T., CLEMENTS, R.G., CHILD, G.I., and DUEVER, J.J. (1975). *Mineral Cycling in a Tropical Moist Forest Ecosystem.* University of Georgia Press, Athens, GA. 248 pp.

GREACEN, E.L. and SANDS, R. (1980). Compaction of forest soils: A review. Aust. J. Soil Res. 18:163-188.

GREAVES, M.P. and MALKONEY, H.P. (1980). Effects on soil microflora. In: *Interactions Between Herbicides and the Soil* (R. J. Hance, ed), pp. 223-253. Academic Press, NY.

GREAVES, M.P., DAVIES, H.A., MARSH, J.A., and WINGFIELD, G.I. (1976). Herbicides and soil micro-organisms. CRC Critical Rev. Microbiol. 5:1-38.

GREEN, D.C. and GRIGAL, D.F. (1980). Nutrient accumulations in jack pine stands on deep and shallow soils over bedrock. For. Sci. 26:325-333.
GREEN, G.M. and SUSSMAN, R.W. (1990). Deforestation history of the eastern rain forests of Madagascar from satellite images. Science 248:212-219.
GREENLAND, D.J. and KOWAL, J.M.L. (1960). Nutrient content of the moist tropical forests of Ghana. Plant Soil 12:154-173.
GRIER, C.C. (1975). Wildfire effects on nutrient distribution and leaching in a coniferous ecosystem. Can. J. For. Res. 5:599-607.
HAMMOND, P.M. (1974). Changes in the British coleopterous fauna. In: *The Changing Flora and Fauna of Britain*, Proceedings of the Symposium of the Systematics Association (D.L. Hawksworth, ed), pp. 323-369. Academic Press, London.
HARMON, M.E. and HUA, C. (1991). Coarse woody debris dynamics in two old-growth ecosystems. Bioscience 41:604-610.
HARMON, M.E., FRANKLIN, J.F., SWANSON, F.T., SOLLINS, P., GREGORY, S.V., LATTIN, J.D., ANDERSON, N.H., CLINE, S.P., AUGMAN, N.G., SEDELL, J.R., LIENDAEMPER, G.W., CROMAC, K., JR., and CUMMINS, K.W. (1986). Ecology of coarse woody debris in temperate ecosystems. Adv. Ecol. Res. 15:133-302.
HARMON, M.E., FERRELL, W.K., and FRANKLIN, J.F. (1990). Effect on carbon storage of conversion of old-growth forests to young forests. Science 247:699-702.
HATCHELL, G. E., RALSTON, G. W. Ralston, and FOIL, R.R. (1970). Soil disturbance in logging. J. For. 68:772-775.
HEEDE, B.H. and KING, R.M. (1990). State-of-the-art timber harvest in an Arizona mixed conifer forest has minimal effect on overland flow and erosion. Hydrol. Sci. J. 35:623-635.
HEILMAN, P.E. (1961). Effects of Nitrogen Fertilization on the Growth and Nitrogen Nutrition of Low-site Douglas-fir Stands, PhD. Dissertation. University of Washington, Seattle, WA.
HELMS, J. A. and HIPKIN, C. (1986). Effects of soil compaction on tree volume in a California ponderosa pine plantation. West J. Appl. For. 1:121-124.
HEWLETT, J.D. and HIBBERT, A.R. (1963). Moisture and energy conditions within a sloping soil mass during drainage. J. Geophys. Res. 68:1081-1087.
HOLLING, C.S. (1973). Resilience and stability of ecological systems. Ann. Rev. Ecol. Syst. 4:1-23.
HURST, G.A. (1988). Vegetation following imazapyr for site preparation. Proc. South. Weed Sci. Soc. 41:201.
HURST, G. A. and BLAKE, P.M. (1987). Plant species composition following hexazinone treatment-site preparation. Proc. South. Weed Sci. Soc. 40:194.
IDE, F. (1967). Effects of forest spraying with DDT on aquatic insects of salmon streams in New Brunswick. J. Fish. Res. Bd. Can. 24:769-805.
JOHNSON, D.W. and TODD, D.E. (1987). Nutrient export by leaching and whole tree harvesting in a loblolly pine and mixed oak forest. Plant Soil 102:99-109.
JOHNSON, D.W., WEST, D.C., TODD, D.E., and MANN, L.K. (1982). Effects of sawlog vs. whole-tree harvesting on nitrogen, phosphorus, potassium and calcium budgets of an upland mixed oak forest. Soil Sci. Soc. Am. J. 46:1304-1309.
JONES, K. (1970). Nitrogen fixation in the phyllosphere of the Douglas-fir *Pseudotsuga douglasii*. Am. Bot. 34:239-244.
KEEVES, A. (1966). Some evidence of loss of productivity with successive rotations of *Pinus radiata* in the south-east of South Australia. Aust. For. 30:51-63.

KINERSON, R.S., RALSTON, C.W., and WELLS, C.G. (1977). Carbon cycling in a loblolly pine plantation. Oecologia 29:1-10.

KODAMA, A.E. and VAN LEAR, D.H. (1980). Prescribed burning and nutrient cycling relationships in young loblolly pine plantations. South. J. Appl. For. 4:118-121.

KRAUSE, H.H., WEETMAN, G.F., and ARP, P.A. (1979). Nutrient cycling in boreal forest ecosystems of North America. In: *Forest Soils and Land Use*, Proceedings of the Fifth North American Forest Soils Conference (C.T. Youngberg, ed), pp. 287-319. Colorado State University Press, Fort Collins, CO.

LAL, R. (1986). Conversion of a tropical rainforest: Agronomic potential and ecological consequences. Adv. Agron. 39:173-259.

LAMB, D. (1990). Exploiting the Tropical Rainforest: An Account of Pulpwood Logging in Papau, New Guinea. UNESCO, Paris.

LEVNO, A. and ROTHACHER, J. (1969). Increases in maximum stream temperature after burning in a small experimental watershed, Res. Note PNW-110. USDA Forest Service, Washington, DC.

LEWIS, W.L., Jr. (1974). Effects of fire on nutrient movement in a South Carolina pine forest. Ecology 55:1120-1127.

LIKENS, G.E., BORMANN, F.H., JOHNSON, N.M., FISHER, D.W., and PIERCE, R.S. (1970). Effects of forest cutting and herbicide treatment on nutrient budgets in the Hubbard Brook ecosystem in New Hampshire. Ecol. Monogr. 40:23-47.

LITTLE, E.L., JR. (1981). Rare and local trees in the eastern deciduous forest biome. In: *Report 2-90*, pp. 89-103. School of Forestry and Wildlife Resources, Virginia Polytechnic Institute and State University, Blacksburg, VA.

LOCKE, G.M.L. (1978). The growing stock of regions. Forestry 51:5-8.

LUGTHART, G.J. (1987). The effect of an insecticide-induced insect reduction on salamander diets in a southern Appalachian headwater stream. Bull. N. Am. Benthol. Soc. 4:57.

MACARTHUR, R.H. (1972). Geographical Ecology. Princeton University Press, Princeton, NJ.

MAKKONEN, O. (1967). Ancient forestry - an historical study, Part I: Facts and information on trees. Acta. For. Fenn. 82:1-76.

MANN, L.K., JOHNSON, D.W., WEST, D.C., COLE, D.W., HORNBECK, J.W., MARTIN, L.W., RIEKERK, H., SMITH. C.T., SWANK, W.T., TRITTON, L.M., and VAN LEAR, D.H. (1988). Effects of whole-tree and stem-only clearcutting on postharvest hydrologic losses, nutrient capital and regrowth. For. Sci. 34:412-428.

MARKS, P.L. (1974). The role of pin cherry (*Prunus pensylvanica*) in the maintenance of stability in northern hardwood ecosystems. Ecol. Monogr. 44:73-88.

MARTIN, J.P. (1963). Influence of pesticide residues on soil microbiological and chemical properties. Residue Rev. 4:98-129.

MASER, C. and TRAPPE, J.M. (1984). The seen and unseen world of the fallen tree, Gen. Tech. Rep. PNW-164. USDA Forest Service, Washington, DC.

MASER, C.S., CLINE, P., CROMACK, K., JR., TRAPPE, J.M., and HANSEN, E. (1988). What we know about large trees that fall to the forest floor. In: *From the Forest to the Sea: A Story of Fallen Trees*, Gen Tech. Rep. PNW-GTR 229 (C. Maser, R.F. Tarrant, J.M. Trappe, and J.F. Franklin, eds), pp. 25-46. USDA Forest Service, Washington, DC.

MAY, T.A., MAY, D.E., and MCCORMACK, M.L., JR. (1982). Plant response to herbicides in clear cut conifer strips. Proc. Northeast Weed Sci. Soc. 36:205-208.

MAYER, R. and ULRICH, B. (1974). Conclusions on the filtering actions of forests from ecosystem analysis. Oceol. Plant. 9:157-168.

MCCOLL, J.G. and GRIGAL, D.F. (1979). Nutrient losses in leaching and erosion by intensive forest harvesting. In: *Proceedings of the Symposium on Intensive Harvesting on Forest Nutrient Cycling*, pp. 231-248. USDA Northeast Forestry Experimental Station, Broomwall, PA.

MCCOMB, W.C. and HURST, G.A. (1987). Herbicide and wildlife in southern forests. In: *Managing Southern Forests for Wildlife and Fish*, Gen Tech. Rep. SO-65 (J.G. Dickson and O. E. Maughan, eds), pp. 28-39. USDA Forest Service, Washington, DC.

MEEHAN, W.R., FARR, W.A., BISHOP, D.M., and PATRICK, J.H. (1969). Some Effects of Clearcutting on Salmon Habitat of Two Southwest Alaska Streams, Res. Pap. PNW-82. USDA Forest Service, Washington, DC.

MICHAEL, J.L. (1980). Long-term impact of aerial application of 2,4,5-T to longleaf pine. Weed Sci. 28:255-257.

MICHAEL, J.L. (1985). Growth of loblolly pine treated with hexazinone, sulfometuron methyl, and metsulfuron methyl for herbaceous weed control. South. J. Appl. For. 9:20-26.

MICHAEL, J.L. and NEARY, D.G. (1989). Fate and movement of herbicides on forest sites in the southern United States of America. I.U.F.R.O. Division I: Herbicides in Forestry, International Meeting, Nancy, France.

MILLER, R.E. and BIGLEY, R.E. (1990). Effects of burning Douglas-fir logging slash on stand development and site productivity. In: *Sustained Productivity of Forest Soils*, Proceedings of the Seventh North American Forest Soils Conference (S.P. Gessel, D.S. Lacate, G.F. Weetman, and R.F. Powers, eds), pp. 362-376. University of British Columbia, Faculty of Forestry, Vancouver, BC.

MILLER, K.W. and WITT, J.S. (1991). Wildlife habitat responses to herbicide site preparation. In: *Proceedings of the 6th Biennial Silvicultural Research Conference*, Gen. Tech. Rep. SE-90 (S. Coleman and D. G. Neary, eds), pp. 795-800. USDA Forest Service, Washington, D. C.

MILLER, H.G., COOPER, J.M., and MILLER, J.D. (1976). Effect of nitrogen supply on nutrients in litterfall and crown leaching in a stand of Corsican pine. J. Appl. Ecol. 13:233-248.

MILLER, K.V., BUSH, P.G., TAYLOR, J., and NEARY, D.G. (1990a). Herbicides and wildlife habitat: A review of 35 years of research. Proc. South. Weed Sci. Soc. 43:202.

MILLER, K.W., BUSH, P.B., TAYLOR, J., and NEARY, D.G. (1990b). Herbicides and Wildlife Habitat: An Annotated Bibliography on the Effects of Herbicides on Wildlife Habitat and Their Uses in Habitat Management, Tech. Publ. R8-TP13. USDA Forest Service, Washington, DC. 152 pp.

MORRIS, L.A. and LOWERY, R.F. (1988). Influence of site preparation on soil conditions affecting stand establishment and tree growth. South. J. Appl. For. 12:170-178.

MORRIS, L.A. and PRITCHETT, W.L. (1982). Nutrient storage and availability in two managed pine flatwoods forests. In: *Proceedings of the Intensive Forest Management Practices Symposium* (C. Coleman, A.C. Mace., Jr., and B.F. Swindel, eds), pp. 17-26. School of Forest Resources and Conservation, University of Florida, Gainesville, FL.

MORRIS, L.A., PRITCHETT, W.L., and SWINDEL, B.F. (1983). Displacement of nutrient into windrows during site preparation of a flatwoods forest. Soil Sci. Soc. Am. J. 47:591-594.

MORRISON, M.L. and MESLOW, E.C. (1983). Impacts of forest herbicides on wildlife: Toxicity and habitat alteration. Trans. 48th N. Am. Wildl. Conf. 48:175-185.

MURRAY, M.D. (1991). The tree that fights cancer. Am. For. July-Aug:52-54.

NEARY, D.G., BUSH, P.B., and DOUGLAS, J.E. (1983). Off-site movement of hexazinone in stormflow and baseflow from forested watersheds. Weed Sci. 31: 543-551.

NEARY, D.G., MORRIS, L.A., and SWINDEL, B.F. (1984). Site preparation and nutrient management in southern pine forests. In: *Forest Soils and Treatment Impacts* (E.L. Stone, ed), pp. 121-143. University of Tennessee Press, Department of Forestry, Wildlife and Fisheries, Knoxville, TN.

NEARY, D.G., BUSH, P.B., DOUGLAS, J.E., and TODD, R.L. (1985). Picloram movement in an Appalachian hardwood forested watershed. J. Environ. Qual. 14:585-592.

NEARY, D.G., BUSH, P.B., and GRANT, M.A. (1986). Water quality of ephemeral forest streams after site preparation with the herbicide hexazinone. For. Ecol. Manage. 14:23-40.

NEARY, D.G., SMITH, J.E., SWINDELL, B.F., and MILLER, K.V. (1990). Effects of forestry herbicides on plant species diversity. In: *Proceedings of the Southwest Weed Science Society*, pp. 266-272. Atlanta, GA.

NEWTON, M. and NORRIS, L.A. (1976). Evaluating short-term and long-term effects of herbicides on non-target forest and range biota. Down Earth 32:18-26.

NEWTON, M.B., HASSAN, E.L., and ZAVITKOUSKI, J. (1968). Role of red alder in western Oregon forest succession. In: *Proceedings of the Symposium on the Biology of Alder* (J.M. Trappe, J.F. Franklin, R.F. Tarrant, and G.M. Hansen, eds), pp. 73-84. USDA PNW Forest Range Experimental Station, Portland, OR.

NOSS, R.F. (1990). What can wilderness do for biodiversity? In: *Preparing to Manage Wilderness in the 21st Century*, Gen. Tech. Rep. SE-60. USDA Forest Service, Washington, DC.

NUTTER, W.L., KNISEL, W., BUSH, P.B., and TAYLOR, J.W. (1990). Use of GLEAMS to predict insecticide losses from pine seed orchards. Environmental Toxicology and Chemistry Society (SETAC). Annual Meeting, Arlington, VA.

NWOBOSHI, L.C. (1983). Potential impacts of some harvesting options on nutrient budgets of a Gmelina pulpwood plantation in Nigeria. In: IUFRO Symposium on Forest Site and Continuous Productivty, Gen. Tech. Rep. PNW-163 (R. Ballard and S.P. Gessel, eds), pp. 212-217. USDA Forest Service, Washington, D. C.

OKUBO, A. (1980). Diffusion and Ecological Problems: Mathematical Models. Springer-Verlag, NY.

OLIVER, C.D. (1981). Forest development in North America following major disturbance. For. Ecol. Manage. 3:153-168.

OVINGTON, J.D. (1962). Quantitative ecology and the woodland ecosystem concept. Adv. Ecol. Res. 1:103-192.

PATRIC, J.H. (1976). Soil erosion in the eastern forest. J. For. 74:671-677.

PEET, R.K. (1981). Changes in biomass and production during secondary forest succession. In: *Forest Succession: Concepts and Management* (D.C. West, H.H. Shugart, and D.B. Botkin, eds), pp. 324-338. Springer-Verlag, NY.

PERCICH, J.A. and LOCKWOOD, J.L. (1978). Interaction of atrazine with soil microorganisms: Population changes and accumulation. Can. J. Microbiol. 24:1145-1152.

PERRY, T.O. (1964). Soil compaction and loblolly pine tree growth. Tree Plant. Notes 67:9.
PIELOU, E.C. (1991). After the Ice Age. The University of Chicago Press, Chicago, IL.
PIERCE, R.S., MARTIN, C.W., REEVES, C.C., LIKENS, G.E., and BORMANN, F.H. (1972). Nutrient loss from clearcuttings in New Hampshire. In: *Proceedings of the National Symposium on Watersheds in Transition* (S.C. Csallany, T.G. McLaughlin, and W.D. Striffler, eds), pp. 285-295. American Water Resources Association, Urbana, IL.
POFF, N.L. and WARD, J.V. (1990). Physical habitat template of lotic systems: Recovery in the context of historical pattern of spatiotemporal heterogeneity. Environ. Manage. 14:629-645.
RACKHAM, O. (1986). The history of the British countryside. J. M. Dent, London (as cited in Warren and Key 1991).
RADTKE, R.E., BYELICH, J., and IRVINE, G. (1983). Endangered birds of fire — the Kirtland's warbler. In: *Yearbook of Agriculture*. U.S. Department of Agriculture, Washington, DC.
REESE, K.P. and RATTI, J.T. (1988). Edge effect: A concept under scrutiny. Trans. N. Am. Wildl. Nat. Resour. Conf. 53:127-136.
REID, W.V. and MILLER, K.R. (1989). Keeping Options Alive: The Scientific Basis for Conserving Biodiversity. World Resources Institute, Washington, DC.
RICHTER, D.D., RALSTON, C.W., and HARMS, W.R. (1982). Prescribed fire: Effects on water quality and forest nutrient cycling. Science 215:661-663.
RIEKERK, H. (1981). Impacts of silviculture on flatwoods runoff, water quality and nutrient budgets. In: *Proceedings of the Symposium on Progress in Wetlands Utilization and Management* (P. McCaffrey, ed). Coordinating Council on Restoration of the Kissimmee River Valley and Taylor Creek-Nubbin Slough Basin, Tallahassee, FL.
RIEKERK, H. (1982). Water quality management in flatwoods of Florida. In: *Impacts of Intensive Forest Management Practices Symposium*, pp. 35-52. University of Florida, Gainesville, FL.
SAMPSON, R.N. (1988). Releaf for global warming. Am. For. 9-14.
SAVIL, P.S. and EVANS, J. (1986). Plantation Silviculture Temperate Regions. Oxford University Press, NY.
SEDJO, R.A. (1990). The global carbon cycle—are forests the missing sink? J. For. 88(10):33-34.
SHARITZ, R.R., BORING, L.R., and VAN LEAR, D.H. (In press). Integrating principles of ecology with natural resource management of southern forests. Ecol. Appl.
SHARPLEY, A.N. and SYERS, J.K. (1979). Phosphorus inputs into a stream draining an agricultural watershed: II amounts and relative significance of runoff types. Water Air Soil Pollut. 9:417-428.
SHEA, S.R.K., GILLEN, J., and KITT, R.J. (1978). Variation in sporangial production of *Phytopthora cinnamoni* on jarrah forest sites with different understory compositions. Aust. For. Res. 8: 219-226.
SHIRT, D.B. (1987). British Red Data Book-2, Insects. Nature Conservancy Council, Peterborough, UK.
SIDLE, R.C. (1991). A conceptual model of changes in root cohesion in response to vegetation management. J. Environ. Qual. 20: 43-52.
SINGER, M.J. and RUST, R.H. (1975). Phosphorus in surface runoff from a deciduous forest. J. Environ. Qual. 4:307-311.
SMITH, D.M. (1986). The Practice of Silviculture, 8th ed. John Wiley & Sons, NY.

SOLIS, D.M., JR. (1983). Summer Habitat Ecology of Spotted Owls in Northwestern California, M.S. Thesis. Humbolt State University, Arcata, CA. 168 pp.

SOLLINS, R., GRIER, C., MCCORISON, F.M., CROMACK, K., JR., and FOGEL, R. (1980). The internal cycles of an old growth Douglas-fir ecosystem in western Oregon. Ecol. Monogr. 50:261-285.

SPEIGHT, M.R. and SPEECHLY, H.T. (1982). Pine shoot moths in S.E. Asia, I: Distribution, biology and impact. Commonwth. For. Rev. 61:121-134.

SPRUGEL, D.G. (1985). Natural disturbance and ecosystem energetics. In: *The Ecology of Natural Disturbance and Patch Dynamics* (S.T.A. Pickett and P.S. White, eds), pp. 335-352. Academic Press, NY.

SQUIRE, R.O., FARRELL, P.W., FLINN, D.W., and AEBERLI, B.C. (1985). Productivity of first and second rotation stands of radiata pine on sandy soils. Aust. For. 48:127-137.

STONE, E. L. (1975). Soil and man's use of land. In: *Forest Soils and Forest Land Management*, (B. Bernier and C.H. Winget, eds), Proceedings of the Fourth North American Forest Soils Conference. Laval University Press, Quebec, Canada.

SUTTON, S.L. and COLLINS, N.M. (1989). Insects and tropical forest conservation. In: *The Conservation of Insects and Their Habitats* (N.M. Collins and J.A. Thomas, eds), pp. 405-424. Academic Press, London.

SWANK, W.T. and HENDERSON, G.S. (1976). Atmospheric input of some cations and anions to forest ecosystems in North Carolina and Tennessee. Water Resour. Res. 12:541-546.

SWIFT, L.W. and MESSER, J.B. (1971). Forest cutting raises stream temperatures of small streams in the southern Appalachians. J. Soil Water Conserv. 26:111-116.

SWINDEL, B.F., CONDE, L.F., and SMITH, J.E. (1987). Index-free diversity orderings: Concept, measurement, and observed response to clearcutting and site preparation. For. Ecol. Manage. 20:195-208.

SWITZER, G.L. and NELSON, L.E. (1973). Maintenance of productivity under short rotations. In: *Proceedings of the International Symposium on Forest Fertilization*. FAO/IUFRO, Paris.

SWITZER, G.L., NELSON, L.E., and BAKER, J.B. (1976). Accumulation and distribution of dry matter in *Aigeiros* poplar plantations. In: *Proceedings of the Symposium on Eastern Cottonwood and Related Species* (B.A. Thielges and S.B. Land, Jr., eds), pp. 359-378. Division of Continuing Education, Louisiana State University, Baton Rouge, LA.

TAMM, C.O. (1969). Site damage by thinning due to removal of organic matter and plant nutrients. In: *Proceedings of the IUFRO Meeting on Thinning and Mechanization*. Stokholm, Sweden.

TEW, D.T., MORRIS, L.A., ALLEN, H.L., and WELLS, C.G. (1986). Estimates of nutrient removal, displacement and loss resulting from harvest and site preparation of a *Pinus taeda* plantation in the Piedmont of North Carolina. For. Ecol. Manage. 15:257-267.

THORUD, D.B. and FRISSEL, S.S. (1976). Time changes in soil density following compaction under an oak forest. Minn. For. Res. Notes 257, School of Forestry, University of Minnesota, St. Paul, MN.

TILMAN, D. (1982). Resource Competition and Community Structure. Princeton University Press, Princeton, NJ.

TURNER, J. and SINGER, M.J. (1976). Nutrient distribution and cycling in a sub-alpine coniferous forest ecosystem. J. Appl. Ecol. 13:295-301.

U.S. FOREST SERVICE. (1988). The South's Fourth Forest: Alternatives for the Future, Rep. FRR 24. USDA Forest Service, Washington, DC.

UNITED STATES DEPARTMENT OF AGRICULTURE (USDA). (1984). Pesticide Background Statements, Volume 1: Herbicides, Agriculture Handbook No. 663. USDA Forest Service, Washington, DC.

UNITED STATES DEPARTMENT OF AGRICULTURE (USDA). (1986a). Pesticide Background Statements, Volume I: Herbicides (supplement), Handbook No. 633. U.S. Government Printing Office, Washington, DC.

UNITED STATES DEPARTMENT OF AGRICULTURE (USDA). (1986b). Pesticide Background Statements, Volume II: Fungicides and Fumigants, Agriculture Handbook No. 661. USDA Forest Service, Washington, DC.

UNITED STATES DEPARTMENT OF AGRICULTURE (USDA). (1987a). Pesticide Background Statements, Volume III: Nursery Pesticides, Agriculture Handbook No. 670. USDA Forest Service, Washington, DC.

UNITED STATES DEPARTMENT OF AGRICULTURE (USDA). (1987b). Human Health Risk Assessment for the Uses of Pesticides in the USDA Forest Service Nurseries, FS412, U.S. Printing Office Document No. 1987-716-000:02106/FS. USDA Forest Service, Washington, DC.

UNITED STATES DEPARTMENT OF AGRICULTURE (USDA). (1988). Draft Environmental Impact Statement: Vegetation Management in the Coastal Plain/Piedmont, Vol. 2 with appendices, Bull. R8-MB 15. USDA Forest Service, Washington, DC.

URSIC, S.J. (1970). Hydrologic Effects of Prescribed Burning and Deadening Upland Hardwoods in Northern Mississippi, Res. Paper SO-54. USDA Forest Service, Washington, DC. 16 pp.

VAN LEAR, D.H., DOUGLAS, J.E., COS, S.K., and AUGSPURGER, M.K. (1985). Sediment and nutrient export in runoff from burned and harvested pine watersheds in the South Carolina Piedmont. J. Environ. Qual. 14:169-174.

VITOUSEK, P.M. (1991). Can planted forests counteract increasing carbon dioxide? J. Environ. Qual. 20:348-354.

VITOUSEK, P.M. and MATSON, P.A. (1985). Disturbance, nitrogen availability, and nitrogen losses in an intensively managed loblolly pine plantation. Ecology 66:1360-1376.

WALLACE, R.R. and HYNES, H.B.N. (1975). The catastrophic drift of stream insects after treatments with methoxychlor. Environ. Pollut. 8:225-268.

WALLACE, R.R. and HYNES, H.B.N. (1981). The effect of chemical treatments against blackfly larvae on the fauna of running waters. In: *Black Flies: The Future for Biological Methods in Integrated Control* (M. Laird, ed), pp. 237-258. Academic Press, NY.

WALLACE, J.B., WEBSTER, J.R., and CUFFNEY, T.F. (1982). Stream detritus dynamics, regulation by invertebrate consumers. Oecologia (Berl.) 53:197-200.

WALLACE, J.B., VOGEL, D.S., and CUFFNEY, T.F. (1986). Recovery of a headwater stream from an insecticide-induced community disturbance. J. N. Am. Benthol. Soc. 5:115-126.

WARREN, M.S. (1987). The ecology and conservation of the health fritillary butterfly, *Mellicta athalia* III Population dynamics and the effect of habitat management. J. Appl. Ecol. 24:499-513.

WARREN, M.S. and KEY, R.S. (1991). Woodlands: Past, present and potential for insects. In: *The Conservation of Insects and Their Habitats*, Proceedings of the 15th Symposium of the Royal Entomological Society of London (N.M. Collins and J.A. Thomas, eds), pp. 155-211. Academic Press, London.

WAUCHOPE, R.D., HORNSBY, A.G., and TAYLOR, J.W. (1991). Pesticide properties data base: A set of parameter values for first-tier comparative water pollution risk analysis. Rev. Environ. Toxicol. 123:1-155.

WEBBER, B. (1978). Potential increases in nutrient requirements of *Pinus radiata* under intensified management. N.Z.J. For. Sci. 8:146-159.

WEED SCIENCE SOCIETY OF AMERICA. (1979). Herbicide Handbook, Fourth Ed. Weed Science Society of America, Champaign, IL. 479 pp.

WEETMAN, G.F. and WEBBER, B. (1972). The influence of wood harvesting on the nutrient status of two spruce stands. Can. J. For. Res. 3:351-369.

WEIDEMANN, E. (1923). Zuwachsruckgang and wuchsstockungen der achte in den mittleren und den unteren hohenlagen der sachischen staats forsten. Tharandt. 181 pp. (USDA Forest Service Translation 302, 1936).

WELLS, C.G. and JORGENSEN, J.R. (1975). Nutrient cycling in loblolly pine plantations. In: *Forest Soils and Forest Land Management*, Proceedings of the Fourth North American Forest Soils Conference (B. Bernier and C.H. Winget, eds), pp. 137-158. Laval University Press, Quebec.

WHITTAKER, R.H., LIKENS, G.E., BORMANN, F.H., EATON, J.S., and SICCAMA, T.G. (1979). The Hubbard Brook ecosystem study: forest nutrient cycling and element behavior. Ecology 60:203-270.

WHYTE, A.G.D. (1973). Productivity of first and second crops of *Pinus radiata* on the moutere soils of Nelson. New Zealand J. For. 18:87-103.

WILLIAMS, T.M., ALM, A.A., and MACE, A.C., JR. (1989). Fifteen-year jack pine growth on full-tree harvesting experiments in northeastern Minnesota. In: *Research Strategies for Long-term Site Productivity*, Proceedings of the IEA/BE A3 Workshop (W.J. Dyck and C.A. Mees, eds), pp. 111-122. New Zealand For. Res. Inst. Bull. 152.

WILSON, E.O. (1989). Threats to biodiversity. Sci. Am. 261(3):108-116.

WOOD, P.J. (1975). The world situation as it will affect the U.K. Scott. For. 29:25-38.

YASUNO, M., FUKUSHIMA, S., HASEGAWA, J., SHIOYAMA, F., and HATAKEYAMA, S. (1982). Changes in benthic fauna and flora after application of temephos to a stream on Mt. Tsukuba. Hydrobiologia 89:205-214.

YOHO, N.S. (1980). Forest management and sediment production in the South—a review. South J. Appl. For. 4:27-36.

CHAPTER 11

ASSESSING ENVIRONMENTAL RISK FROM GENETICALLY ENGINEERED MICROORGANISMS AND PRODUCTS CONTAINING RECOMBINANT DNA

David R. Orvos
University Center for Environmental and Hazardous Materials Studies
Virginia Polytechnic Institute and State University
Blacksburg, Virginia

INTRODUCTION

During the 1970s, several techniques were developed that allowed the *in vitro* fragmentation and rejoining of molecules of deoxyribonucleic acid (DNA) and the placement of these molecules into living cells. Use of these recombinant DNA techniques led to the development of biotechnology, broadly defined as the application of biological organisms and their products to commercial processes (Office of Technology Assessment [OTA] 1988). Among the types of organisms now being used for biotechnological applications, bacteria have been used more than others because of their short generation times, lack of a defined nucleus, and relatively "simple" genetics as compared to eukaryotes. However, other organisms, including fungi, algae, plants, and animals, are also being used as genetic engineering hosts. Genetic engineering of microorganisms is being used to perform specific, commercially desired tasks including producing pesticides and pharmaceuticals, fixing nitrogen, leaching ores, recovering oil, degrading hazardous wastes, and constructing biological control agents for pests or diseases. Biotechnological products will, one hopes, benefit human society.

Current environmental applications of genetically engineered microorganisms (GEMs) have been limited to relatively small scale trials because of a lack of knowledge about their impact on ecosystems and an increased public awareness over the release of genetically manipulated organisms. Critics of these relatively new techniques often compare environmental release to exotic species release, such as the starling and Dutch elm disease pathogen. However, such comparisons are not valid since exotic species represent an exotic genome while GEMs may represent indigenous genomes altered only to a small extent (OTA 1988). To date, no adverse consequences have resulted from the small scale release of GEMs (Lindow et al. 1989; OTA 1988; author's review of National Biotechnology Impact

---

1. Corresponding Author: David R. Orvos, Environmental Research Laboratory, SmithKline Beecham Corporation, P. O. Box 1539, L-25, King of Prussia, PA 19406.

2. Key Words: GEMs, recombinant DNA, bacteria, biotechnology, environmental effects.

Assessment Program news reports database); however, the majority of these releases have entailed the use of microbes that have had parts of their genome removed or have had innocuous insertions that pose no obvious environmental threats. At some point in time, GEMs with capabilities of altering basic ecosystem processes, such as nitrogen fixation and nutrient cycling, will be released. This review addresses the environmental release of GEMs and other nonliving products containing or derived from recombinant DNA (rDNA), addresses their potential environmental effects, and attempts to construct a risk assessment paradigm to predict the environmental fate and effects of a GEM or other product containing rDNA.

FACTORS FOR CONSIDERATION

Several factors must be considered before GEMs or products containing rDNA are released in large quantities into the environment. Survival of an organism under laboratory conditions often does not ensure survival in natural ecosystems. Conversely, the ability to contain and control species in the laboratory does not preclude their establishment, survival, and adverse environmental effects outside the laboratory (Cairns and Pratt 1986). Biotechnological products, especially the engineered organisms themselves, present unique regulatory problems since, unlike toxic chemicals, GEMs have the potential to increase in quantity in the environment at the site of release or following transport to some unintended and unexpected location. Similarly, rDNA in a process waste stream has the potential to be incorporated into indigenous microbes during or after the waste treatment process.

GEMs may potentially affect the environment through any one of three ways: (1) by transferring DNA from or into the GEM, thus altering the genetic composition of the GEM or indigenous microbial populations; (2) by affecting the structural integrity of communities, both microbial and macrobial; and (3) by affecting functional processes, including geochemical cycling of nutrients.

Genetic Considerations

Bacteria possess genetic mechanisms for the dispersal and alteration of their DNA; such mechanisms naturally enhance bacterial survival in the environment (Grant and Long 1981). Four basic mechanisms for exchanging genetic material are known: (1) transformation, the uptake of naked DNA as plasmids in the case of Gram-negative bacteria or as linear DNA fragments by Gram-positive bacteria; (2) transduction, the incorporation of bacterial DNA in bacteriophage particles and their transfer to recipient bacteria; (3) conjugation, the transfer of DNA by direct cell-to-cell contact; and (4) fusion and recombination, a mechanism that is limited to actinomycetes.

Most, if not all, bacteria in the environment exchange genetic information with nearby microbes (Slater 1985). Several steps must be followed when evaluating the risk from potential transfer of the genome from a GEM or rDNA product to indigenous bacteria. First, molecular techniques must confirm that gene transfer has indeed occurred. Such techniques may include DNA:DNA or DNA:RNA hybridization and sequencing of the DNA region in question. Second, if a transfer has indeed occurred, information must be obtained on whether the newly created genetic information will be sustained and expressed in the bacterium that acquired it. Third, if genetic information is maintained and expressed, the effects that such expression may have on other biota must be properly evaluated (OTA 1988). Several criteria, including gene transfer frequency and genetic distance among species, can be used in executing an assessment, along with consideration of the construction of the gene itself and the vector on which the gene resides (OTA 1988).

While studies of potential genetic exchange mechanisms among microorganisms have been undertaken, few have examined these mechanisms outside the laboratory. Transfer frequencies on filter membranes are useful, but they fail to provide information concerning selective pressures. In addition, DNA is rarely found in the naked, available state in environmental samples; rather, it is either degraded by exogenous nucleases or sorbed to soil or sediment particles and, therefore, unavailable for transfer (Ogram et al. 1988; Romanowski et al. 1991). Such sorption is regulated by soil type, clay and organic material content, ionic strength, pH, and DNA length (Ogram et al. 1988).

Plasmid DNA transfer, via conjugation, in the environment has been studied in both terrestrial and aquatic systems. Plasmids (autonomously replicating, extrachromosomal, circular pieces of DNA) are ubiquitous in bacteria. Many environmental isolates have been shown to be recipients of plasmids. Genthner et al. (1988) tested 68 bacterial isolates from aquatic sources and found 26 of these able to receive plasmids from *Pseudomonas aeruginosa* donors. Schilf and Klingmuller (1983), using plasmid pRD1 carried by *Escherichia coli*, discovered that 1.3% of soil bacteria and 17.3% of aquatic bacteria could receive the plasmid from a donor. Evidence for plasmid selection or exchange in the environment is also based upon high incidence of plasmids in polluted areas (Hada and Sizemore 1981), coexistence of identical plasmids in different strains, and data from *in situ* transfer experiments (Grimes et al. 1988). While plasmids are known to affect the growth of microorganisms, Cruz-Cruz et al. (1988) found that bacteria containing large (100 megadalton) plasmids may survive *in situ* equally as well as those with smaller plasmids. They also suggest that genetic alteration of indigenous microorganisms is made possible by either conjugation or transformation in aquatic habitats. Bacterial conjugation has been observed in both non-sterile and sterile soils during extensive studies by Stotzky (e.g., Weinberg and Stotzky 1972). Finally, transconjugants of *Enterobacter cloacae* were found in the digestive tracts and fecal pellets of the variegated cutworm (*Peridroma saucia*) by Armstrong et al. (1991), who concluded that recombinant DNA transfer may occur at rare frequencies. These findings indicate that higher trophic levels must be part of any risk assessment paradigm.

Transfer of plasmids also appears to be influenced by environmental parameters. Stotzky and Krasovsky (1981) found that recombination was affected markedly by pH, with the survival of both donors and recipients increasing at neutrality. Moisture content of soil does not appear to be influential if it is between 20 and 100% of the total holding capacity of the soil (Trevors and Oddie 1986). In addition to transfer in soil and aquatic systems, plasmids may be transferred on plants (Lacy et al. 1984).

The DNA transfers studied from GEMs already released appear to be innocuous; this is expected since antibiotic resistance from the marker plasmid is often the only parameter monitored. Nevertheless, it is conceivable that plasmid transfer may affect important functional characteristics of the organism. Rafii and Crawford (1988) found that several conjugative plasmids were transferred in sterile soil from the donor *Streptomyces lividans* to several *Streptomyces* strains isolated from environmental samples. In a concurrent study, Wang and Crawford (1988) found that a similar *Streptomyces* strain caused significantly higher soil carbon mineralization rates in non-sterile soil amended with lignocellulose. It is believed that the effect was caused by a foreign DNA insert. The transfer of such an insert could alter the function of other organisms. Finally, Glick et al. (1986) found that plasmid DNA introduction into *Azotobacter vinelandii* via transformation reduced its capacity for nitrogen fixation, mean cell size, and siderophore production.

Plasmids may not be suitable for environmental applications since they are frequently passed from one organism to another (Slater 1985). However, because of ease of manipulation and potential usefulness, they cannot simply be neglected; any risk attributed to them must be evaluated and predicted. Walter et al. (1987a) found that no single mating technique was superior in predicting the frequency of plasmid transfer and developed a new technique, a combination of several older ones, that predicts the frequency with which plasmids move from one cell to another. Another group (Knudsen et al. 1988) at the Corvallis Environmental Research Laboratory of the USEPA developed a predictive computer model that accurately forecasts transconjugant populations in the rhizosphere and on leaf surfaces. This and other future models may eventually lead to accurate assessments of plasmid transfer prior to environmental release.

DNA transfer via transduction has been documented in environmental samples from both soil and water. Zeph et al. (1988) found that transduction of *Escherichia coli* by bacteriophage P1 occurred in soil, both sterile and non-sterile; however, they did not observe transduction of indigenous organisms. It appears that a minimum number of bacteria are needed to ensure replication of the phage (Wiggins and Alexander 1985). This limit minimizes viral damage in the environment because of the relatively low populations of individual bacteria (Wiggins and Alexander 1985). Transduction in aquatic samples has also been demonstrated by Saye et al. (1987), who used *P. aeruginosa* containing plasmid Rms149, although the presence of natural microbial communities significantly decreased transduction frequency.

Transformation, especially in *Bacillus subtilis*, has been shown in non-sterile soil (Graham and Istock 1978, 1979). However, Greaves and Wilson (1970) found that DNA or RNA added to non-sterile soil was rapidly degraded, thus causing an increase in indigenous non-transformed populations. Both of these studies have implications for evaluating rDNA in process waste streams. Although the GEMs may have been destroyed by sterilization, the rDNA may still be viable for the transformation process.

Transfer of genetic information is a viable concern when designing a risk assessment scheme to predict potential environmental consequences. Recent technologies such as suicide vectors (Bej et al. 1988; Knudsen and Karlstrom 1991) may eventually minimize movement of recombinant genes. Even more important than the actual transfer of genetic information is the expression of that genetic information. If such expression is to be manifested, the bacterium must survive after it is released. Alexander (1985) surmises that if a GEM is to have an effect, good or bad, it must survive. The question of survival is central to both the pro and con arguments regarding the efficacy and risk of GEMs.

Survival

A great deal of information has been acquired regarding survival of single species of bacteria, especially *Rhizobium* (Stacey 1985). However, relatively few data have been obtained concerning the release of non-root colonizing bacteria into an environment that may or may not have originally contained the bacteria being released. The data collected from introductions of *Rhizobium* (and other bacteria) are useful in comprehending biotic and abiotic factors that affect microbial survival.

Abiotic factors that affect microbial growth in the environment include substrate type, nutrient availability, temperature, desiccation, salinity, and pH (Grant and Long 1981; Stacey 1985). Among the most important for a newly released GEM are nutrient availability, temperature, and pH. Deficiencies or extreme fluctuations in any of these parameters may severely compromise GEM integrity. Even if the GEM encounters conditions favorable for its establishment, it must contend with predation by phages, protozoans, or other bacteria. Predation has been shown to adversely affect *Rhizobium* populations (Stacey 1985). In addition, deleterious enzymes or antibiotics produced by indigenous microbes may cause elimination of the introduced species. Effects of many of these factors can be reduced by increasing the concentrations of bacteria being applied to the area of interest. With this rationale, though, comes an increased chance of producing runoff that may transport the GEM to an area not originally intended.

It has been demonstrated that exogenous bacteria, engineered or not, do not often survive for extended periods of time because of environmental pressures exerted on them. Indeed, if the bacteria being introduced have been derived from indigenous species, they will have to out compete those species already firmly established in their niche. To make these introduced microbes predominate, strong selective pressure will have to be applied (Stacey 1985). Alexander (1985) summarizes this by concluding that bacteria die when they fail to overcome any of the biotic or abiotic factors that they encounter.

Although survival has been examined for many species of engineered bacteria, few studies have endeavored to examine bacterial populations below the level of detection found with popular enumeration techniques. Scanferlato et al. (1990) found that engineered *Erwinia carotovora* do indeed decline to extinction after prolonged periods in microcosms. This is probably true for all other engineered microbes released to date since the external pressures from interspecific competition, predation, and nutrient availability are immense. Postma et al. (1991) examined the importance of physical parameters, such as altered cell surface properties and particle association, for survival of *Rhizobium* and determined that no clear relationship existed.

Trevors et al. (1990) used vertical microcosms to assess downward migration of genetically engineered *P. fluorescens* and found that cells could be detected even after three flushes of groundwater and 10 incubation days. This finding substantiated the opinion that vertical movement of GEMs to groundwater must be considered for risk assessment.

Predicting survival will be an important parameter in risk assessment. Walter et al. (1987b) developed a technique that allows determination of bacterial survival in soil extracts. This technique appears to be useful for an industrial assay in initial data generation. The USEPA group found that engineered strains survived longer in sterile soil than in non-sterile soil. This is not surprising since a lack of competition would encourage survival. In all of the GEM soil survival studies to date, bacteria have always declined in concentration over time, most likely from the enormous competition they encounter.

Few data exist for GEM releases into aquatic systems. Like their terrestrial counterparts, bacteria decline when released into aquatic microcosms (Burton et al. 1987). Scanferlato et al. (1989) found that engineered *E. carotovora* declined at the same rate as its wildtype parent using a microcosm design that allowed GEM detection over a 64-day period. Another microcosm study by Awong et al. (1990) found no significant differences in decline between engineered and non-engineered *P. putida* over 20 days.

While documentation exists for survival of GEMs in soil, few studies have been executed to measure aerial dispersion of microorganisms from plants and soil; little research has been undertaken to elucidate the conditions, such as wind speed, plant leaf condition and age, inoculum age, and moisture, that may influence dispersal of GEMs into unanticipated areas. Because of their small mass, bacteria and viral particles require extensive periods of time to settle out of the atmosphere; therefore, they may be transported over long distances (Barnthouse and Palumbo 1986). Bacteria have traveled hundreds of kilometers on air currents; however, bacteria carried by plant tissue or airborne soil particles settle far more rapidly (Barnthouse and Palumbo 1986). In addition, the method used to apply the bacteria in the environment has been shown to affect survival and transport after application (Marthi et al. 1990).

Bacterial plant pathogens disseminate in aerosols produced by raindrop impacts, irrigation systems, and mechanical manipulation (Harrison 1980) and are capable of inducing pathogenesis after movement. Lindemann and Upper (1985) observed that bacteria re-entrain on plant surfaces and soil after being disseminated in aerosols; in addition, these researchers discovered that plant canopies are excellent sources of these aerosols when leaves are dry on sunny days. More recently, two studies have documented that bacteria altered by natural processes do indeed aerially disperse from a site of introduction. Marshall et al. (1988) found that *Escherichia coli*, containing a Tn*5* derivative of the plasmid ColE1, dispersed from the sites of introduction, a laboratory office and a barn, with maximal dispersion being observed in the first 30 minutes following release. While rapid aerosol die-off was observed, it is conceivable that many of the bacteria attained a viable, but non-culturable status (Colwell et al. 1985) that was not detected. Such bacteria could increase in concentration if favorable conditions were encountered. Marshall and colleagues also learned that, in several instances, *E. coli* containing ColE1 and Tn*5* demonstrated enhanced survival, apparently related to a growth advantage from the transposon. In a more environmentally realistic study, Lindow et al. (1988) determined that viable *P. syringae* cells experienced an exponential decrease in concentration as distance from aerosol application site increased. While survival rate of bacteria was 10 times higher on the intended plants than on adjacent plants, such dispersion was demonstrated even with relatively calm winds (0.73 to 1.5 m/s) that were not favorable to produce dissemination. Organisms on the soil surface appeared to die rapidly, as expected, since *P. syringae* is an epiphyte.

In a study of the first true environmental releases of a recombinant bacterium, Seidler and Hern (1988) of the USEPA reported that low numbers of *P. syringae* and *P. fluorescens* left test plots even in the presence of minimal wind conditions; they also found that strong vertical mixing occurred at three spray releases. Even though 99.99% of the introduced bacteria remained on the soil and plants for which they were intended, some 1.2×10^6 colony forming units (CFU) of bacteria migrated into a 15-m buffer zone during the course of one experiment; bacteria could still be found around the site 2 days after the test. In addition, a greater drift of bio-aerosol particles occurred in an experiment conducted at Tulelake, California, as compared to experiments carried out at Brentwood, California. Seidler and Hern believe that a larger sprayed area and a greater source strength of bacterial inoculum may have significantly affected aerosol drift during the one release. Clearly, additional investigation is warranted to determine which factors contribute to the aerial dispersion of GEMs during and after application.

A major problem for ascertaining GEM survival is limitations in the enumeration techniques available to the investigator. Currently, indirect methods such as the plate count are limited in that no one medium will support the growth of all bacterial species present in a particular sample (Mills and Bell 1986). Studies performed by the author have proven the plate count to be an inexpensive, yet reliable, method of determining concentrations of GEMs and indigenous bacteria (Orvos et al. 1990). This conclusion is also supported by Olsen and Bakken (1987). In addition, use of selective media allows enumeration of communities of interest. More recent studies by the USEPA have found that use of the drop plate technique for enumeration is as efficient as spread plating but requires less plates and, hence, less effort (Donegan et al. 1991). Direct counting of bacteria using simple or fluorescent (Daley 1979) stains has contributed little information since it is impossible to detect the organism of interest from the remainder of the sample. Use of fluorescent antibodies, specific for the organism in question, has not been as useful as initially hoped because of low sensitivity and presence of contaminating substances in environmental samples (Levin et al. 1987). However, refinements in the technique will undoubtedly make it even more useful.

Molecular techniques for detection of GEMs show a great deal of promise but are more complicated and expensive than conventional technologies. Methods have been developed to ascertain the amount of dissolved environmental DNA (DeFlaun et al. 1986) as well as DNA from bacterial cells (Holben et al. 1988; Steffan et al. 1988). Recent modifications of these techniques allow their use with minimal ($\approx$1 g) substrate sample sizes (Tsai and Olson 1991). Use of this DNA in hybridization studies allows detection of GEMs by genotype rather than phenotype, as well as allowing detection of nonculturable microbes (Barkay and Sayler 1988). Application of these techniques has allowed detection of as few as 10 cells/g soil (Fredrickson et al. 1988). However, use of probes in the past usually resulted in low sensitivity (Barkay and Sayler 1988; Levin et al. 1987). Application of the polymerase chain reaction (PCR) for use in environmental samples by Steffan and Atlas (1988) has shown promise for GEM detection. Steffan and Atlas (1988) reported the detection of 1 cell/g sediment using *P. cepacia* containing 15 to 20 copies of a 1.3 kilobase (kb) gene. Nonetheless, use of the PCR technique with GEMs containing only one gene per organism has not been especially productive because of environmental contamination and low initial DNA quantities. Hence, further methods development is indicated before the technique can be exploited to its maximum efficiency.

While use of these technologies will eventually allow rapid and sensitive detection, adaptation of existing techniques such as most probable number and selective enrichment allowed Scanferlato et al. (1990) to detect as few as 1 to 10 target cells/10 g soil. Engineered *Erwinia carotovora*, added initially at only 10^4 cells/g, could be detected for 45 days using this combination technique.

A major problem when considering GEM survival in developing a risk assessment paradigm is that the GEM may indeed be present but undetectable. Colwell et al. (1985) coined the phrase "viable, but non-culturable" for such a situation. The study of Colwell and her colleagues found that, in both microcosms and field tests, *Vibrio cholerae* were present in higher numbers when determined by immunofluorescent microscopy than by plate counts. Thus, even when plate counts yielded negative information concerning survival, bacteria were indeed present.

Potential Ecosystem Effects

Two major biotechnology product effects must be considered when determining impact to the environment. One is that such products may cause alterations in community structure and the other is that community function may be affected. Ecosystems have two fundamental properties: (1) they are composed of structures, and (2) these structures facilitate specific functions (Cairns and Pratt 1986). Since structure and function are closely related, an alteration in one may soon affect the other. A variety of studies will be required to estimate probable effects resulting from the controlled or uncontrolled release of GEMs (Cairns and Pratt 1986; Chakrabarty et al. 1984). Complex environmental effects resulting from displacement of native bacterial populations may occur, especially when GEMs are derived from common free-living taxa. Displacement of bacterial functional groups may produce an alteration of detrital processing and nutrient cycling that could have broad consequences, especially if introduced organisms have high fitness and are transferred among ecosystems (Cairns and Pratt 1986).

A more recent example of an unforeseen effect was observed in Louisiana where a field test was conducted with soybean seeds coated with *Bradyrhizobium japonica* genetically modified for improved nitrogen fixation and marker genes. This test was for only one season, and all plants and seeds were then incinerated. The company involved ceased monitoring and involvement with the field trial. The USEPA, which had given a consent order for the release, also terminated involvement when the consent order expired. The field was then plowed and resown with soybeans. The plowing erased the distinction between the test and control plots, and the engineered *Rhizobia* were spread over a 4-acre area. Subsequent monitoring revealed that the engineered *Rhizobia* were out-competing the indigenous *Rhizobia*. The implications of these observations are not clear at this time, and the Louisiana Department of Agriculture is suggesting another planting of soybeans to limit the spread of the genetically engineered strain. One of the major considerations with this case is that a microbe with an extensive historical database was used in a well planned and thoroughly reviewed experiment and an unpredicted result was still observed.

Many ecologists believe that bacteria crucial to nutrient processing are similar for virtually all comparable ecosystems, while others believe that a great deal of functional redundancy is present among a diverse array of species. Bacteria decomposing detritus produce the majority of energy in most aquatic ecosystems; the mineralization of this material is the most important role of bacteria (Fenchel and Blackburn 1979). Regardless, even though the probability of adverse effects from GEM release to these microorganisms may be small, the failure to evaluate any potential risks could be disastrous since functional effects would have severe consequences if a crucial parameter was affected or if the assimilative capacity of the ecosystem was exceeded. At the microbial level, most imported and exported materials carry a portion of the resident biota (Pimentel 1985), thus providing the opportunity for introduction of GEMs into new habitats.

It is important to note that no consensus exists as to what parameters in the environment should be monitored, even when dealing with chemical impacts. The number of parameters that may be monitored for assessing biotechnology risk are numerous and include nutrient and enzyme concentrations, production/respiration ratios, oxidation-reduction potentials, and pH shifts. No agreement exists on what is a "normal" value for a particular parameter under a particular set of circumstances (Cairns and Pratt 1986). Indeed, far more research is indicated in this area for both chemicals and GEMs.

It has been shown that GEMs are able to produce functional effects that are not observed in the parent organism. Glick et al. (1986) found that the introduction of plasmid DNA into *A. vinelandii* caused a decrease in nitrogen fixation, mean cell size, and siderophore synthesis. Glick and his colleagues hypothesized that the plasmid introduction imparted a metabolic load on the bacterium that impaired its ability to function. Wang and Crawford (1988) found that a recombinant strain of *S. lividans* significantly affected soil organic carbon turnover rates. They noted that the organism displayed an increased mineralization rate, particularly in soils amended with lignocellulose; these rates remained significantly higher over the 30-day duration of the test.

Because of the uncertainty of the impact from GEM releases, it is likely that small, easily contained microcosms will form the basis for preliminary assessments of ecological effects (A.D. Little, Inc. 1984). Such studies have proven useful for examining a number of ecological problems, although results must be interpreted with caution (King 1980). The small size and contained nature of these microcosms make possible effective dose-response studies for possible dissemination of recombinant DNA. Small or large inoculum levels of donor and recipient bacteria may be used with safety since the laboratory environments may be autoclaved. Further, the small size of the chambers makes introducing various environmental stresses possible; such stresses may affect the frequency of DNA exchange. Microcosm construction, optimization, and limitations have been reviewed elsewhere (King 1980).

BIOTECHNOLOGY RISK ASSESSMENT

Risk is defined as the "potential for adverse consequences of an event or activity" (Fiksel and Covello 1986), and *risk assessment* is defined as "use of scientific data to estimate the effects of exposure to hazardous materials or conditions" (OTA 1988). The risk assessment process for genetically engineered microorganism introduction, according to the OTA (1988) is the following: (1) risk identification, (2) risk-source characterization, (3) exposure assessment, (4) dose-response assessment, and (5) risk estimation. This scheme was derived from those used for chemicals, but is not as applicable when applied to GEMs or other rDNA-containing products because, as documented in the OTA report, protocols did not and currently do not exist for predicting kinds of risk or the magnitude of potential risk. Several differences exist between biotechnology and chemical risk assessment. First, GEMs have the potential to increase in concentration after their release. This complicates risk assessment strategy development. Second, sensitive detection methodologies, often parts-per-trillion, exist for many toxic substances. Sensitive techniques for GEMs and the rDNAs they contain, such as the polymerase chain reaction, are still being refined. Third, end points for measuring the magnitude, or even actuality, of risk have not been generally agreed upon for biotechnology products. Risk cannot adequately be assessed unless end points are defined and agreed upon as being pertinent. Finally, toxic chemicals are evaluated using such concepts as maximum acceptable toxicant concentration, lethal concentration, and no-observable-effects concentration. These concepts are difficult to apply to biotechnology because of the present poor database and lack of observable effects.

The Coordinated Framework for Regulation of Biotechnology (OTA 1986) was issued by the Executive Office of the President, Office of Science and Technology Policy, to announce a policy of federal agencies that are responsible for the review and regulation of biotechnology. Agencies included in the Coordinated Framework include the Food and Drug Administration, the USEPA, the Occupational Safety and Health Administration, the Department of Agriculture, the National Institutes of Health, and the National Science Foundation. Utilizing existing statutes, this framework stipulates which federal agency is responsible for review of a particular product at a particular time. The strengths and weaknesses of the Coordinated Framework have been discussed elsewhere in the literature (Cohrssen 1988; Jones 1988).

Other alternative assessment strategies have been proposed. Alexander (1985) suggested that, while the risk from a GEM release may be minimal, the consequences would be enormous if the event were to occur. Alexander also examined six factors important in ascertaining the probability of an adverse effect, including the nature of the rDNA and GEM survival. A committee of the Ecological Society of America (Tiedje et al. 1989) reviewed the ecological principles that must be considered in GEM release and gave a balanced assessment of the risks of such a release. A conclusion of this report is "The absence of problems at an early stage suggests that the screening mechanisms are working correctly, but should not be interpreted to mean that the introduction of all genetically engineered organisms is inherently safe." Other articles have given different views of assessment strategies and even the question the need for such a strategy (Dean-Ross 1986; Hardy and Glass 1985; Levin et al. 1987; National Academy of Sciences 1987).

Important Factors for Consideration

GEM Construction and Gene Transfer. Evaluation of GEM construction should include assessment of the sequence of any DNA inserted into the genome, the origin of this DNA, and the intended function of the DNA. The assessment strategy proposed later in this chapter assigns a potential hazard ranking to different categories of DNA origin and function (Comeaux et al. 1990). The location of DNA insertion in the microbe is also of importance since DNA fragments inserted into plasmids will disseminate more easily than if they were placed into chromosomes (Comeaux et al. 1990).

The potential for gene transfer is an important parameter in risk assessment strategy development. Several factors must be considered when assessing risk due to potential genomic transfer from a GEM to indigenous bacteria. First, conventional and molecular hybridization techniques must be used to confirm that gene transfer has indeed occurred and at what frequency. Second, if transfer has occurred, it must be ascertained if novel genetic information is sustained and expressed in the bacterium that just acquired it. Third, if such new genetic information is maintained and expressed, effects that such expression may have on other biota must be evaluated (OTA 1988). Several criteria, including gene transfer frequency and genetic distance between donor and recipient species, can be used (OTA 1988).

Assessment of GEM Survival and Transport. If the GEM is to be released for commercially desired purposes, then it must survive long enough to produce a desired effect. Because of the oligotrophic nature of environmental substrates and the extreme interspecific competition that is found in them, GEMs usually decline in density after their release. This period of decline may take from days to months, depending upon the particular GEM and the environment into which it is being placed (Kluepfel et al. 1990; Marshall et al. 1988; Orvos et al. 1990; Scanferlato et al. 1989). During that time, the GEM may be transported through air, water, or soil. Therefore, accurate detection methodologies must continue to be developed since organisms can only be detected to the threshold of the method being used. This may be as little as one cell in 1 g of substrate or as many as 10 thousand cells (Scanferlato et al. 1990). To complicate the situation, data exist that indicate the presence of viable but non-culturable cells (Colwell et al. 1985). Finally, perturbation of a system, for example by a thermal stress, increases the length of survival of engineered organisms by partial elimination of indigenous competitors and predators (Scanferlato 1990). These situations must be considered in any assessment strategy.

The use of microcosms is advantageous in evaluating survival and potential gene transfer prior to small scale field trials. Microcosms, or environmental surrogates, can be reflective of a variety of environmental conditions if properly designed, engineered, and utilized. Microcosms offer the advantages of small size; replicability; lack of confounding factors; realistic interactions between physical, chemical, and biological processes; simultaneous estimation of fate; and an estimation of ecological effects. Using microcosms in biotechnology risk assessment offers many advantages, including making a preliminary determination of survival, structural and functional effects, and potential DNA transfer before the actual environmental release of the microbe.

While subject to limitations (Cairns and Pratt 1986; King 1980), microcosms appear to give reliable data for weeks (Orvos et al. 1990; Scanferlato et al. 1989) or even a year (Kluepfel et al. 1990). Microcosms may range in size from small 10-ml containers to large 200-L ones and include ecological compartments such as trophic levels and functional groups (Cairns 1986). It is believed that most risk assessments can be accomplished in microcosm systems before conducting them in more realistic and expensive settings (Cairns 1986).

Assessment of Functional Effects. Effects of GEMs on ecosystem function are difficult to predict and assess. Although the mobility of microbes is low, they are readily dispersed physically and/or biologically and have immense potential for gene transfer. As discussed previously, no adverse functional effects have been observed to date. However, depending on the intended use of the GEM, it is plausible that one will be created that may affect functional processes. How to determine this effect and ascertain its risk remains somewhat ambiguous. According to the OTA (1988), impacts that involve the disruption of ecosystem processes are the least likely to occur but are the most potentially serious to the environment. GEMs capable of nitrogen fixation, cellulose degradation enhancement, and hazardous material degradation may have deleterious effects in areas of unintended release.

Such adverse effects must be predicted and detected early in the laboratory testing phases of risk assessment so that development costs can be refined and environmental protection ensured. However, how to predict such effects accurately remains unclear, as does selection of end points that are useful in such prediction. Our research group has found that end points such as enzyme activities, production, respiration, and nutrients are not useful because of their lack of sensitivity. However, Doyle et al. (1989) did observe differences in soil enzymes and carbon dioxide evolution in microcosms containing *P. putida* engineered to degrade 2,4-dichlorophenoxyacetic acid and uninoculated microcosms. Doyle and his colleagues believe the effects are not ecologically significant. Finally, as discussed above, Wang and Crawford (1988) found that altered *Streptomyces* strains mineralized carbon at rates significantly higher in non-sterile soil amended with lignocellulose, probably because of a foreign DNA insert. Therefore, selection of end points to be monitored must depend on the construction and intended use of the GEM.

Proposed Strategy for Risk Assessment

A five-step strategy for risk assessment is proposed. It should be noted that this is not an all-inclusive framework for all biotechnology product releases, but rather is intended as a basic framework for GEM release that will be refined as data become available. This strategy is for environmental consequences, not human health risks.

Step 1: Evaluate Risks Associated with rDNA Origin, Location, and Function

A. rDNA Origin (4 Categories [#1 is minimal risk])
   1. Deletion of a gene from a wildtype genome
   2. Enhancement or change in regulation of gene already present in wildtype genome
   3. Addition to a genome of genetic construction not present in wildtype genome but derived from genomes in the same habitat
   4. Addition to a genome of genetic construction not present in wildtype genome or derived from genomes not in same habitat

B. rDNA Location (3 categories [#1 is minimal risk])
   1. No DNA insertions are present
   2. rDNA insertion is in chromosome
   3. rDNA insertion is in a plasmid

C. rDNA Function (3 categories [#1 is minimal risk])
   1. DNA does not encode for an obvious competitive advantage
   2. DNA confers possible competitive advantage
   3. DNA possibly confers detrimental phenotype

Step 2: Assess GEM Survival Under Various Conditions

A. Survival Assessment—Assess survival in environmentally realistic microcosms simulating soil, sediment, water, and *in planta* environments of both target and non-target release areas (5 categories [#1 is minimal risk])
   1. GEM declines to extinction, if suitable method is available to determine this (e.g. Scanferlato et al. 1990)
   2. GEM declines to sub-detectable concentrations within a reasonable period of time (determined by regulatory officials)

3. GEM declines initially, but then maintains a stable concentration (below that of the inoculum concentration)
4. GEM does not decline in concentration, but does not increase
5. GEM increases in concentration after release

B. Survival Assessment in Perturbed Microcosms—Perturbations may include chemical, thermal, moisture, and radiation (UV) stresses; categories as above

Step 3: Evaluate Gene Transfer Under Various Conditions
Transfer in different substrates must be ascertained as well as the implications of such transfer. No categories are assigned, since all microorganisms transfer genetic information via several mechanisms. However, the implication of such transfer must be assessed by the following:

A. Determining the construction of the GEM. Plasmid-borne traits are more likely to exchange genetic information than are chromosome-borne inserts (Comeaux et al. 1990)
B. Ascertaining if transferred DNA is expressed in the recipient
C. Determining the frequency of transfer using a variety of methods and substrates (including soil, water, and sediment from areas of intended and unintended release)

Step 4: Evaluate Possible Structural/Functional Effects
This is the most ambiguous category since lack of effect upon a particular end point may indicate a poor end point choice rather than a lack of functional effect. Some possible end points for consideration follow, with no rankings assigned.

A. Structural End Points
   1. Evaluate potential for displacement of indigenous bacterial populations
      a. Bacterial structural groups - effect on genera/species dominant at specific habitats (e.g. Orvos et al. 1990)
      b. Bacterial functional groups - effects on functional groups (Scanferlato et al. 1989) present at specific habitats
   2. Effects upon total biomass
   3. Effects upon ATP content
B. Functional End Points
   1. Effects upon selected microbial functional groups (e.g. Scanferlato et al. 1989)
   2. Effects upon biostimulation (algal productivity)
   3. Effects upon microbial enzyme activities including dehydrogenases or electron transport system enzymes, phosphatases, and arylsulfatases
   4. Nutrient dynamics
   5. Production/biomass ratios

Step 5: Conduct Small Scale Field Trials
If the results of the aforementioned studies indicate minimal risk of gene transfer, excessive survival/colonization, and functional/structural effects,

then small scale field trials must be executed to validate these findings. All applicable federal and state guidelines must be followed.

Other Considerations

Once environmental risk has been estimated, a quantification of the uncertainty in that estimate must be made (Cairns 1986; USEPA 1989). Too often, estimates of risk are made as if the estimates are absolute. However, uncertainty is inherently present in each step of the assessment process and must be adequately quantified. A risk assessment that overstates the risk of an engineered organism that may be of potential benefit to society may be as bad as a risk assessment that understates risk to that same society. Finally, hazard communication must be completed as with any other risk assessment. This has been overlooked in the past and may arouse suspicion among the general public when not done properly.

CONCLUSIONS

This chapter briefly reviews some important considerations in developing a risk assessment strategy for genetically engineered microorganism release. These include survival, gene transfer, and ecological effects. Using these variables, a five-step risk assessment strategy is presented for consideration. The strategy is being refined so that classes of risk (minimal, moderate, not acceptable) can be derived from the scaling score.

ACKNOWLEDGEMENTS

While research leading to this article was supported by the National Science Foundation (BSR-8705445) and the U.S. Environmental Protection Agency (R-813805-01-0) to the University Center for Environmental and Hazardous Materials Studies, Virginia Polytechnic Institute and State University, these views do not reflect those of any federal agency or industrial organization. The author has been supported, in part, by the Society for Environmental Toxicology and Chemistry Pre-Doctoral Fellowship sponsored by the Procter & Gamble Company and the Cunningham Research Fellowship from Virginia Polytechnic Institute and State University. I am grateful to Darla Donald and Andrea Orvos for editorial assistance.

REFERENCES

A.D. LITTLE, INC. (1984). Risk Analysis Methods for Environmental Applications of Biotechnology. Cambridge, MA.

ALEXANDER, M. (1985). Ecological consequences: reducing the uncertainties. Issues Sci. Tech. Spring:57-68.

ARMSTRONG, J.L., WOOD, N.D., and PORTEOUS, L.A. (1991). Transconjugation between bacteria in the digestive tract of the cutworm (*Peridroma saucia*). Appl. Environ. Microbiol. 56:1492-1493.

AWONG, J., BITTON, G., and CHAUDHRY, G.R. (1990). Microcosm for assessing survival of genetically engineered microorganisms in aquatic environments. Appl. Environ. Microbiol. 56:977-983.

BARKAY, T. and SAYLER, G.S. (1988). Gene probes as a tool for the detection of specific genomes in the environment. In: *Aquatic Toxicology and Hazard Assessment: 10th volume, STP971.* (W.J. Adams, G.A. Chapman and W.G. Landis, eds), pp. 29-36. American Society for Testing and Materials, Philadelphia, PA.

BARNTHOUSE, L.W. and PALUMBO, A.V. (1986). Assessing the transport and fate of bioengineered microorganisms in the environment. In: *Biotechnology Risk Assessment* (J. Fiksel and V.T. Covello, eds), pp. 109-128. Pergamon Press, NY.

BEJ, A.K., PERLIN, M.H., and ATLAS, R.M. (1988). Model suicide vector for containment of genetically engineered microorganisms. Appl. Environ. Microbiol. 54:2472-2477.

BURTON, G.A., GUNNISON, D., and LANZA, G.R. (1987). Survival of pathogenic bacteria in various freshwater sediments. Appl. Environ. Microbiol. 53:633-638.

CAIRNS, J., JR. (1986). Multispecies toxicity testing: A new information base for hazard evaluation. Curr. Prac. Environ. Sci. Eng. 2:37-49.

CAIRNS, J., JR. and PRATT, J.R. (1986). Ecological consequence assessment: effects of bioengineered organisms. In: *Biotechnology Risk Assessment* (J. Fiksel and V.T. Covello, eds), pp. 88-108. Pergamon Press, NY.

CHAKRABARTY, A.M., KARNS, J.S., LILBANE, J.J., and CHATTERJEE, D.K. (1984). Selective evolution of genes for enhanced degradation of persistent, toxic chemicals. In: *Genetic Manipulation: Impact on Man and Society* (W. Arber, K. Illmensee, W.J. Peacock and P. Starlinger, eds), pp 43-54. Cambridge Press, NY.

COHRSSEN, J.J. (1988). United States biotechnology policy. American Biotechnology Laboratory. January:22-27.

COLWELL, R.R., BRAYTON, P.R., GRIMES, D.J., ROSZAK, D.B., HUQ, S.A., and PALMER, L.M. (1985). Viable but non-culturable *Vibrio cholerae* and related pathogens in the environment: Implications for release of genetically engineered organisms. Bio/Technology 3:817-820.

COMEAUX, J.L., POORANAMPILLAI, C.D., LACY, G.H., and STROMBERG, V.K. (1990). Transfer of genes to other populations and analysis of associated potential risks. In: *Risk Assessment in Agricultural Biotechnology* (J.J. Marois and G. Bruening, eds), pp. 132-145. Cooperative Extension Service, University of California Press, Berkley, CA.

CRUZ-CRUZ, N.E., TORANZOS, G.A., AHEARN, D.G., and HAZEN, T.C. (1988). In situ survival of plasmid-bearing and plasmidless *Pseudomonas aeruginosa* in pristine tropical waters. Appl. Environ. Microbiol. 54:2574-2577.

DALEY, R.J. (1979). Direct epifluorescence enumeration of native aquatic bacteria: uses, limitations, and comparative accuracy. In: *Native Aquatic Bacteria: Enumeration, Activity, and Ecology, STP695* (J.W. Costerton and R.R. Colwell, eds), pp. 29-45. American Society for Testing and Materials, Philadelphia, PA.

DEAN-ROSS, D. (1986). Release of genetically engineered organisms: hazard assessment. ASM News 52:572-575.

DeFLAUN, M.F., PAUL, J.H., and DAVIS, D. (1986). Simplified method for dissolved DNA determination in aquatic environments. Appl. Environ. Microbiol. 52:654-659.

DONEGAN, K., MATYAC, C, SEIDLER, R., and PORTEOUS, A. (1991). Evaluation of methods for sampling, recovery, and enumeration of bacteria applied to the phylloplane. Appl. Environ. Microbiol. 57:51-56.

DOYLE, J., SHORT, K., and STOTZKY, G. (1989). Effects of *Pseudomonas putida* PPO301(pRO103), genetically engineered to degrade 2,4-dichlorophenoxyacetic acid, on microbe-mediated ecological processes in soil. Abstracts of the 89th Annual Meeting of the American Society for Microbiology, p. 353, Q-140.

FENCHEL, T. and BLACKBURN, T.H. (1979). Bacteria and Mineral Cycling. Academic Press, NY. 225 pp.

FIKSEL, J.R. and COVELLO, V.T. (1986). The suitability and applicability of risk assessment methods for environmental applications of biotechnology. In: *Biotechnology Risk Assessment: Issues and Methods for Environmental Introductions* (J.R. Fiksel and V.T. Covello, eds), pp. 1-34. Pergamon Press, NY.

FREDRICKSON, J.K., BEZDICEK, D.F., BROCKMAN, F.J., and LI, S.W. (1988). Enumeration of Tn5 mutant bacteria in soil by using a most-probable-number DNA hybridization procedure and antibiotic resistance. Appl. Environ. Microbiol. 54:446-453.

GENTHNER, F.J., CHATTERJEE, P., BARKAY, T., and A.W. BOURQUIN. (1988). Capacity of aquatic bacteria to act as recipients of plasmid bacteria. Appl. Environ. Microbiol. 54:115-117.

GLICK, B.R., BROOKS, H.E., and PASTERNAK, J.J. (1986). Physiological effects of plasmid DNA transformation on *Azotobacter vinelandii*. Can. J. Microbiol. 32:145-148.

GRAHAM, J.B. and ISTOCK, C.A. (1978). Genetic exchange in *Bacillus subtillis* in soil. Mol. Gen. Genet. 166:287-290.

GRAHAM, J.B. and ISTOCK, C.A. (1979). Gene exchange and natural selection cause *Bacillus subtillis* to evolve in soil culture. Science 204:637-639.

GRANT, W.D. and LONG, P.E. (1981). Environmental Microbiology. Halsted Press, Glasgow, Scotland. 215 pp.

GREAVES, M.P. and WILSON, M.J. (1970). The degradation of nucleic acids and montmorillonite-nucleic acid complexes by soil microorganisms. Soil Biol. Biochem. 2:257-268.

GRIMES, D.J., SOMERVILLE, C.C., STRAUBE, W., ROSZAK, D.B., ORTIZ-CONDE, B.A., MACDONNELL, M.T., and COLWELL, R.R. (1988). In: *Aquatic Toxicology and Hazard Assessment: Tenth Volume, STP971* (W.J. Adams, G.A. Chapman, and W.G. Landis, eds), pp. 37-42. American Society for Testing and Materials, Philadelphia, PA.

HADA, H.S. and SIZEMORE, R.K. (1981). Incidence of plasmids in marine *Vibrio* spp. isolated from an oil field in the northwestern Gulf of Mexico. Appl. Environ. Microbiol. 41:199-202.

HARDY, R.W.F. and GLASS, D.J. (1985). Our investment: what is at stake? Issues Sci. Tech., Spring.

HARRISON, M.D. (1980). Aerial dissemination of bacterial plant pathogens. Ann. NY Acad. Sci. 353:94-104.

HOLBEN, W.E., JANSSON, J.K., CHELM, B.K., and TIEDJE, J.M. (1988). DNA probe method for the detection of specific microorganisms in the soil bacterial community. Appl. Environ. Microbiol. 54:703-711.

JONES, D.D. (1988). USDA safety review of biotechnology research in agriculture. American Biotechnology Laboratory. January:30-37.

KING, D.L. (1980). Some cautions in applying results from aquatic microcosms. In: *Microcosms in Ecological Research* (J.P. Giesy, ed), pp. 164-191. National Technical Information Service, Springfield, VA.

KLUEPFEL, D.A., KLINE, E.L., MUELLER, J., DRAHOS, D.J. BARRY, G., and HEMMING, B.C. (1990). Evaluation of the risks associated with the release of soil-borne genetically engineered bacteria into the environment. Abstracts of the 199th Annual Meeting of the American Chemical Society, Agrochemicals Division, No. 155.

KNUDSEN, S.M. and KARLSTROM, O.H. (1991). Development of efficient suicide mechanisms for biological containment of bacteria. Appl. Environ. Microbiol. 57:85-92.

KNUDSEN, G.R., WALTER, M.V., PORTEOUS, L.A., PRINCE, V.J., ARMSTRONG, J.L., and SEIDLER, R.J. (1988). Predictive model of conjugative plasmid transfer in the rhizosphere and phyllosphere. Appl. Environ. Microbiol. 54:343-347.

LACY, G.H., STROMBERG, V.K., and CANNON, N.P. (1984). *Erwinia amylovora* mutants and in planta derived transconjugants resistant to oxytetracycline. Can. J. Plant Path. 6:33-39.

LEVIN, M.A., SEIDLER, R., BORQUIN, A.W., FOWLE, J.R., and BARKAY, T. (1987). EPA developing methods to assess environmental release. Biotechnology 5:38-45.

LINDEMANN, J. and UPPER, C.D. (1985). Aerial dispersal of epiphytic bacteria over bean plants. Appl. Environ. Microbiol. 50:1229-1232.

LINDOW, S.E., KNUDSEN, G.R., SEIDLER, R.J., WALTER, M.V., LAMBOU, V.W., AMY, P.S., SCHMEDDING, D., PRINCE, V., and HERN, S. (1988). Aerial dispersal and epiphytic survival of *Pseudomonas syringae* during a pretest for the release of genetically engineered strains into the environment. Appl. Environ. Microbiol. 54:1557-1563.

LINDOW, S.E., PANOPOULOS, N.J., and MCFARLAND, B.L. (1989). Genetic engineering of bacteria from managed and natural habitats. Science 244:1300-1307.

MARSHALL, B., FLYNN, P., KAMELY, D., and LEVY, S.B. (1988). Survival of *Escherichia coli* with and without ColE1:Tn5 after aerosol dispersal in a laboratory and a farm environment. Appl. Environ. Microbiol. 54:1776-1783.

MARTHI, B., FIELAND, V.P., WALTER, M., and SEIDLER, R.J. (1990). Survival of bacteria during aerosolization. Appl. Environ. Microbiol. 56:3463-3467.

MILLS, A.L. and BELL, P.E. (1986). Determination of individual organisms and their activities in situ. In: *Microbial Autecology* (R.L. Tate, ed), pp. 27-60. Wiley, New York.

NATIONAL ACADEMY OF SCIENCES. (1987). Introduction of Recombinant DNA-engineered Organisms into the Environment: Key Issues. National Academy Press, Washington, DC.

OFFICE OF TECHNOLOGY ASSESSMENT [OTA]. (1986). Coordinated framework for regulation of biotechnology. Federal Register 51:23301-23350.

OFFICE OF TECHNOLOGY ASSESSMENT . (1988). New Developments in Biotechnology - Field-Testing Engineered Organisms: Genetic and Ecological Issues, OTA-BA-350. Washington, DC. 150 pp.

OGRAM, A., SAYLER, G.S., GUSTIN, D., and LEWIS, R.J. (1988). DNA adsorption to soils and sediments. Environ. Sci. Tech. 22:982-984.

OLSEN, R.A. and BAKKEN, L.R. (1987). Viability of soil bacteria: optimization of plate-counting technique and comparison between total counts and plate counts within different size groups. Microbiol. Ecol. 13:59-74.

ORVOS, D.R., LACY, G.H., and CAIRNS, J., JR. (1990). Genetically engineered *Erwinia carotovora*: Survival, intraspecific competition, and effects upon selected bacterial genera. Appl. Environ. Microbiol. 56:1689-1694.

PIMENTEL, D. (1985). Using genetic engineering for biological control: reducing ecological risks. In: *Engineered Organisms in the Environment: Scientific Issues* (H.O. Halvorson, D. Pramer, and M. Rogul, eds), pp. 129-140. American Society for Microbiology, Washington, DC.

POSTMA, J., HOK-A-HIN, C.H., SCHOTMAN, J.M.T., WIJFFELMAN, C.A., and VAN VEEN, J.A. (1991). Population dynamics of *Rhizobium leguminosarum* Tn5 mutants with altered cell surface properties introduced into sterile and nonsterile soils. Appl. Environ. Microbiol. 57:649-654.

RAFII, F. and CRAWFORD, D.L. (1988). Transfer of conjugative plasmids and mobilization of a nonconjugative plasmid between *Streptomyces* strains on agar and in soil. Appl. Environ. Microbiol. 54:1334-1340.

ROMANOWSKI, G., LORENZ, R.G., and WACKERNAGEL, W. (1991). Adsorption of plasmid DNA to mineral surfaces and protection against DNase I. Appl. Environ. Microbiol. 57:1057-1061.

SAYE, D.J., OGUNSEITAN, O., SAYLER, G.S., and MILLER, R.V. (1987). Potential for transfer of plasmids in a natural freshwater environment: Effect of plasmid donor concentration and a natural microbial community on transduction in *Pseudomonas aeruginosa*. Appl. Environ. Microbiol. 53:987-995.

SCANFERLATO, V.S. (1990). Environmental Risk Assessment for Toxic Chemicals and Genetically-engineered Microorganisms: A Microcosm Approach. Ph.D. Dissertation, Virginia Polytechnic Institute and State University, Blacksburg, VA. 121 pp.

SCANFERLATO, V.S., ORVOS, D.R., CAIRNS, J., JR., and LACY, G.H. (1989). Genetically engineered *Erwinia carotovora* in aquatic microcosms: Survival and effects on functional groups of indigenous bacteria. Appl. Environ. Microbiol. 55:1477-1482.

SCANFERLATO, V.S., ORVOS, D.R., LACY, G.H., and CAIRNS, J., JR. (1990). Enumerating low densities of genetically-engineered *Erwinia carotovora* in soil. Lett. Appl. Microbiol. 10:55-59.

SCHILF, W. and KLINGMULLER, W. (1983). Experiments with *E. coli* on the dispersal of plasmids in environmental samples. Recombinant DNA Tech. Bull. 6:101-102.

SEIDLER, R.J. and HERN, S. (1988). Special Report: The Release of Ice Minus Recombinant Bacteria at California Test Sites. USEPA, CERL, Corvallis, OR. 83 pp.

SLATER, J.H. (1985). Gene transfer in microbial communities. In: *Engineered Organisms in the Environment: Scientific Issues* (H.O. Halvorson, D. Pramer and M. Rogul, eds), pp. 89-98. American Society for Microbiology, Washington, DC.

STACEY, G. (1985). The *Rhizobium* experience. In: *Engineered Organisms in the Environment: Scientific Issues* (H.O. Halvorson, D. Pramer and M. Rogul, eds), pp. 109-121. American Society for Microbiology, Washington, DC.

STEFFAN, R.J. and ATLAS, R.M. (1988). DNA amplification to enhance detection of genetically engineered bacteria in environmental samples. Appl. Environ. Microbiol. 54:2185-2191.

STEFFAN, R.J., GOKSOYR, J., and ATLAS, R.M. (1988). Recovery of DNA from soil and sediments. Appl. Environ. Microbiol. 54:2908-2915.

STOTZKY, G. and KRASOVSKY, V.N. (1981). Ecological factors that affect the survival, establishment, growth, and genetic recombination of microbes in natural habitats. In: *Molecular Biology, Pathogenicity, and Ecology of Bacterial Plasmids* (S.B. Levy, R.C. Clowes and E.L. Koenig, eds), pp. 31-42. Plenum Press, NY.

TIEDJE, J.M., COLWELL, R.K., GROSSMAN, Y.L., HODSON, R.E., LENSKI, R.E., MACK, R.N., and REGAL, P.J. (1989). The planned introduction of genetically engineered organisms: Ecological considerations and recommendations. Ecology 70:298-315.

TREVORS, J.T. and ODDIE, K.M. (1986). R-plasmid transfer in soil and water. Can. J. Microbiol. 32:610-613.
TREVORS, J.T., VAN ELSAS, J.D., VAN OVERBEEK, L.S., and STARODUB, M. (1990). Transport of a genetically engineered *Pseudomonas fluorescens* strain through a soil microcosm. Appl. Environ. Microbiol. 56:401-408.
TSAI, Y. and OLSON, B.H. (1991). Rapid method for direct extraction of DNA from soil and sediments. Appl. Environ. Microbiol. 57:1070-1074.
USEPA. (1989). Ecological Assessment of Hazardous Waste Sites: A Field and Laboratory Reference, EPA/600/3-89/013. U.S. Environmental Protection Agency, Washington, DC.
WALTER, M.V., PORTEOUS, A., and SEIDLER, R.J. (1987a). Measuring genetic stability in bacteria of potential use in genetic engineering. Appl. Environ. Microbiol. 53:105-109.
WALTER, M.V., BARBOUR, K., MCDOWELL, M., and SEIDLER, R.J. (1987b). A method to evaluate survival of genetically engineered bacteria in soil extracts. Curr. Microbiol. 15:193-197.
WANG, Z. and CRAWFORD, D.L. (1988). Effects of wildtype and recombinant *Streptomyces* on mineralization of organic carbon. Abstracts of First International Conference on the Release of Genetically Engineered Microorganisms. No. 12, p. 13.
WEINBERG, S.R. and STOTZKY, G. (1972). Conjugation and genetic recombination of *Escherichia coil* in soil. Soil Biol. Biochem. 4:171-180.
WIGGINS, B.A. and Alexander, M.A. 1985. Minimum bacterial density for bacteriophage replication: Implications for significance of bacteriophages in natural ecosystems. Appl. Environ. Microbiol. 49:19-23.
ZEPH, L.R., ONAGA, M.A., and STOTZKY, G. (1988). Transduction of *E. coli* by bacteriophage P1 in soil. Appl. Environ. Microbiol. 54:1731-1737.

CHAPTER 12

ECOLOGICAL RISKS OF HIGHWAYS

Robert B. Atkinson and John Cairns, Jr.
University Center for Environmental and Hazardous Materials Studies
Virginia Polytechnic Institute and State University
Blacksburg, Virginia

Risks associated with highway construction and operation include direct and secondary impacts associated with both construction and use of highways. Direct impacts include, but are not limited to, habitat fragmentation, genetic risks, species loss and species invasion, air pollution, runoff, release of toxic substances, and altered ecosystem stress regimes. Secondary impacts include any environmental degradation that occurs as a result of improved access to what were minimally disturbed habitats. Loss of environmental services is likely to be most severe where secondary development is concentrated. Many areas of the country are experiencing dramatic growth, e.g., Florida and Washington, DC, and provide testing grounds for growth management practices for secondary impact control.

The effects of highways on wetlands are special cases. Mitigation efforts currently in use are controversial and expensive, yet research to evaluate the effectiveness of these efforts has not been a priority. Compensation for wetland loss is particularly controversial, and a variety of environmental risks are associated with this practice. Several researchers have described the phenomenon of cumulative impacts to wetlands, which present landscape-level risks and jurisdictional problems.

Effective advanced planning of direct impacts, incorporation of a landscape-level perspective, and a more realistic consideration of secondary impacts are necessary to reduce environmental risks. Consideration of environmental risks is needed for calculation of true highway construction costs, so that current plans to increase federal highway funding might provide incentives for use of effective mitigation measures.

INTRODUCTION

In colonial times, transportation depended on waterways, canals, and development of the steamboat. By the end of the 1800s, railroad networks

---

1. Corresponding Author: Robert B. Atkinson, University Center for Environmental and Hazardous Materials Studies, Virginia Polytechnic Institute and State University, Blacksburg, VA 24061-0415.

2. Key Words: corridors, environmental risks, habitat, secondary impacts, wetlands.

had been encouraged by the Civil War and replaced waterways in importance. Highway construction and improvements followed the 60-year railroad era, and interstate highways were built after World War II. Highway transportation began to gain prominence in the 1920s and has continued to grow. Since 1960, a 450% increase in miles of interstate highway has occurred nationwide (Council on Environmental Quality [CEQ] 1989). This growth has outpaced both gross national product and population, and transportation needs are not projected to change through the year 2020 (Greene et al. 1988).

In developed countries, the extent of highways is related to gross national product and life styles of the society that highways are designed to support. Highway planning and construction is, therefore, a function of economic and political needs. Since highway construction technology preceded the development of ecological models capable of predicting environmental consequences, opportunities to avoid environmental degradation were minimal. Concerns for environmental impacts had little effect on policy before legislation mandated protection efforts (CEQ 1989; Cohn 1982).

In 1893, the U.S. Department of Agriculture established the Office of Road Inquiry to study road construction methods. Planning and research were encouraged by the Federal Aid Highway Act of 1944. However, not until the 1960s were environmental risks given serious consideration. Metropolitan areas with populations over 50,000 were required to utilize "a continuing, comprehensive and cooperative process" including environmental effects, according to the Federal Aid Highway Act of 1962 (P.L. 87-866, 23 USC 134). Section 102 of the National Environmental Policy Act of 1969 (P.L. 91-190, 42 USC 4321-4327) required federal agencies to include environmental design as part of an interdisciplinary approach to planning and decision making and required consultation with the CEQ. The 1970 Federal Aid Highway Act (P.L. 91-605, 23 USC 109) required that states comply with detailed environmental guidelines in order to be eligible for federal funds. With a large portion of the responsibility delegated to states, federal control was maintained by development of the *Process Guidelines for the Development of Environmental Action Plans* under the Federal Aid Highway Program. Three components of the policy required (1) that social, economic, and environmental effects be identified and that studies be performed early enough to permit analysis and consideration, while alternatives are being formulated and evaluated, (2) that other agencies and the public be involved in the system planning and project development early enough to influence studies and final decisions, and (3) that appropriate consideration be given to responsible alternatives, including the alternative of not building. Prior to these laws, highway construction decisions were based entirely on economic and engineering considerations with little or no input from other agencies or individuals (Cohn 1982).

Though some of the necessary legislation for reducing environmental risks has been enacted, other impediments remain. Environmental concerns have historically received less attention than economic and political priorities and are considered antithetical to some goals of highway planners and constructors. According to Suhrbier and Deakin (1988), transportation professionals treat environmental concerns as "constraints, restrictions, and problems...", causing a major problem in environmental protection. Cohn (1982) describes most state highway departments as being deficient in personnel who are qualified in environmental quality. The Transportation Research Board (TRB 1984) stated that "the financial commitment to highway research also appears surprisingly low compared to research funds invested by other sectors of the economy." Even low technology industries such as textiles, mining, and building materials spend more than eight times as much of their budget on research. Research expenditures have ranged from 0.25 to 0.17% of total highway program expenditures (TRB 1984).

This chapter outlines several sources of environmental risk associated with highway construction and use and concludes with advocating (1) effective advanced planning for direct impacts, (2) incorporation of a landscape-level perspective, and (3) a more realistic consideration of secondary impacts. This discussion also encourages consideration of environmental risks in calculating true costs of highway construction, so that current plans to increase federal funding of highways might reward the use of environmentally sensitive construction techniques and alternatives.

DIRECT IMPACTS

Risks to Habitat

Highways are linear aspects of the landscape and, as such, may disect ecosystems. Limiting the ability of a species to travel either to or from an area is a situation analogous to an island existence (Diamond and May 1976). The basic theory of island biogeography was elucidated by MacArthur and Wilson (1967). According to the theory, island size is of great importance to the number of species that will inhabit the island. Immigration and extinction rates intersect to define the equilibrium species number for an island. Smaller islands exhibit lower immigration rates and higher extinction rates, which lowers the equilibrium species number. Ecosystems that are fragmented create small islands capable of supporting fewer species (Lovejoy and Oren 1981).

Vegetation communities can be fragmented and isolated from other similar communities by highways. Seed dispersal among communities is generally by wind and animals and may be limited by highway placement. In the case of the maple's winged fruit, *Acer* spp., less than 1% of the seeds were transported beyond several hundred meters, and the frequency of these species was lowered (Johnson et al. 1981). During 20 years of urban sprawl, 16,000 ha surrounding Milwaukee, Wisconsin, were urbanized and led to conversion of fields and forests to woodlots. Though plant species diversity remained high, edge species dominated and serious declines in elm, *Ulmus* spp., and American beech, *Fagus grandifolia*, were noted (Levenson 1981).

Matthiae and Stearns (1981) found greater mammalian species diversity on larger forest fragments following urbanization and agricultural development in Wisconsin. Exceptions included such species as rabbits, which are considered edge inhabitants. Settlement and agricultural development in the area since 1835 is responsible for the regional extinction of wolverine, elk, lynx, timber wolf, moose, porcupine, fisher, cougar, coyote, bobcat, and others (Matthiae and Stearns 1981). Animal populations may also be at risk from selective pressures associated with highways. Impact-intolerant allelic forms could be lost if juvenile migration is genetically based and is selected against by highway traffic. Non-migrants will be selected for and the population may eventually be at risk.

Limits to the dispersal of vegetation, snails, tortoises, frogs, toads, snakes, and mammals by highways are somewhat intuitive. For example, the threatened black bear migrates along the Wekiva River in Florida, which may be at risk from the Northwest Beltway (Gwynn et al. 1989). Although one might assume that the capacity for flight would allow birds to overcome highway-induced isolation mechanisms and maintain population levels regardless of degree of habitat fragmentation, bird isolation may result from use of small forest fragments as home territorial limits. Anderson et al. (1977) studied transmission line corridors and found that these created natural territorial boundaries that affected bird population distributions. Bird species diversity declines with decreasing fragment size. Some neotropical species and some forest interior species were absent in small to medium-sized fragments, but present in an extensive forest in that region (Whitcomb et al. 1981).

Natural corridors provide links that can increase effective habitat areas (Noss 1987). When such corridors are disrupted, demes and metapopulations may become isolated (Wallace 1981) and species may be extirpated (Murphy 1988). Small populations are often subject to sudden local extinction by natural disaster and inbreeding (Gilpin and Soule 1986). Normal populations may have a variety of mechanisms that effectively limit inbreeding (Ralls et al. 1986), which may not operate when population sizes reach critically low levels. Inbreeding can result in increased expression of deleterious alleles, lower genetic variation in the population, and a diminished capacity of the population to survive environmental variations (Wallace 1981). The minimal size of animal populations that is viable can be calculated (Lacava and Hughes 1984; Reed et al. 1986) and can vary from species to species. However, minimal areas selected for a representative suite of species may risk the entire community due to the interdependence of species in communities (Ehrlich and Ehrlich 1990). Many conservation biologists believe that habitat fragmentation is the "most serious threat to biological diversity and is the primary cause of the present extinction crisis" (Wilcox and Murphy 1985).

Risks of Highway Corridors

Vegetation along highways may differ greatly from that of adjacent ecosystems. Natural corridors facilitate migration of certain native species, but migration of r-selected (e.g., rapid colonizing) and non-native species may be facilitated by highways that act as corridors. New records of some aquatic plants (e.g., a rush, *Juncus arborvitus*) are limited to highway ditches that appear to be a dispersal mechanism (Porter and Wieboldt, unpublished[a]). Spread of kudzu vine, *Pueraria lobata*, and honeysuckle, *Lonicera* spp., along roads has been implicated as a risk to endangered species, including the noonday snail (Norse et al. 1986). The U.S. Park Service has actively attempted to limit accidental automobile conveyance of gypsy moths, *Porthetria dispar*, along highways. This species is defoliating forests in the Blue Ridge Mountains of Virginia and is progressing southward. Jackson (1979) summarizes much of the available literature documenting range expansion of r-selected barn swallows, *Hirundo rustica*, along interstates. Replacement of wooden bridges with cement ones has provided improved nesting sites and helped these birds colonize 300 miles in 23 years (Jackson 1979). An example of plant invasion following disturbance is provided by phragmites, *Phragmites communis*. An excellent competitor in disturbed sites such as roadsides, mud waves, and created wetlands, phragmites grows in near monotypic stands providing minimal ecological value. Breakdowns in natural isolation may result in the introduction of harmful alleles or displacement of native species.

Air Pollution Risks

Highway use is a major energy consumer and produces large amounts of emissions. Transportation is responsible for more than one-fourth of all energy consumed in the United States, including almost two-thirds of all oil—more oil than all developing countries combined use for all purposes (Greene et al. 1988). Automobile emissions account for 70% of the carbon monoxide, 35% of hydrocarbons, and 40% of nitrogen oxides released in the United States (CEQ 1989). Other components of automobile emissions include odor, sulphur oxides, other organic compounds, and particulate matter that may include metals. These emissions are known to contribute significantly to acid rain (Organisation for Economic Co-Operation and Development [OECD] 1986; Suhrbier and Deaken 1988) and to the greenhouse effect (Ehrlich 1989; Suhrbier and Deaken 1988). Photochemical oxidation of nitrogen oxides in combination with reactive hydrocarbons results in ozone formation. In areas such as Los Angeles, California, where atmospheric inversions are frequent occurrences, concentrations of these emissions frequently exceed "safe" levels for human health (OECD 1986).

Although the risks of air pollution to human health (e.g., ozone effects on the human respiratory system) and economic issues (e.g., risks of acid rain to agriculture) are usually recognized, effects on other organisms may not be considered. In addition to ozone, sulphur and nitrogen oxides can be toxic to both woody and herbaceous plant species (OECD 1986). While

[a]Porter, D.M. and Wieboldt, T.F. (In preparation), The Endangered and Threatened Vascular Flora of Virginia. Virginia Polytechnic Institute and State University, Blacksburg, VA.

most air pollutant emissions have been reduced, ozone and carbon monoxide emissions continue to increase (Suhrbier and Deaken 1988). Toxic concentrations that would place ecological systems at risk are generally not known.

Randall et al. (1978) studied the contribution of several atmospheric pollutants to storm water quality near commuter routes into Washington, DC. Significant impacts to surface water quality were found following both rain and snow precipitation events. Contribution to the problem of surface water enrichment is also a possibility. Nitrogen oxides can increase nitrogen concentrations in these systems (Moore, personal communication[b]).

Risks Associated with Runoff

Acute runoff associated with highways may occur during construction. No erosion controls were used before the 1970s. Even though numerous erosion control measures are now available, many highway construction projects still lead to significant water quality problems. Causes include selection of inadequate or inappropriate erosion control measure(s), flawed installation, and inadequate maintenance. Sediment and nutrient sources are often difficult to determine since development indirectly associated with highway construction may contribute to this pollution. Phosphorus loading is often associated with sediment runoff and eutrophic effects that result can be extreme, often adversely affecting oxygen content of water. Affected systems may include any area the sediments eventually reach. Though many aquatic systems have evolved under conditions of sediment loading from natural sources, loading rates may exceed tolerance ranges, resulting in significant negative ecological consequences.

Chronic runoff from highway surfaces includes deicing salts, heavy metals, and oil, which are transported into surface and groundwater (Pye et al. 1983). Over 10 million tons of salt are used annually to deice U.S. roads. Additional salt runoff from uncovered storage piles has been noted (CEQ 1989). Significant concentrations of several trace metals have been found in detention basins receiving highway runoff (Wigington et al. 1983), but capacity for long-term retention of metals was not determined. Routine applications of chemicals including salt, herbicides, and fertilizers may accumulate to harmful levels in sediments.

Risks Associated with Transport of Hazardous Materials

Hazardous substances are transported via land, sea, and air, but the focus here is on highway carriers that include both commercial and non-commercial transport. Abkowitz et al. (1989) describe the amount of hazardous materials transported in the United States as "significant and growing;" hazardous materials comprise over 25% of all freight shipments. Trucks

---

[b]Dr. Kenneth Moore, Professor of Marine Science, Virginia Institute of Marine Science, Gloucester Point, VA. November 1990.

carry over 60% of the hazardous materials transported domestically. Hazardous materials range from chemically pure shipments by container to various mixtures of wastes, generally contained in 55-gallon drums. Frequency of incident and severity of damages have been described as inversely related (Abkowitz et al. 1989). Therefore, environmental risk assessment for hazardous materials needs to focus on both probability and severity, rather than generation of a single number. Furthermore, human risk assessment for spills has outpaced that for ecological structure and function. Long-term research at spill sites is necessary to correct this deficiency.

Growth in the chemical industry in the 1950s and 1960s outpaced development of regulations in the United States and the world. When first enacted, the Resource Conservation and Recovery Act (RCRA) led to massive toxic substance disposal by firms attempting to avoid regulation. Later, loopholes in the law allowed evasion by the most flagrant polluters and RCRA was subsequently revised (Wynne 1987). However, the idea of "cradle- to-grave" management (i.e., from waste generation to waste disposal) of hazardous materials has historically failed during the transportation phase (Cashman 1986). Inconsistencies in state regulations governing highway transport of hazardous materials worsens the problem. State regulations of hazardous material transport range from strict and expensive to virtual nonexistence (Cashman 1986). Dawson and Mercer (1986) describe the hazardous waste transport industry as "a large conglomerate of small operators who enter and exit the market with impunity. Local haulers experience little regulatory control and have but small assets at risk." The difficulties associated with meeting disparate state regulations, coupled with the potential for tremendous profits, makes illegal operation appealing. The feared, unscrupulous operator or midnight dumper who illegally disposes of wastes, with or without the knowledge of the waste generator, truly exists (Cashman 1986; Dawson and Mercer 1986). Statistics on the pervasive nature of illegal and unreported discharges and spills are inaccessible. However, reported incidents involving hazardous materials are quite frequent. A nine-county, somewhat urbanized region of Southeastern Virginia experienced over 400 reported incidents in just 2 years (1989 and 1990; Holloway, personal correspondence[c]).

Even toxicants placed on unpaved roads may not remain there. An interesting example is the 71 pounds of dioxin mixed with waste oil and sprayed on roads in Missouri to keep dust down (Erbaugh and Hinck 1984; Howard 1984). Dioxin degrades slowly in the environment and bioaccumulates. Windborne dust alone has transported this chemical up to 150 yards from "treated" roads, and hundreds of sites have been exposed (Howard 1984). Some investigators have recommended that the government purchase property where dioxin occurs (e.g., Times Beach, Missouri) as an economical

---

[c]James Holloway, Chief, Technology Hazards Branch, Department of Emergency Services, Commonwealth of Virginia, Richmond, VA. November 1990.

alternative to clean up. Costs for cleanup at a single site were estimated at from \$3.5 to \$4 million (Exner et al. 1984). Although more recent interpretations indicate dioxin may be less toxic than once predicted (Bailey 1992), the point of this paragraph is that disposal of wastes on roads generally has risks far in excess of the benefits in most circumstances.

Hazardous substances are typically defined by human, rather than ecological, terms, e.g., materials in quantities and forms that the U.S. Department of Transportation has found may pose an unreasonable risk to health and safety or to property when transported. Likewise, most studies and popular press attention are devoted to public health aspects of hazardous materials transport. Although state controls vary throughout the country, a clear trend has been established—carriers are routed away from population centers and towards rural areas (Abkowitz et al. 1989). These rural areas are less likely to have sophisticated emergency response programs and are, therefore, subject to greater potential environmental risk. Furthermore, evaluation of risk associated with hazardous materials is dependent upon accurate flow data, i.e., types and amounts of toxics transported (Glickman 1984), which appears inadequate at this time.

Removal of Ecosystem Stresses

Some ecosystems are stress dependent, and removing natural stresses may place these systems at risk. Fire is recognized as an essential component of many natural systems. Though usually caused by lightning, other natural causes include volcanic activity, spontaneous combustion, and sparks from falling rocks. Fire has a variety of effects on ecosystems, ranging from altered soil chemistry to synchronized flowering time (Barbour et al. 1980). Many ecosystems have evolved beyond fire tolerance and are now actually fire dependent (Vogl 1980). Highways decrease the spread of fire (Norse et al. 1986), and the impact of fire is difficult to assess. An endangered species, Kirtland's warbler (*Dendroica kirtlandii*) depends on young jack pine forests that are fire dependent (Vogl 1980). The Oregon silverspot butterfly (*Speyaria zerene hippolyta*), an endangered species, requires successional stages that occur after fire (Norse et al. 1986). Build up of organic matter resulting from reduced fire frequency could result in elevated combustion temperatures above that of fire-adapted species tolerance levels. The negative effects associated with the absence of fire may be undetectable for long periods of time.

Flooding provides an important disturbance regime in many systems. Culverts used in fills for highway crossings may not be adequate for surface flow rates, and they generally fail to allow for reduced subsurface flow in the valley-axial direction. Alteration of frequency, timing, depth, or duration of flooding is likely to alter wetlands and associated aquatic systems. These systems are generally stress dependent and may undergo an alternative or expedited successionary progression. Large fills are also more likely to affect adjacent systems by erosion and deposition or by mud waves. Bridges can reduce these and many other environmental risks, including wetlands loss.

SECONDARY IMPACTS

Secondary impacts are environmental degradation resulting from development following improved access. A new highway through a pristine area may facilitate development and hasten environmental degradation through habitat loss, disturbance during construction, toxic runoff, auto emissions, etc. Enlargement of an existing highway may increase access and encourage additional development.

In developed countries, urban sprawl is facilitated by highways that support condominiums, hotels, and shopping malls. Forests initially fragmented by colonization and conversion to agricultural and silvicultural activities become more isolated (Murphy 1988). In developing countries, poor planning of road systems has caused much damage to tropical forests (Wilson 1988), probably as a result of increased access. Attempts to minimize tropical species loss have led to debate regarding the wisdom of preserving a single, large fragment or several smaller fragments (acronymed SLOSS) of forests (Diamond and May 1976; Simberloff and Abele 1982; and others). The literature supports the concept that small areas can only provide habitat for edge species, while larger tracts of land also accommodate interior species. Governmental policy affecting land use might encourage protection for larger, minimally disturbed areas.

Land use issues are highlighted and new policies developed in areas of concentrated population growth around the United States. Examples from two such areas, Florida and metropolitan Washington, DC, are discussed below.

Florida

By any measure—population, manufacturing sector, etc.—growth is occurring in Florida at a staggering rate. While most people moving to Florida are attracted by clean beaches, pristine natural areas, and the vacation atmosphere, these are the very resources threatened by growth. As a result of growth and increased tourism, Floridians currently enjoy unusually low state taxes. However, the increasing population curve might intersect a curve representing declining natural resources, e.g., freshwater (salt water intrusion threatens Floridian aquifers). Regardless of the factors responsible, Florida state legislators have been forced to address quality of life issues associated with dramatic growth. New highways, bridges, interchanges with previously limited access, and substantial improvements to existing highways all increase access, thereby affecting land use patterns (McCue 1988). Responsibility for growth management is acknowledged by the Florida Department of Highways, and new initiatives include developing and promoting access control policies for the state highway system (McCue 1988).

Washington, DC

The watershed of the nation's largest estuary, the Chesapeake Bay, is expected to have an additional 2.6 million people by the year 2020. This area includes part of Virginia, Maryland, Pennsylvania, Delaware, and the District of Columbia. In planning for the anticipated growth, a panel of the Chesapeake Executive Council supported high density housing and reported:

> More than any single factor, we were concerned with low density sprawl. Sprawl is an ineffective use of the land, difficult to service with infrastructure and transportation, requiring extensive use of automobiles, and consuming large land areas (Year 2020 Panel 1988).

Expected acreages of land used for roads ranged from 125,714 for low density development to 51,640 for high density development over the 30-year period from 1990 to 2020. The panel strongly recommended rigorous zoning ordinances and encouraged development in selected areas by controlling highway construction.

Although strongly supported for short-term economic gains (Suhrbier and Deakin 1988), highway construction projects may face pressure to address fully the environmental consequences of potential secondary development. A bypass through Washington, DC, was proposed to facilitate passage through the city and enhance public safety. This proposal was supported by the Washington Trade Council and the development community in general. A number of environmental organizations noted environmental concerns related to primary impacts of the project and inadequacies of the environmental impact statement (EIS). A lack of non-construction alternatives was also cited. However, civic associations, local governments, many regional governments, numerous legislators, as well as environmental organizations expressed serious reservations regarding potential secondary impacts and the effect on growth management of the proposed project. Two groups were particularly effective in communicating the enviromental risks of the project. A commission involved in restoration of the Chesapeake Bay articulated these concerns in the form of a resolution. The Chesapeake Bay Foundation, a privately funded environmental group, responded to the proposed bypass in a 70-page report. As of this writing, the proposal has been withdrawn and alternatives are being sought.

HIGHWAYS AND WETLANDS

Background

Wetlands were once considered wastelands in the United States, and less than half of the country's original wetland acreage remains (Tiner 1984). If wetland hydrology (the controlling factor in wetland ecology) is altered by activities such as filling for highway construction or for development, wetland functions and values may be lost. Wetland functions include ecosystem maintenance processes (e.g., nutrient cycling and microbial respiration), while wetland values are recognized attributes with socio-economic benefits (e.g., flood abatement and water quality enhancement).

Normal wetland function is a result of interaction between wetland vegetation, soils, and hydrology. Vegetation is described as hydrophytic, and wetland species possess morphological and/or metabolic adaptations to survive in inundated conditions. Wetland soils are termed hydric and, like the vegetation, are formed in response to flooded conditions. The soils are usually low in oxygen (Ponnamperuma 1972) and rich in organic matter (Megonigal and Day 1988). The presence of "hydrology" on site refers to some minimal amount of flooding needed for the establishment of hydrophytic vegetation and hydric soils. In addition to bringing about the existence of a wetland, hydrology is also responsible for many of the differences found among wetlands (Carter 1986; Odum 1978).

An array of values are associated with a diversity of wetland types, and they are reviewed in the literature, including community profiles (e.g., Josselyn 1983; Sharitz and Gibbons 1982; Wharton et al. 1982), conferences (Fisk 1989; Kusler and Riexinger 1986; Sharitz and Gibbons 1989), summary reports (Sather and Smith 1984), and texts (Greeson et al. 1978; Hook et al. 1988; Mitsch and Gosselink 1986). Wildlife values were among the first to gain public attention (Shaw and Fredine 1956). Additional wetland values were studied during and since the environmental movement of the 1960s. Wetland contribution to aquatic and terrestrial food chains has been assessed (Wetzel 1983; Wharton et al. 1982). Water quality values include nutrient abatement (Brinson et al. 1984; Richardson 1989), amelioration of some toxic substances (Gambrell and Patrick 1978; Nichols 1988), groundwater recharge and discharge (Hollands 1986), and sediment stabilization (Broome 1990; Garbisch et al. 1975). Values associated with flooding include storm water management (Adams and Dove 1984; Athanas 1988) and modification of flood peak (Hollands et al. 1987; Maltby 1987). In spite of considerable research, most wetland scientists agree that functional assessment of wetlands on an individual basis remains a difficult task.

Wetland Legislation

Initial regulation of wetlands adjacent to navigable waterways was provided by Section 10 of the Rivers and Harbors Act of 1899, administered by the U.S. Army Corps of Engineers (Corps). Protection in the form of consideration of Corps environmental impacts began in 1968 following increased public concern regarding lost wetland values. The next year, the National Environmental Policy Act of 1969 (NEPA; 42 U.S.C. 4321 et. seq.) was enacted. The NEPA required federal agencies to consider environmental impacts in all federal projects and to complete an EIS if "major" impacts were involved. To afford greater protection for wetlands, President Carter signed Executive Order 11990 in May 1977, and the Federal Highways Administration (FHWA) began developing wetland functional assessment methodologies to implement the order (DesJardins 1986).

Section 404 of the Clean Water Act of 1977 (CWA; 33 U.S.C. Sections 1251-1376) arguably provides the most protection for wetlands. The CWA further increased the role of the Corps in wetland regulation, since the courts would later interpret "waters of the U.S." to be much more inclusive than "navigable waters" (Salveson 1990). The Corps is believed by many to be a reluctant regulator, partly as a result of its long history of wetland destruction resulting from ongoing dredge and fill operations (Salveson 1990; Zaleha 1988). As a check on the increased powers of wetland regulation given to the Corps, Congress gave the U.S. Environmental Protection Agency (USEPA) regulatory oversight powers under Section 404 of CWA. USEPA has issued the 404 (b) (1) guidelines, which the Corps must follow in issuing wetland permits. Veto power over permitting done by the Corps was also granted USEPA, and comment opportunity on permits was granted to the U.S. Fish and Wildlife Service (USFWS) and the National Marine Fisheries Service (NMFS). Nearly two decades of confrontation has followed establishment of Section 404 of the CWA, yet this remains the primary tool of wetland protection throughout the United States. Over most of this time, the Corps and USEPA have even defined wetlands differently, but they finally agreed on a wetland delineation manual in 1989 (Federal Interagency Committee for Wetland Delineation 1989).

In practice, Section 404 of CWA is less than complete in its mitigation of wetland impacts. Loopholes, such as automatically issued general permits, allow some wetland impacts to go virtually unregulated. Some aspects of bridge construction and "minor" road crossings are covered by general permits, and are not afforded the same agency review process. Understaffing in USEPA and Corps field offices leads to many unpermitted, illegal fills still undetected; when after-the-fact fills are detected, penalties have been light (Salveson 1990; Zaleha 1988). Mitigation is defined by the CEQ as having five parts:

(a) avoiding the impact altogether by not taking a certain action or parts of an action;
(b) minimizing impacts by limiting the degree or magnitude of the action and its implementation;
(c) rectifying the impact by repairing, rehabilitating, or restoring the affected environment;
(d) reducing or eliminating the impact over time by preservation and maintenance operations during the life of the action; and
(e) compensating for the impact by replacing or providing substitute resources or environments. (40 CFR Part 1508.20(a-e)).

When permits are obtained, the mitigation process may be truncated. While USEPA, USFWS, and NMFS interpret these as separate steps, the Corps has allowed unsequenced mitigation. The Corps apparently believes that compensation can replace all wetland functions and values, but USEPA, USFWS, and NMFS disagree (Salveson 1990).

Local highway department policy towards wetland regulations is largely controlled by the FWHA. FHWA policy requires that mitigation measures be incorporated into proposed actions. A reasonable public expenditure must be made for mitigating impacts in order to be eligible for federal funding. Privately owned wetlands mitigation policy and procedures have been described by FHWA (23 CFR 777). Local highway departments must obtain wetland permits from their local Corps district engineer. Permit applications may receive comments from USFWS, NMFS, and USEPA recommending aggressive mitigation; however, FHWA retains considerable control over mitigation through its interpretation of reasonable public expenditure and by deciding which mitigation measures are eligible for federal highway funding.

Compensation for Wetland Loss

The most controversial of all highway and wetland-related activities is Part e of CEQ's definition of mitigation, termed *compensation*. Although the term is explicitly defined in Part e, confusing synonyms have arisen, including *mitigation*, *compensatory mitigation*, *wetland creation*, *artificial wetland creation*, *replacement wetland creation*, *restoration*, *enhancement*, etc. Compensatory measures considered by FHWA include "preservation, restoration, enhancement, creation, and replacement" (Isaacson 1988); however, this discussion focuses on compensation through wetland creation, i.e., construction of new wetlands.

Recently, highway departments have embraced the concept of mitigation via creation. Permits for highway construction form the bulk of wetland fill permits in the Southeastern United States (Deitz, personal communication[d]). Some literature is available to assist those who build highways in this country in complying with wetland regulations. These include training manuals (Erickson et al. 1984), the 1988 FHWA procedural guidelines for compliance with NEPA and Section 404 of CWA, methodologies for assessing wetland function (Adamus and Stockwell 1983; Adamus et al. 1987; DesJardins 1986), and methodologies for constructing wetlands (Garbisch 1986). However, a lack of sufficient training in wetland design has been cited as a serious cause of failed compensation attempts (Kusler 1990).

Regardless of how wetland regulations are interpreted or enforced, highway projects resulting in wetland fills are inevitable. The question becomes one of risk. To what extent should highway departments perform Parts a-d (avoiding and minimizing impacts) before the remaining acreage filled is replaced through compensation? Clearly, regulatory agencies need to know the probability that compensation will successfully replace the structure and function of the lost wetland. However, compensation is being widely used (Wolf et al. 1986) with public funds and without: (a) sufficient monitoring to analyze success or recommend improvements, and (b) ex-

[d]Randy Deitz, Committee on Public Works and Transportation, U.S. Fish and Wildlife Service, Washington, DC. March 1991.

perimental design to develop improvements (Butts 1988; Kusler 1990; Larson 1988; Race 1985; Atkinson et al. submitted[e]).

Some studies have begun to address compensation effectiveness (Fisk 1989; Hillsboro Community College 1974-to date; Kusler and Brooks 1988; Larson and Neill 1987; Swanson 1979). A USEPA-funded publication (Kusler and Kentula 1990) provides wetland restoration and creation examples from around the United States. No evaluation technique for created wetland success enjoys widespread approval (LaRoe 1988). Success in wetland creation can be defined as achieving the goals set for the new wetland (Quammen 1988); therefore, an absence of either stated or implied goals limits the evaluation. When explicitly stated, goals range from replacement of vegetative cover over the site to replacement of all natural wetland functions (Kusler 1990). If revegetation is used as the primary criterion for success, some researchers have found few projects to be successful. However, in at least one case, revegetation by *Phragmites communis*, generally considered to be a pest (Pierce 1988), was described as very successful (e.g., C.E. Maguire, Inc. 1985). In another study, Reimold and Thompson (1988) found serious problems in all five sites studied, but somehow concluded "based on our review, mitigation [by wetland creation] is a viable concept." Most papers reviewing compensation efforts express doubt that constructed wetlands perform most normal wetland functions as well as the systems they were designed to replace (Butts 1988; D'Avanso 1987; Quammen 1988; Race 1985; Race and Christie 1982; Walker 1988).

While fully successful compensation attempts seem difficult to find, many researchers have identified potential causes of failure. Some problems are of an administrative/policy nature. For example, monitoring of compliance with terms of the permit has been weak (Kusler 1990). Others are more indicative of the early stage in the development of the science. Regardless of these shortcomings, the environmental risks of compensation by wetland construction must be addressed.

Wetlands are characterized by hydrology, vegetation, and soils and these are major sources of compensation failure. Hydrology is of great importance because of its controlling role in wetland processes (Mitsch and Gosselink 1986). Problems with hydrology include inadequate elevation and hydroperiod (Broome 1990; Lewis 1990), inadequate supply of dissolved nutrients (Brinson and Lee 1989), excessive wave energy (Lewis 1990), and failure to consider hydrologic variations, such as those described in Odum (1978) and Weller (1988).

---

[e]Atkinson, R.B., Perry, J.E., Smith, E., and Cairns, J., Jr. Use of constructed wetlands delineation and weighted averages as a component of assessment. Virginia Polytechnic Institute and State University, Blacksburg, VA.

Wetland classification is most often based on vegetation, which is critical to many wetland functions and values. Problems with vegetation in compensation wetlands include low survivorship of planted material (Quammen 1988); invasion of exotic species, such as *Lythrum salicaria* (Kusler 1990; Pierce 1988) and *Phragmites communis*; introduction of non-native species; lack of acclimation for nursery-grown plant material (Lewis 1990); poor planting techniques (Josselyn et al. 1990); grazing by animals (Pierce 1988); invasion by upland species; and incomplete colonization resulting in bare areas (personal observation).

Wetland soils are essential to some wetland functions and values, many of which result from the fact that wetlands are the primary reducing ecosystem on the landscape (Gambrell and Patrick 1978). Soil problems include severe sedimentation (Kusler 1990), erosion (personal observation), lack of proper soil microflora (Kruczynski 1990), low percent organic matter (Walker 1988), high bulk density, and length of time needed for hydric soil development.

Many ecological concerns are within the realm of landscape ecology. Compensation wetlands are sometimes constructed off-site, that is, the constructed wetland is not adjacent to the impacted wetland. In other cases, wetland losses are compensated for by a wetland bank in which acres lost are traded for acres constructed at another site. In either case, wetland functions dependent on location within the landscape are likely to be forfeited, and such tradeoffs cannot be supported on a scientific basis (Kusler 1990). Compensation sites are sometimes isolated by steep banks that affect habitat functions, but design should allow wetlands to act as corridors (Willard and Hiller 1990). In other cases, wetlands are created within loops, or otherwise surrounded by highways. Such areas clearly present risks to animals, not to mention motorists, attracted to these sites (Harris 1988). Habitat functions lost when a new road fill crosses over an otherwise pristine wetland system are not fully mitigated by wetlands created between lanes or adjacent to the highway.

Ecological risks have evolved with the practice of compensation. Some permit requests now include, implicitly or explicitly, replacement of one type of wetland with a different type. In-kind compensation (constructing the same wetland type as that lost) and out-of-kind compensation (constructing some other type of wetland) occupy opposite ends of the spectrum. At one end, the same species assemblage, successional stage, biomass, soil characteristics, and hydrologic regime are replaced; at the other end of the spectrum, an emergent wetland may replace a forested wetland, a shallow pond could replace an emergent wetland, a nontidal system replace a tidal wetland, or a degraded, lowlying upland replace a wetland. Many of the associated ecological risks of out-of-kind compensation have been characterized (Brinson and Lee 1989; Kruczynski 1990; Kusler 1990). Compensation for losses of certain wetland types are nearly always out-of-kind. For example, forested wetland construction techniques generally involve planting saplings or recolonization and must rely on community development over time for establishment of a forested wetland. Development of forested wetland structure, functions, and values may take decades to occur, if at all (Brinson and Lee 1989; Atkinson et al. submitted[e]); however, contractors are seldom responsible for survivorship beyond one year after planting. Wetlands are components of the landscape and evolve in response to chance events occurring over time. Uncertainty regarding the effects of random events on the development of the compensation wetland are always a threat to a project (Kusler 1990; Willard and Hiller 1990).

Areas adjacent to the compensation site may be placed at risk. Adjacent wetlands may be at risk due to hydrologic alteration. Grading to lower than appropriate elevations or diverting a stream could lower the water table in the adjacent system. The highway fill may dam water and increase inundation times, which will alter wetland productivity (Odum 1978) and often kills most floodplain plant species (impacts such as these are seldom included in the mitigation process). Vegetation risks to adjacent systems include invasion by exotic species and introduction of non-native allelic forms. Soil-related risks include sedimentation from erosion in the compensation wetland, microbial oxidation of natural wetland organic matter if draining occurs (Terry 1986), and introduction of toxic materials (if present in soils spread over a site).

In addition to the risk that the compensation wetland may not adequately replace that filled, three other landscape components may be placed at risk: (1) a donor system, perhaps a forested upland, may be used as the wetland construction site (upland areas have functions and values that may be sacrificed for the compensation wetland; Kruczynski 1990), (2) natural wetlands are often used as a source for plant material for the compensation site, which places these systems at great risk (Race 1985), and (3) wetland acreage in excess of that directly filled may be negatively affected. Alteration of hydrology by highways has been shown to cause effects ranging from eutrophication to plant invasions (Mitsch and Gosselink 1986).

Risks of Cumulative Impacts to Wetlands
Cumulative impact was defined by CEQ (40 CFR Part 1508.7) in 1978 as environmental impacts resulting from an incremental impact, when added to other past, present, and reasonably forseeable future actions regardless of which agency (governmental or non-governmental) or person is responsible. Muir et al. (1990) describe cumulative impacts on wetlands as a major source of wetland loss and functional degradation. Weller et al. (1988) described cumulative effects on wetlands as the result of piecemeal losses of wetland system or type that ultimately affects the functioning of

---

[e]Atkinson, R.B., Perry, J.E., Smith, E., and Cairns, J., Jr. Use of constructed wetlands delineation and weighted averages as a component of assessment. Virginia Polytechnic Institute and State University, Blacksburg, VA.

the system significantly. Discussions of cumulative impacts and methods for dealing with associated wetland losses can be found in the literature (Gosselink and Lee 1989; Gosselink et al. 1990; McCort et al. 1988). Gosselink and Lee (1988) borrowed from island biogeography and the SLOSS debate in determining the goals of cumulative impact assessment for bottomland hardwood wetlands.

(1) Preserve *large blocks* of bottomland/upland forest appropriately interspersed with smaller tracts.
(2) Preserve *continuity* among patches.
(3) Maintain *contiguity* between bottomland forest and stream, and between bottomland forest and upland forest.

Muir et al. (1990) reviewed 16 statutes and regulations that require cumulative impact consideration and described all these measures as ineffective. Reasons for the failure included economic, political, regulatory agency, and scientific factors. Economic and political causes for failure might be considered as successes in circles with opposing goals. Regulatory agencies contribute to the failure of cumulative impact consideration partially because they are responsible for only part of the ecosystem and are limited geographically (Cairns 1990). Scientific causes include spatial and temporal complexities associated with landscape-level processes (Gosselink and Lee 1989). Most of the available information necessary for consideration of cumulative impacts during the Section 404 permit review process is simply not used by review agencies. Even if agencies (e.g., USFWS, NMFS, and USEPA) had sufficient capacity to consider these impacts, the political cost might be excessive. For example, Stakhiv (1988) focuses on developer concerns regarding economic impacts to development interests.

Cumulative impacts include, by CEQ definition, both direct and indirect impacts. Interestingly, CEQ described indirect effects as caused by the action and are later in time or farther removed in distance, but are still reasonably forsceable. Cumulative impacts resulting from highways have a direct effect on the wetland filled, and perhaps on adjacent wetlands via altered hydrology, mud waves, toxic runoff, etc. But in addition, it is reasonably forseeable that the access to undeveloped areas derived from the new highway leads to environmental impacts from development. Piecemeal loss of wetland acreage follows as individual permits for wetland fills are issued for housing, shopping malls, schools, etc. in addition to unpermitted fills. Without regard for cumulative impacts, the functional integrity of wetlands and adjacent waters may be destroyed. Indirect impacts of highways are not limited to wetland and other aquatic systems; in fact, terrestrial and aquatic systems may also be at risk to secondary impacts.

CONCLUSIONS

Some sources of environmental risk have been addressed in this discussion but more remain. Consideration of the risks associated with highways appears to be quite inadequate throughout most of the United States. Effective advanced planning for direct impacts, a landscape-level perspective, and realistic predictions of secondary impacts should be given the most serious consideration and should include assistance from the scientific community. In the special case of wetlands construction, additional research is mandatory, including (a) sufficient monitoring to analyze success and to recommend improvements and (b) incorporation of experimental design to develop improvements. Current proposals by the Bush administration, the U.S. Senate, and the U.S. House of Representatives all call for substantial increases in highway transportation expenditures. It is essential that highway planning include reasonable costs of maintaining environmental services by considering environmental risks.

Advanced planning for highway construction would benefit from a check list of environmental risks to be considered while modifications can still be effective. Advisory groups should be established and called upon, perhaps using the Florida Environmental Advisory Group (EAG) as an example. The Orlando-Orange Expressway Authority (OOCEA) of Orlando, Florida, responded to confrontations with environmental groups by establishing the EAG and a special consultant. The role of the consultant is to facilitate and expedite exchange of concerns between the EAG and OOCEA. The group, entirely separate from the OOCEA, is primarily concerned with selection of alignments, development of mitigation plans, and maintenance of mitigation sites. In addition to monthly meetings, the EAG visits alignment sites in particularly sensitive areas (Gwynn et al. 1989).

The importance of a landscape-level perspective for adequate mitigation of adverse environmental effects cannot be overstated. Agencies with environmental review responsibilities need to access available databases (i.e., geographic information system) and incorporate landscape-level concerns into management decisions. Such tools would improve information for selection of highway alignments and support decisions regarding use of bridges instead of fill and box culverts. These agencies typically lack sufficient resources to address many other significant environmental risks. Incorporation of landscape-level concepts has aided the development of cumulative impact assessment in wetlands, which also needs to be incorporated into agency review of proposed activities.

Secondary impacts of highways pose grave environmental risks. Examples from fast-growing urbanized areas indicate that aggressive zoning regulations accompanied by early highway planning may control and direct growth and limit urban sprawl. Information generated by detailed responses to proposals should be obtained by those who seek to apply good land use management to growth problems they may face, e.g., response of the Chesapeake Bay Foundation to the proposed Washington bypass. Risks of secondary development to undeveloped, silvicultural, and agricultural areas should be taken into account when highway routing and access via intersections and exits are planned. The risk to valuable ecosystems (i.e., wetlands loss to secondary development) may be very high and should be incorporated into costs and mitigation planning of the new highway or the enlargement of an existing highway.

Highway construction is often justified by demonstration of a cost/benefit analysis. In the past, expertise necessary for adequate consideration of environmental costs of highway construction was not available. Currently, environmental personnel on highway department staffs can consult environmental experts, such as wetland scientists, when additional expertise is necessary. Although many environmental risks are difficult to quantify, qualitative concerns must be taken into consideration, whether they are a result of primary or secondary impacts. More responsibility for direct and indirect environmental risks of highways should be taken by transportation departments and legislators.

ACKNOWLEDGMENTS

Financial support was provided by the du Pont Educational Aid Program, Mobil Foundation, Inc., and the Virginia Academy of Science. We would like to express gratitude for constructive criticism of the manuscript from Dr. C.M. Sellers, Barbara Niederlehner, Stephanie Hill, Karen Holl, and Clint Sellers. Thanks go to Dr. J.E. Perry for conversations regarding wetlands and to Dr. Bruce Wallace for genetics discussions. The editorial assistance of Darla Donald is gratefully acknowledged.

REFERENCES

ABKOWITZ, M., LIST, G., and RADWAN, A.E. (1989). Critical issues in safe transport of hazardous materials. J. Transport. Eng. 115(6):608-629.

ADAMS, L.W. and DOVE, L.E. (1984). Urban Wetlands for Stormwater Control and Wildlife Enhancement. National Institute for Urban Wildlife, Columbia, MD. 15 pp.

ADAMUS, P.R. and STOCKWELL, L.T. (1983). A Method for Wetland Functional Assessment: Volume 1, FHWA-IP-82-83; Volume 2, FHWA-IP-82-24. Federal Highway Administration, Washington, DC. 181 pp.

ADAMUS, P.R., CLAIRAIN, E.J., SMITH, R.D., and YOUNG, R.D. (1987). Wetland Evaluation Technique (WET); Volume 2: Methodology, Operational Draft TR-Y-87. U.S. Army Engineers Waterways Experiment Station, Vicksburg, MS.

ANDERSON, S.H., MANN, K., and SHUGART, H.H., Jr. (1977). The effect of transmission-line corridors on bird populations. Am. Midl. Nat. 97(1):216-221.

ATHANAS, C. (1988). The creation and use of wetlands for stormwater treatment. In: *Proceedings of the National Wetland Symposium: Wetland Hydrology* (J.A. Kusler and G. Brooks, eds), pp. 224-238. Association of State Wetland Managers, Berne, NY.

BAILEY, J. (1992). Dueling studies: How two industries created a fresh spin on the dioxin debate. The Wall Street Journal CCXIX(35):1, 4.

BARBOUR, M.G., BURK, J.H., and PITTS, W.D. (1980). Terrestrial Plant Ecology. Benjamin/Cummings Publishing Company, Inc., Menlo Park, CA. 604 pp.

BRINSON, M.M. and LEE, L.C. (1989). In-kind mitigation for wetland loss: Statement of ecological issues and evaluation of examples. In: *Freshwater Wetlands and Wildlife* (R.R. Sharitz and J.W. Gibbons, eds), pp. 1069-1085. U.S.DOE, Office of Scientific and Technical Information, Oak Ridge, TN.

BRINSON, M.M., BRADSHAW, H.D., and KANE, E.S. (1984). Nutrient assimilation capacity of an alluvial floodplain swamp. J. Appl. Ecol. 21:1041-1057.

BROOME, S.W. (1990). Creation and restoration of tidal wetlands of the Southeastern United States. In: *Wetland Creation and Restoration: The Status of the Science* (J.A. Kusler and M.E. Kentula, eds), pp. 37-72. Island Press, Washington, DC.

BUTTS, M.P. (1988). Status of wetlands creation/mitigation projects on state highway projects in Connecticut. In: *Proceedings of the National Wetlands Symposium: Mitigation of Impacts and Losses* (J.A. Kusler, M.L. Quammen, and G. Brooks, eds), pp. 185-187. Association of State Wetland Managers, Berne, NY.

CAIRNS, J., Jr. (1990). Gauging the cumulative effects of developmental activities on complex ecosystems. In: *Ecological Processes and Cumulative Impacts: Illustrated by Bottomland Hardwood Wetland Ecosystems* (J.G. Gosselink, L.C. Lee, and T.A. Muir, eds), pp. 239-256. Lewis Publishers, Chelsea, MI.

CARTER, V. (1986). An overview of hydrologic concerns related to wetlands in the United States. Can. J. Bot. 64:364-374.

CASHMAN, J.R. (1986). Transportation: Chem-Freight, Inc. In: *Management of Hazardous Waste Treatment/Storage/Disposal Facilities* (J.R. Cashman, ed), pp. 245-252. Technomic Publishing Company, Inc., Lancaster, PA.

C.E. MAGUIRE, INC. (1985). Wetland Replacement Evaluation. U.S. Army Corps of Engineers, Norfolk District, Virginia Beach, VA.

COHN, L.F. (1982). Environmental action planning for transportation project development. Transport. Quart. 36(4):503-525.

COUNCIL ON ENVIRONMENTAL QUALITY (1989). Environmental Trends. Superintendent of Documents, U.S. Government Printing Office, Washington, DC. 152 pp.

D'AVANSO C. (1987). Vegetation in freshwater replacement wetlands in the Northeast. In: *Mitigating Freshwater Wetland Alterations in the Glaciated Northeastern United States: An Assessment of the Science Base* (J.S. Larson and C. Neill, eds), Pub. No. 87-1, pp. 53-81. University of Massachusetts, Amherst, MA.

DAWSON, G.W. and MERCER, B.W. (1986). Hazardous Waste Management. John Wiley and Sons, NY. 532 pp.

DESJARDINS, C. (1986). Implementation of Executive Order 11990-the real world. In: *Proceedings: National Wetland Assessment Symposium* (J.A. Kusler and P. Riexinger, eds), pp. 80-83. Association of State Wetland Managers, Chester, VT.

DIAMOND, J.M. and MAY, R.M. (1976). Island biogeography and the design of natural reserves. In: *Theoretical Ecology* (R.M. May, ed), pp. 163-186. Blackwell Scientific Publishers, Oxford.

EHRLICH, P.R. (1989). Carbon dioxide and the human predicament: An overview. In: *Atmospheric Carbon Dioxide and Global Warming, Proceedings of the First Presidential Symposium on World Issues* (J. Cairns, Jr. and P.F. Zweifel, eds), pp. 5-11. Virginia Polytechnic Institute and State University, Blacksburg, VA.

EHRLICH, P.R. and EHRLICH, A.H. (1990). The Population Bomb. Simon and Schuster, NY. 320 pp.

ERBAUGH, M.D. and HINCK, L.G. (1984). Health, safety, and personnel protection for sites contaminated with 2,3,7,8-tetrachlorodibenzo-p-dioxin. In: *Hazardous Material Spills Conference Proceedings* (J. Ludwigson, ed), pp. 223-229. Government Institutes Inc., Rockville, MD.

ERICKSON, P.A., CAMOUGIS, G., HOLCOMB, B.P., SUURBALLE, N.C., and MCCROHON, T.A. (1984). Effects of Highway Development on Wetland Areas, Instructors Guide. U.S. Department of Transportation, Federal Highway Administration, Washington, DC. 84 pp.

EXNER, J.H., ERICKSON, D.G., CIBULSKIS, R., and KEFFER, W.D. (1984). Disposal, treatment, and mitigation options for Missouri dioxin sites. In: *1984 Hazardous Spills Conference* (J. Ludwigson, ed), pp. 245-251. Government Institutes Inc., Rockville, MD.

FEDERAL INTERAGENCY COMMITTEE FOR WETLAND DELINEATION (1989). Federal Manual for Identifying and Delineating Jurisdictional Wetlands. Cooperative Technical Publication, U.S. Army Corps of Engineers, U.S. Environmental Protection Agency, U.S. Fish and Wildlife Service, and U.S.D.A. Soil Conservation Service, Washington, DC. 76 pp.

FISK, D.W. (1989). Proceedings of the Symposium, Wetlands: Concerns and Successes, Tampa, Florida. American Water Resources Assocociation, Bethesda, MD. 568 pp.

GAMBRELL, R.P. and PATRICK, W.H., Jr. (1978). Chemical and microbiological properties of anaerobic soils and sediments. In: *Plant Life in Anaerobic Environments* (D.D. Hook and R.M.M. Crawford, eds), pp. 375-423. Ann Arbor Science, Ann Arbor, MI.

GARBISCH, E.W., Jr. (1986). Highways and Wetlands, Compensating Wetland Losses. FHWA-IP-82-22, U.S. Department of Transportation, Federal Highway Administration, Washington, DC. 60 pp.

GARBISCH, E.W., Jr., WOLLER, P.B., and MCCALLUM, R.J. (1975). Salt marsh establishment and development. TM-52, U.S. Army Corps of Engineers, CERRE-EC, Fort Belvior, VA. 110 pp.

GILPIN, M.E. AND SOULE, M.E. (1986). Minimum viable populations: Processes of species extinction. In: *Conservation Biology* (M.E. Soule, ed), pp. 19-34. Sinauer Associates, Sunderland, MA.

GLICKMAN, T. (1984). Data requirements for hazardous materials transportation risks analysis. In: *1984 Hazardous Spills Conference* (J. Ludwigson, ed), pp. 322-326. Government Institutes Inc., Rockville, MD.

GOSSELINK, J.G. and LEE, L.C. (1988). Cumulative impact assessment principles. In: *Proceedings of the National Wetlands Symposium: Mitigation of Impacts and Losses* (J.A. Kusler, M.L. Quammen, and G. Brooks, eds), pp. 196-203. Association of State Wetland Managers, Berne, NY.

GOSSELINK, J.G. and LEE, L.C. (1989). Cumulative impact assessment in bottomland hardwood forests. Wetlands 9: Special Issue. 174 pp.

GOSSELINK, J.G., LEE, L.C., and MUIR, T.A. (1990). Ecological Processes and Cumulative Impacts. Lewis Publishers, Inc., Chelsea, MI. 708 pp.

GREENE, D.L., SPERLING, D., and MCNUTT, B. (1988). Transportation energy to the year 2020. In: *A Look Ahead, Year 2020: Proceedings of the Conference on Long-Range Trends and Requirements for the Nation's Highway and Public Transit Systems*, pp. 207-231. Transportation Research Board, National Research Council, Washington, DC.

GRESSON, P.E., CLARK, J.R., and CLARK, J.E. (1978). Wetlands Functions and Values: The State of Our Understanding, Tech. Pub. Series No. TPS79-2. American Water Resources Association, Minneapolis, MN.

GWYNN, D.W., DITTMEIER, J.M. and SCOTT, H. (1989). Environmental advice for new expressways. Transport. Quart. 33(2):183-192.

HARRIS, L.D. (1988). The faunal significance of fragmentation of Southeastern bottomland hardwood forests. In: *Proceedings of the Symposium: The Forested Wetlands of the Southern United States* (D.D. Hook and R. Lea, eds), pp. 126-134. U.S. Department of Agriculture, Forest Service, Washington, DC.

HILLSBORO COMMUNITY COLLEGE (1974 to date). Proceedings of the Annual Conference on the Restoration and Creation of Wetlands. Hillsboro Community College, Tampa, FL.

HOLLANDS, G.G. (1986). Assessing the relationship of groundwater and wetlands. In: *Proceedings of the National Wetland Assessment Symposium* (J.A. Kusler and P. Riexinger, eds), pp. 55-57. Association of State Wetland Managers, Chester, VT.

HOLLANDS, G.G., HOLLIS, G.E., and LARSON, J.S. (1987). Science base for freshwater wetland mitigation in the glaciated Northeastern United States: hydrology. In: *Mitigating Freshwater Wetland Alterations in the Glaciated Northeastern United States: An Assessment of the Science Base* (J.S. Larson and C. Neill, eds), Pub. No. 87-1, pp. 131-143. University of Massachusetts, Amherst, MA.

HOOK, D.D., MCKEE, W.H., SMITH, H.K., GREGORY, J., BURRELL, V.G. Jr., DEVOE, M.R., SOJKA, R.E., GILBERT, S., BANKS, R., STOLZY, L.H., BROOKS, C., MATTHEWS, T.D., and SHEAR, T.H. (1988). The Ecology and Management of Wetlands, Vol. 2: Management, Use and Value of Wetlands. Timber Press, Portland, OR. 394 pp.

HOWARD, D.M. (1984). Planning for immediate removals at Eastern Missouri Dioxin sites. In: *1984 Hazardous Spills Conference* (J. Ludwigson, ed), pp. 240-244. Government Institutes Inc., Rockville, MD.

ISAACSON, L. (1988). The federal highway program and wetland mitigation. In: *Proceedings of the National Wetlands Symposium: Mitigation of Impacts and Losses* (J.A. Kusler, M.L. Quammen, and G. Brooks, eds), pp. 43-45. Association of State Wetland Managers, Berne, NY.

JACKSON, J.A. (1979). Highways and wildlife—some challenges and opportunities for management. In: *The Mitigation Symposium: A National Workshop on Mitigating Losses of Fish and Wildlife Habitats* (G.A. Swanson, technical coordinator), General Technical Report RM-65, pp. 566-571. Fort Collins, CO.

JOHNSON, W.C., SHARPE, D.M., DEANGELIS, D.L., FIELDS, D.E., and OLSON, R.J. (1981). Modeling seed dispersal and forest island dynamics. In: *Forest Island Dynamics in Man-Dominated Landscapes* (R.L. Burgess and D.L. Sharpe, eds), pp. 215-239. Springer-Verlag, NY.

JOSSELYN, M. (1983). The ecology of San Francisco Bay tidal marshes: A community profile. U.S. Fish and Wildlife Service, Division of Biological Services, FWS/OBS-83/23. Washington, DC. 102 pp.

JOSSELYN, M., ZEDLER, J.B., and GRISWOLD, T. (1990). Wetland mitigation along the Pacific coast. In: *Wetland Creation and Restoration: The Status of the Science* (J.A. Kusler and M.E. Kentula, eds), pp. 3-36. Island Press, Washington, DC.

KRUCZYNSKI, W.L. (1990). Options to be considered in preparation and evaluation of mitigation plans. In: *Wetland Creation and Restoration: The Status of the Science* (J.A. Kusler and M.E. Kentula, eds), pp. 555-570. Island Press, Washington, DC.

KUSLER, J.A. (1990). Views on scientific issues relating to the restoration and creation of wetlands. In: *Issues in Wetland Protection: Background Papers Pre-*

pared for the National Wetlands Policy Forum (G. Bingham, E.H. Clark, L.V. Haygood, and M. Leslie, eds), pp. 217-230. The Conservation Foundation, Washington, DC.

KUSLER, J.A. and G. BROOKS. (1988). Proceedings of the National Wetland Symposium: Wetland Hydrology. Association of State Wetland Managers, Berne, NY. 339 pp.

KUSLER, J.A. and KENTULA, M.E. (1990). Wetland Creation and Restoration, the Status of the Science. Island Press, Washington, DC. 591 pp.

KUSLER, J.A. and RIEXINGER, P. (1986). Proceedings of the National Wetland Assessment Symposium. Association of State Wetland Managers, Chester, VT. 331 pp.

LACAVA, J. and HUGHES, J. (1984). Determining minimum viable population levels. Wildl. Soc. Bull. 12(4):370-376.

LAROE, E.T. (1988). Mitigative research needs: summary. In: *Proceedings of the National Wetland Symposium: Mitigation of Impacts and Losses* (J.A. Kusler, M.L. Quammen, and G. Brooks, eds), pp. 431-435. Association of State Wetland Managers, Berne, NY.

LARSON, J.S. (1988). Adequacy of the science base for wetland restoration and creation. Third International Wetlands Conference. University of Rennes, Rennes, France.

LARSON, J.S. and NEILL, C. (1987). Mitigating Freshwater Wetland Alterations in the Glaciated Northeastern United States: An Assessment of the Science Base, Publ. No. 87-1. The Environmental Institute, University of Massachusetts, Amherst, MA. 143 pp.

LEVENSON, J.B. (1981). Woodlots as biogeographic islands in Southeastern Wisconsin. In: *Forest Island Dynamics in Man-Dominated Landscapes* (R.L. Burgess and D.L. Sharpe, eds), pp. 13-39. Springer-Verlag, NY.

LEWIS, R.R. (1990). Creation and restoration of coastal plain wetlands in Florida. In: *Wetland Creation and Restoration: The Status of the Science* (J.A. Kusler and M.E. Kentula, eds), pp. 73-101. Island Press, Washington, DC.

LOVEJOY, T.E. and OREN, D.C. (1981). The minimum critical size of ecosystems. In: *Forest Island Dynamics in Man-Dominated Landscapes* (R.L. Burgess and D.L. Sharpe, eds), pp. 7-12. Springer-Verlag, NY.

MACARTHUR, R.H. and WILSON, E.O. (1967). The Theory of Island Biogeography. Princeton University Press, Princeton, NJ. 203 pp.

MALTBY, E. (1987). Soils science base for freshwater wetland mitigation in the Northeastern United States. In: *Mitigating Freshwater Wetland Alterations in the Glaciated Northeastern United States: An Assessment of the Science Base*, Publ. No. 87-1 (J.S. Larson and C. Neill, eds), pp. 17-52. University of Massachusetts, Amherst, MA.

MATTHIAE, P.E. and STEARNS, F. (1981). Mammals in forest islands in southeastern Wisconsin. In: *Forest Island Dynamics in Man-Dominated Landscapes* (R.L. Burgess and D.L. Sharpe, eds), pp. 55-66. Springer-Verlag, NY.

MCCORT, W.D., LEE, L.C., and WEIN, G.R. (1988). Mitigating for large-scale wetland loss: A realistic endeavor? In: *Proceedings of the National Wetland Symposium: Mitigation of Impacts and Losses* (J.A. Kusler, M.L. Quammen, and G. Brooks, eds), pp. 359-367. Association of State Wetland Managers, Berne, NY.

MCCUE, P.J. (1988). Florida's challenge: managing explosive growth. In: *A Look Ahead, Year 2020: Proceedings of the Conference on Long-Range Trends and Requirements for the Nation's Highway and Public Transit Systems*, pp. 394-413. Transportation Research Board, National Research Council, Washington, DC.

MEGONIGAL, J.P. and DAY, F.P., Jr. (1988). Organic matter dynamics in four seasonally flooded forest communities of the Dismal Swamp. Am. J. Bot. 75(9):1334-1343.

MITSCH, W.J. and GOSSELINK, J.G. (1986). Wetlands. Van Nostrand Reinhold, NY. 539 pp.

MUIR, T.A., RHOADES, C., and GOSSELINK, J.G. (1990). Federal statutes and programs relating to cumulative impacts in wetlands. In: *Ecological Processes and Cumulative Impacts: Illustrated by Bottomland Hardwood Wetland Ecosystems* (J.G. Gosselink, L.C. Lee, and T.A. Muir, eds), pp. 223-236. Lewis Publishers, Chelsea, MI.

MURPHY, D.D. (1988). Challenges to biological diversity in urban areas. In: *Biodiversity* (E.O. Wilson, ed), pp. 70-76. National Academy Press, Washington, DC.

NICHOLS, A.B. (1988). A vital role for wetlands. J. Water Pollut. Control Fed. 6(7):1215-1221.

NORSE, E.A., ROSENBAUM, K.L., WILCOVE, D.S., WILCOX, B.A., ROMME, W.H., JOHNSTON, J.W., and STOUT, M.L. (1986). Conserving Biological Diversity in our National Forests. Global Printing, Inc. Alexandria, VA. 118 pp.

NOSS, R.F. (1987). Corridors in real landscapes: a reply to Simberloff and Cox. Conserv. Biol. 1(2):159-164.

ODUM, E.P. (1978). The value of wetlands: A hierarchical approach. In: *Wetlands Functions and Values: The State of Our Understanding* (P.E. Greeson, J.R. Clark, and J.E. Clark, eds), Tech. Publ. Series No. TPS79-2, pp. 16-25. American Water Resources Association, Minneapolis, MN.

ORGANISATION FOR ECONOMIC CO-OPERATION AND DEVELOPMENT (OECD) (1986). Environmental Effects of Automotive Transport. The OECD Publications Service, Paris, France. 172 pp.

PIERCE, G.J. (1988). Succession and stability in freshwater marshes in the Northeastern United States. In: *Proceedings of the National Wetland Symposium: Mitigation of Impacts and Losses* (J.A. Kusler, M.L. Quammen, and G. Brooks, eds), pp. 323-324. Association of State Wetland Managers, Berne, NY.

PONNAMPERUMA, F.N. (1972). The chemistry of submerged soils. Adv. Agron. 24:29-92.

PYE, V.I., PATRICK, R., and QUARLES, J. (1983). Groundwater Contamination in the United States. University of Pennsylvania Press, Philadelphia, PA. 315 pp.

QUAMMEN, M.L. (1988). Measuring the success of wetlands mitigation. In: *Proceedings of the National Wetlands Symposium: Mitigation of Impacts and Losses* (J.A. Kusler, M.L. Quammen, and G. Brooks, eds), pp. 242-245. Association of State Wetland Managers, Berne, NY.

RACE, M.S. (1985). Critique of present wetlands mitigation policies in the United States based on an analysis of past restoration projects in San Francisco Bay. Environ. Manage. 9(1):71-82.

RACE, M.S. and CHRISTIE, D.R. (1982). Coastal zone development: mitigation, marsh creation and decision-making. Environ. Manage. 6(4):317-328.

RALLS, K., HARVEY, P.H., and LYLES, A.M. (1986). Inbreeding in natural populations of birds and mammals. In: *Conservation Biology* (M.E. Soule, ed), pp. 35-56. Sinauer Associates, Sunderland, MA.

RANDALL, C.W., HELSEL, D.R., GRIZZARD, T.J., and HOEHN, R.C. (1978). The impact of atmospheric contaminants on storm water quality in an urban area. Prog. Water Tech. 10(5- 6):417-431.

REED, J.M., DOERR, P.D., and WALTERS, J.R. (1986). Determining minimum population sizes for birds and mammals. Wildl. Soc. Bull. 14:255-261.

REIMOLD, R.J. and THOMPSON, D.A. (1988). Wetland mitigation effectiveness. In: *Proceedings of the National Wetlands Symposium: Mitigation of Impacts and Losses* (J.A. Kusler, M.L. Quammen, and G. Brooks, eds), pp. 259-262. Association of State Wetland Managers, Berne, NY.

RICHARDSON, C.J. (1989). Freshwater wetlands: transformers, filters, or sinks? In: *Freshwater Wetlands and Wildlife, 1989* (R.R. Sharitz and J.W. Gibbons, eds), pp. 25-46. U.S. DOE Office of Scientific and Technical Information, Oak Ridge, TN.

SALVESON, D. (1990). Wetlands: Mitigating and Regulating Development Impacts. The Urban Land Institute, Washington, DC. 117 pp.

SATHER, J.H. and SMITH R.D. (1984). Proceedings of the National Wetland Assessment Workshop, FWS/OBS-84-12. Washington, DC. 100 pp.

SHARITZ, R.R. and GIBBONS, J.W. (1982). The Ecology of Southeastern Shrub Bogs (Pocosins) and Carolina Bays: A Community Profile, FWS/OBS-82/04. U.S. Fish and Wildlife Service, Division of Biological Services, Washington, DC. 93 pp.

SHARITZ, R.R. and GIBBONS, J.W. (1989). Freshwater Wetlands and Wildlife: Proceedings of a Symposium. U.S. Department of Energy, Washington, DC. 1265 pp.

SHAW, S.P. and FREDINE, C.G. (1956). Wetlands of the United States, their extent and their value to waterfowl and other wildlife, Circ. 39. U.S. Fish and Wildlife Service, Washington, DC. 67 pp.

SIMBERLOFF, D. and ABELE, L.G. (1982). Refuge design and island biogeographic theory. Am. Nat. 120(1):41-50.

STAKHIV, E.Z. (1988). An Evaluation Paradigm for Cumulative Impact Analysis. U.S. Army Corps of Engineers, Institute for Water Resources, Fort Belvior, VA. 69 pp.

SUHRBIER, J.H. and DEAKIN, E.A. (1988). Environmental considerations in a 2020 transportation plan. In: *A Look Ahead, Year 2020: Proceedings of the Conference on Long-Range Trends and Requirements for the Nation's Highway and Public Transit Systems*, pp. 232-259. Transportation Research Board, National Research Council, Washington, DC.

SWANSON, G.A. (1979). The Mitigation Symposium: A National Workshop on Mitigating Losses of Fish and Wildlife Habitats, Gen. Tech. Rept. RM-65. U.S. Department of Agriculture, Rocky Mountain Forest and Range Experiment Station, U.S. Forest Service, Fort Collins, CO. 696 pp.

TERRY, R.E. (1986). Nitrogen transformations in Histosols. In: *The Role of Organic Matter in Modern Agriculture* (Y. Chen and Y. Avnimelech, eds), pp. 55-70. Martinus Nijhoff Publishers, Dordrecht, The Netherlands.

TINER, R.W. (1984). Wetlands of the United States: current status and recent trends. U.S. Fish and Wildlife Service, Habitat Resources, Newton Corner, MA. 59 pp.

TRANSPORTATION RESEARCH BOARD. (1984). America's Highways, Accelerating the Search for Change, TE 1 N38 n. 202. Transportation Research Board, National Research Council, Washington, DC. 169 pp.

VOGL, R.J. (1980). The ecological factors that produce perturbation-dependent ecosystems. In: *The Recovery Process in Damaged Ecosystems* (J. Cairns, Jr., ed), pp. 63-94. Ann Arbor Science Publishers Inc., Ann Arbor, MI.

WALKER, S. (1988). A review of past mitigation projects in Florida. In: *Proceedings of the National Wetlands Symposium: Mitigation of Impacts and Losses*

(J.A. Kusler, M.L. Quammen, and G. Brooks, eds), pp. 256-258. Association of State Wetland Managers, Berne, NY.

WALLACE, B. (1981). Basic Population Genetics. Columbia University Press, NY.

WELLER, M.W. (1988). The influence of hydrologic maxima and minima on wildlife habitat and production values of wetlands. In: *Proceedings of the National Wetland Symposium: Wetland Hydrology* (J.A. Kusler and G. Brooks, eds), pp. 55-60. Association of State Wetland Managers, Berne, NY.

WELLER, M.W., ZEDLER, J.B., and SATHER, J.H. (1988). Research needs for better mitigation: Future directions. In: *Proceedings of the National Wetlands Symposium: Mitigation of Impacts and Losses* (J.A. Kusler, M.L. Quammen, and G. Brooks, eds), pp. 428-430. Association of State Wetland Managers, Berne, NY.

WETZEL, R.G. (1983). Limnology. Saunders College Publishing, Philadelphia, PA. 858 pp.

WHARTON, C.H., KITCHENS, W.M., PENDLETON, E.C., and SIPE, T.W. (1982). The Ecology of Bottomland Hardwood Swamps of the Southeast: A Community Profile, FWS/OBS-81/37. U.S. Fish and Wildlife Service, Biological Service Program, Washington, DC. 133 pp.

WHITCOMB, R.F., ROBBINS, C.S., LYNCH, J.F., WHITCOMB, B.L., KLIMKIEWICZ, M.K., and BYSTRAK, D. (1981). Effects of forest fragmentation on avifauna of the eastern deciduous forest. In: *Forest Island Dynamics in Man-Dominated Landscapes* (R.L. Burgess and D.L. Sharpe, eds), pp. 125-205. Springer-Verlag, NY.

WIGINGTON, P.J., Jr., RANDALL, C.W., and GRIZZARD, T.J. (1983). Accumulation of selected trace metals in soils of urban runoff detention basins. Water Resour. Bull., 19(5):709- 717.

WILCOX, B.A. and MURPHY, D.D. (1985). Conservation strategy: the effects of fragmentation on extinction. Am. Nat. 125(6):879-887.

WILLARD, D.E. and HILLER, A.K. (1990). Wetland dynamics: Considerations for restored and created wetlands. In: *Wetland Creation and Restoration: The Status of the Science* (J.A. Kusler and M.E. Kentula, eds), pp. 459-466. Island Press, Washington, DC.

WILSON, E.O. (1988). The current state of biological diversity. In: *Biodiversity* (E.O. Wilson, ed), pp. 3-18. National Academy Press, Washington, DC.

WOLF, R.B., LEE, L.C., and SHARITZ, R.R. (1986). Wetland creation and restoration in the United States from 1970 to 1985: an annotated bibliography. Wetlands: Special Issue.

WYNNE, B. (1987). Hazardous wastes risk management and environmental regulation. In: *Risk Management and Hazardous Waste, Implementation and the Dialectics of Credibility* (B. Wynne, ed), pp. 23-44. Springer-Verlag, NY.

YEAR 2020 PANEL. (1988). Population Growth and Development in the Chesapeake Bay Watershed to the Year 2020. The Report to the Year 2020 Panel, Washington, DC. 52 pp.

ZALEHA, D.B. (1988). A guide to federal wetlands protection under Section 404 of the Clean Water Act. Anadromous Fish Law Memo, National Resources Law Institute, Lewis and Clark Law School, Portland, OR. 46:1-34.

CHAPTER 13

USE OF EXPERIMENTAL STREAM MESOCOSMS TO ASSESS ECOSYSTEM RISK: A CASE STUDY WITH A CATIONIC SURFACTANT

Scott E. Belanger
Environmental Toxicology and Stream Ecosystem Research Section
Environmental Safety Department
The Procter and Gamble Company
Cincinnati, Ohio

INTRODUCTION

One consequence of the need for validation in aquatic risk assessment has been the increasing use of large scale experimental test systems such as mesocosms. Mesocosms are units that contain essential ecosystem structural and functional components, multiple trophic levels, and are large enough to sample repeatedly without altering the system. Odum (1984) and Sanders (1982) give features of such systems. The use of a mesocosm may be necessary in some situations to "negate the presumption of risk" as a result of unrealistic responses observed in sensitive single species aquatic toxicity tests (Touart 1988).

Sanders (1985) provides a lucid discussion of the definitions associated with the term *field validation*. In the recent history of aquatic toxicology, validation has become a loosely applied term, used to support various forms of extrapolating single species data for prediction of risks of chemical exposure to ecosystems (Cairns et al. 1988). For validation to be a useful concept to aquatic toxicologists, Cairns (1988) argues that crisp definitions of the term must be applied along with strong experimental evidence. Sanders (1985) states that scientific field validation is a comparison of the overall results of a method, toxicity test, or model with field data in order to determine the degree of correspondence. Validation requires the use of pre-selected acceptability criteria to facilitate decision making.

MESOCOSM EXPERIMENTS AND ENVIRONMENTAL REALISM

A suite of problems exists in using single species data to protect and predict effects of chemicals on ecosystems. Among these are the sensitivity of the

1. Corresponding Author: Scott E. Belanger, Environmental Toxicology and Stream Ecosystem Research Section, Environmental Safety Department, The Procter and Gamble Company, Cincinnati, OH 45217.

2. Key Words: cationic, mesocosm, risk assessment, stream, surfactant, toxicity.

taxa tested (Cairns 1986), the delivery of the test chemical to the organism (i.e., the appropriate bioavailable form) (Versteeg and Woltering 1990), choice of life stages to be tested (Suter et al. 1987), and the ability to assess interactions among taxa within and between different trophic levels. Such considerations have led to testing of some chemicals, such as pesticides, under the Federal Insecticide, Fungicide, and Rodenticide Act in pond mesocosms (see Touart 1988). Mesocosm testing in pond ecosystems is designed to negate or prove the presumption of ecosystem risk associated with the use of pesticides in the real world. This is accomplished by providing a test environment that is more dependable and predictive than traditional single species bioassays.

The importance of the different limitations of single species tests to predict ecosystem level risk will vary depending on the chemical. For example, selenium, which is moderately toxic, exerts its most profound effects on centrarchid fish (e.g., bass and bluegill) through the food chain (i.e., bioaccumulation). Selenium testing in stream mesocosms at the United States Environmental Protection Agency (USEPA) Monticello facility confirmed this fact (Hedtke et al. 1990). Some materials for which aquatic toxicity data are desired may be difficult to evaluate using single species assays due to complex chemistry, unrealistic bioavailability, or inappropriate dose form. Evaluations of toxicity for readily degradable metabolizable or rapidly hydrolyzing materials present problems with common exposure methods such as static renewal or pulsed flow through (e.g., proportional dilutors).

Recently, mesocosm testing has been invoked for evaluating chemical fate, effects, and exposure for a variety of reasons. These have included recognition that:

(a) single species tests cannot assess biological interactions such as predation and competition;
(b) single species tests are unable to evaluate fundamental ecosystem processes such as nutrient cycling or spiralling, and energy flow;
(c) the chosen single species used in a risk assessment may not represent an important link in a functioning ecosystem;
(d) mesocosms can appropriately evaluate key elements of system assimilative capacity such as the influence of water quality (e.g., organic carbon, alkalinity, fluctuating environmental conditions) on test chemical availability;
(e) mesocosms can be used to determine ecosystem structural and functional redundancy (here redundancy refers to the ability of organisms to fill ecological niches or roles previously held by organisms no longer existing in the ecosystem as a result of natural or artificial perturbations; Cairns and Dickson 1977).

Because mesocosm tests should be (and usually are) designed to be comprehensive in nature, they can be used to provide a more comprehensive and accurate assessment and understanding of the safety profile of chemicals that display relatively small safety margins based upon single species testing. Therefore, a mesocosm test is triggered by a fairly substantial data base for the chemical of interest. Acute and chronic toxicity of the chemical to selected algae, invertebrates, and fish should be known in order to identify organisms and environmental compartments on which to focus. Physical and chemical characteristics that may impact the environmental fate of the material require quantification (e.g., solubility, sorptivity, volatility, etc.) to determine the relevant compartments to investigate. Also, an understanding of the expected exposure based upon knowledge of the chemical fate, use, biodegradability, and treatability of the material should precede mesocosm testing. Given the above baseline data, if the margin between predicted biological effect levels and environmental exposure is small or the uncertainty in the data base is high (such as relevance of key single species assays to the predicted exposure), then a mesocosm test may be warranted.

Mesocosms have been used when the social costs (e.g., potential of widespread ecological harm) or monetary risks associated with the chemical use are high. Relative to this latter reason, the cost of repairing damaged systems may be far greater than the cost of ensuring that use of a chemical will be safe. Finally, an oft-unspoken reason for performing mesocosm tests is to discover the fundamental ecological processes of stressed systems (Odum 1984).

The following key elements should be present in order to use mesocosm tests successfully to predict ecosystem risk or to further the understanding of stressed systems.

(a) Mesocosms should reflect essential elements of ecosystem structure and function. Nutrient cycling and energy flow (e.g., primary and secondary productivity, ecosystem respiration) in the mesocosm should be representative of real world systems. Mesocosms are more than a pool of species with interactions that are evaluated using a very limited number of end points. The whole (ecosystem) is greater than the sum of the individual parts (species). A weakness of single species tests used to predict long-term ecosystem level processes is the ability to choose end points that clearly relate to those processes. Finally, structural and functional components of mesocosms and the end points measured should be representative of the type of system being protected.

(b) The system should support a diverse assemblage of aquatic life so that a wide range of organism sensitivities can be evaluated. Sufficiently detailed taxonomic understanding should be employed (i.e., to the lowest practical taxonomic level) to make this assessment possible.

(c) Multiple trophic interactions and pathways should be present. In many instances, it may be desirable to include a fish component, especially if fish are uniquely sensitive to the chemical. If fish are not used in the test system, a clear ecological rationale that would lead to a fish trophic level should be present (e.g., knowledge of invertebrate biomass and production). At the minimum, microbial heterotrophic (bacteria, fungi, protozoa) and autotrophic (algal) communities must be present along with macroinvertebrates (e.g., insects, molluscs).

(d) A mesocosm test should include enough exposure concentrations to establish exposure(dose)-response relationships. Modelling these relationships may be difficult because of the complexity of the test system, but, without a sufficient range of exposures, extrapolation to protect ecosystems would be difficult. An additional point is the trade-off that occurs between pseudo-replication of test systems (Hurlburt 1984), the intensity and cost of sampling effort (made higher in direct proportion to the amount of true replication), and developing adequate dose-response information. Psuedo-replicated streams are vulnerable to unknown confounding variables, but these are unlikely to affect streams in a dose-response manner.

(e) The mesocosm system must be progressively sampled enough such that temporal trends or ecological trajectories can be established. However, if the system is oversampled, effects of chemical exposure may become obscured as the system responds to sampling disturbance.

(f) Mesocosm tests can only be successful if each test is designed using an *a priori* set of hypotheses that are appropriately evaluated (e.g., trends expected in stressed ecosystems [Odum 1984]).

THE PROCTER & GAMBLE EXPERIMENTAL STREAM FACILITY PROGRAM

Procter & Gamble, as a consumer products company, markets a diverse array of products including soap and detergents. Large volumes of these products and ingredients are initially disposed of "down the drain," thereby reaching municipal wastewater treatment facilities (Woltering and Bishop 1989). The portion that does not biodegrade, adsorb to solids, or volatilize may reach surface water as a component of treated wastewater effluent. In order to ensure the environmental safety of such materials, the detergent industry has been broadly involved in environmental risk assessment.

Predictions of ecosystem risk from single species toxicity data, along with predictions of environmental exposure, form the basis for decisions on chemical manufacture made by regulatory bodies (e.g., USEPA 1984). However, predictions necessarily involve an element of uncertainty. Uncertainty may arise from the relationship of single species data, for example, to predict multiple trophic interactions present in real world systems. The use of mesocosm tests in ecosystem risk assessment is spurred by the following concepts, most of which are not unique to consumer products.

(a) There is a need to reduce uncertainty in extrapolating single species laboratory data to the field.

(b) Mesocosms allow scientists to integrate chemical fate, exposure, and effects in a single test unit with sufficient complexity to approximate natural ecosystems.

(c) Understanding mesocosm structure and function can guide and develop the science of effects assessment and extrapolation.

Clearly, mesocosm testing will not replace the need for single species laboratory data, which comprise early tiers of the risk assessment process and are required for global regulatory filings, nor confirmatory field data (i.e., long-term monitoring).

For chemicals that have high safety margins due to inherently low toxicity or low levels of environmental exposure, single species tests may be entirely adequate to establish safety. Mesocosm tests are conducted only where the need is clear because mesocosm tests are expensive and time intensive studies.

Experimental Stream Facility Structure and Operation

The Procter & Gamble Experimental Stream Facility (ESF) was conceived in the mid-1980s (Barnum and Woltering 1989) as a special tool to evaluate various consumer product ingredients. The facility is designed to incorporate relevant aspects of the environmental release of consumer product chemicals to surface waters. Streams are one type of system that is of interest to the consumer product industry because of the volume of municipal wastewater discharged into aquatic environments (Rapaport 1988). The ESF experimental streams receive river water continuously supplied from a high quality stream and a treated wastewater at a fixed v:v percentage. A test chemical is added at selected concentrations to stream channels. The ESF is located near the Lower East Fork of the Little Miami River (LEFR) which is designated as an exceptional warmwater fishery by the Ohio EPA (1987). It is adjacent to the Clermont County (Ohio) Lower East Fork Wastewater Treatment Plant (WWTP). Additional background can be found in Belanger et al. (in press).

Conceptually, the ESF was designed and sited at a location to ensure that the key elements of a mesocosm would be present as discussed earlier. In order for the ESF to be a valid test system, it must incorporate essential elements of ecosystem structure and function as well as support diverse aquatic assemblages with multiple trophic linkages. The experimental design incorporates exposure-response determinations based on rigorous sampling directed by trends expected in stressed ecosystems.

The ESF is a partially environmentally controlled facility that combines aspects of computer control over key parameters with elements of natural variability of the LEFR and WWTP. Photoperiod (duration and timing), flow rate, and stream physical structure are controlled; whereas, natural colonization of biota and water quality reflect LEFR conditions (see Barnum and Woltering 1989; Belanger et al., in press; Cuffney et al. 1990; McCormick et al. 1991). Figure 1 is a diagrammatic representation of the ESF environment.

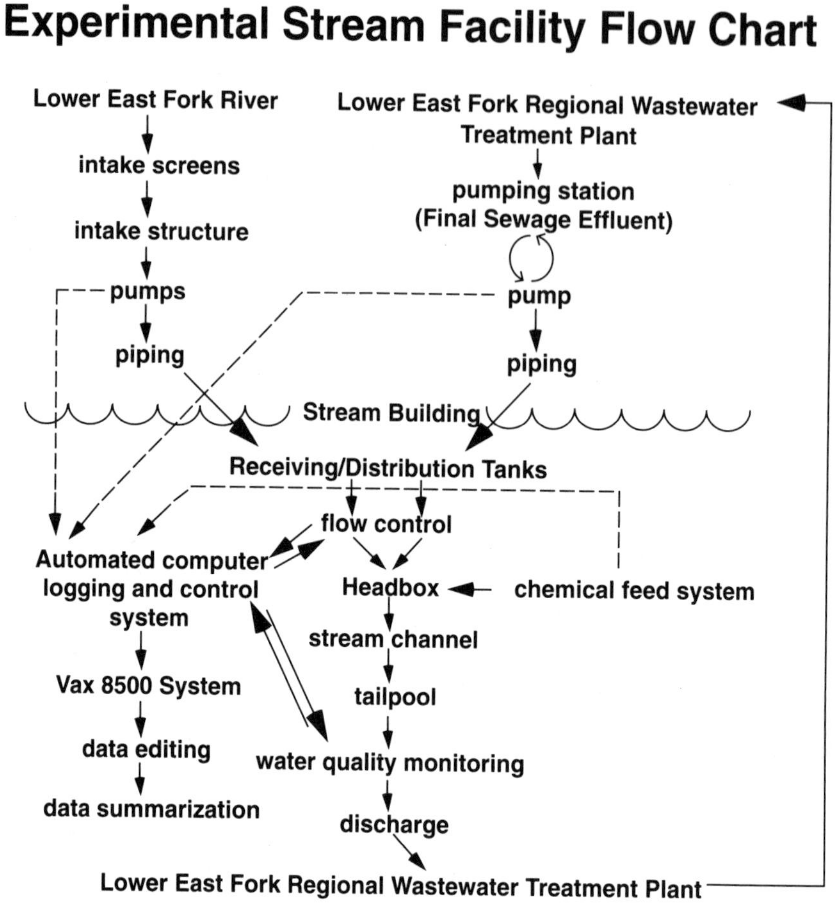

Figure 1. Flow diagram of the Experimental Stream Facility (ESF) environment. Water from the Lower East Fork of the Little Miami River (LEFR) is accessed by a shoreline intake structure that distributes water to the ESF. Water is returned to the wastewater treatment facility after use in the ESF.

The ESF environment is controlled by an on-site automated computer data acquisition and logging system. Computer control exists for announcing catastrophic system failures by an alarm system, flow to each stream and for each water source, and overhead lighting. Water quality (pH, dissolved oxygen, conductivity, temperature) at each stream and ESF water source and light sensors located throughout the building track environmental conditions. Each stream is equipped with separately controlled test chemical feed systems. Test concentration chemicals are tracked using specific analytical techniques.

Stream channels consist of four sections: a headbox (142 L) region (prior to which river and wastewater and chemical are thoroughly mixed); a 4.3-m long microbial study region (1% slope, 0.29 m wide, paved with clay tile); a flared region that leads to a 4.3-m long invertebrate study region (5% slope, 0.53 m wide, containing invertebrate trays described below), and a 1.2-m long tail pool (568 L; sensors for water quality are located here). The ESF is fed LEFR water for use in experimental stream channels (up to eight are available, although only four are discussed in the following sections). Flow to each channel is 167 L/min with 10% by volume as treated final wastewater. A complete discussion of these components is given in Belanger et al. (in press) for periphyton and in Cuffney et al. (1990) for invertebrates. The interior of the ESF with stream channels, monitoring stations, and water distribution networks are shown in Figure 2.

Figure 2. The Experimental Stream Facility building interior features eight stream channels, large volume headtanks that receive treated sewage effluent or river water, flow and lighting control, and environmental monitoring.

Experimental Stream Facility Experimental Design

The ESF program is given in Figure 3. Biotic components (algae, heterotrophs, protozoa, invertebrates) are evaluated at the individual, population, and community levels using traditional and non-traditional methods (see Appendix). Each stream receives one of a logarithmic chemical exposure series. The choice of end points used in the program reflects the state of the science and a desire to develop complementary data that could provide scientific checks and balances (i.e., reduce the likelihood of false positive and negative findings). We actively embrace the use of guiding ecological principles, such as those given in Sanders (1982), Odum (1984), and Kimball and Levin (1985), to evaluate ecosystem risk. Such *a priori* hypothesis testing is being developed for current ESF programs.

Because of the complexity of mesocosm data, it is necessary to follow an ecologically based end point analysis scheme. Mesocosm tests are, or should be, more than a physical structure in which to generate some singular "litmus test" as a surrogate for all ecosystem effects end points (a common approach used by many investigators is to apply a single measure upon which to base a statement of entire ecosystem health). Alternatively, we endorse the following generic scheme to collate, integrate, and understand findings from mesocosm tests.

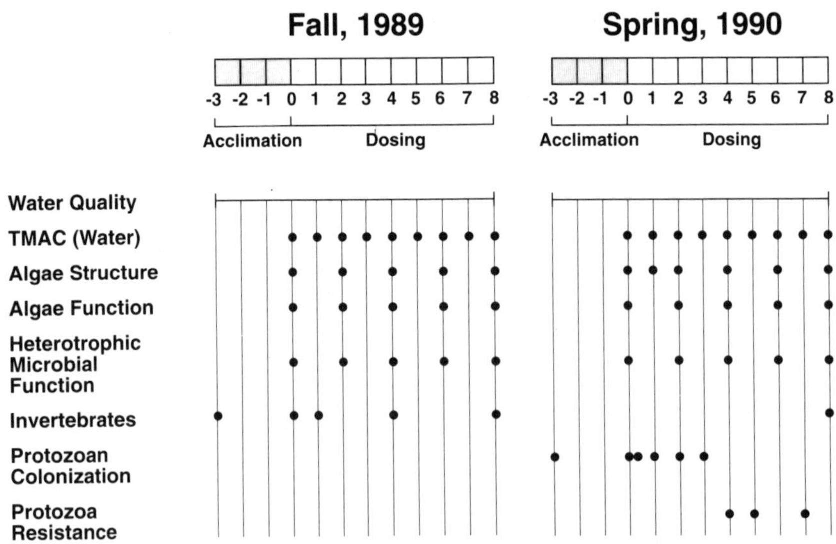

Figure 3. Experimental sampling and phases of the C_{12}TMAC studies in fall, 1989 (August 21-November 9). and spring, 1990 (March 27-June 11). Selected water quality parameters were continuously monitored. TMAC was evaluated weekly. Algal and heterotrophic periphyton structure and function, invertebrate community responses, and protozoan colonization on barren substrates and mature substrates (resistance) were evaluated throughout the exposure (dosing) period.

If a change in an end point is detected (relative to the control) for a sample during an ESF study, a systematic and objective evaluation scheme is applied to determine if a true ecosystem-level adverse effect for the chemical is achieved. To be considered an adverse effect, an end point must be both statistically and ecologically significant. Community no-observed-effect-concentrations (NOEC) for heterotrophic periphyton, algal periphyton, protozoans, and invertebrates are then defined based on these findings, which are then available for integration into a single NOEC for the ecosystem. Finally, the environmental risk is determined by comparing the ecosystem NOEC with expected environmental concentrations. This scheme provides assurance against producing "false positive" or "false negatives" for chemical treatment effects.

If a change in an end point is detected for a parameter during an ESF study, the following series of questions are posed to determine if a true ecosystem-level adverse effect exists for the test chemical.

(a) Is the change statistically significant? Samples are sufficiently replicated within each stream during a study to allow for proper hypothesis testing to determine if possible adverse responses observed in a dosed channel are significantly different from the undosed condition (control). This approach minimizes the possible interpretation of "effects" due to natural variation inherent in all biological communities. If streams or test systems are pseudo-replicated, it should be recognized that the interpretation of collected information is not confined to determining specific treatment level effects compared to a control. This is only the first step in the analysis. Because it is unknown at what specific exposure concentration effects may be observed, as well as the relative importance or consistency of the chosen end points, the analysis proceeds with the questions below.

(b) Is the observed effect ecologically consistent with other related or complementary end points evaluated within a given stream channel? This is the step that benefits from having the array of end points evaluated in a *coordinated* and *integrated* manner. The integrated approach allows ratios of end points that have ecological relevance (e.g., photosynthesis/heterotrophic activity ratio) to be evaluated on each experimental unit; these can then be statistically tested. Coordinated sampling allows periphyton and invertebrate end points to be compared simultaneously. For instance, if algal biomass declines, but photosynthetic production remains at the same level, then one should also evaluate the possible effect on a higher trophic level (e.g., invertebrates) to determine the real ecosystem implications of this change.

Coordination of sampling, both temporally and spatially, to evaluate periphytic and invertebrate responses allows direct comparisons of effects of the test substance on these communities. Further, we evaluate communities developed on several relevant substrates (e.g., clay tiles, rock cobble, and polyurethane foam) to investigate the ecological consistency of responses between these communities.

In addition to integrated and coordinated sampling strategies, the array of end points are selected to allow for internal checks for ecological consistency (e.g., algal biomass using chlorophyll, ash free dry weight, and biovolume).

Step 2 of this end point analysis goes beyond an evaluation of individual end points within the context of others. Instead, a community level evaluation is made. Beyond this point, then, individual end points are no longer considered.

(c) Is the community effect a function of test chemical dosing? That is, is it consistent between streams in a dose-response fashion? There may be effects observed for a given stream that may not be directly related to chemical toxicity. The dose-response experimental design for the ESF project requires that any true treatment effect would have significant community effects in more than one stream, and at levels consistent with the amount of added chemical.

(d) Is the effect consistent over the exposure period, and is it progressive? Sampling over an 8-week dosing period attempts to ascertain the importance of transient or developmental changes in the communities that may not be consistent with long-term treatment effects. For instance, temporal analyses provide a means to evaluate the natural development of periphytic communities through time and allow a proper interpretation of transient or progressive community changes that will not ultimately result in adverse ecosystem consequences (e.g., algal growth and bacterial adaptation to test chemical or biodegradation).

If the answers to all the questions (a-d) above are "yes," we conclude that the observed effects are real and ecologically relevant. The above process can result in defensible community NOECs for heterotrophic periphyton, algal periphyton, protozoans, and invertebrates, which will be used for additional evaluation.

ESF TMAC STUDIES AS A CASE EXAMPLE

Experimental Design

During fall 1989 and spring 1990, two 8-week experimental evaluations of the cationic surfactant C_{12}-trimethyl ammonium chloride (TMAC) were undertaken at the ESF. The chemical used in this case has an alkyl chain 12 carbons long and belongs to the chemical class known as monoalkyl quaternary ammonium compounds (MAQs). TMAC is highly water soluble (>1000 mg/L), possesses low bioaccumulation potential, and is highly biodegraded in wastewater treatment (95% in activated sludge) (Woltering and Bishop 1989). TMAC was chosen because of its well demonstrated environmental profile (fate and effects) to which ESF mesocosm responses could be readily compared, including laboratory and field data.

In order for the integration of the chosen biological and chemical end points into community or ecosystem NOECs to be meaningful, stream channels must be sufficiently chemically and biologically characterized (Fig. 3). Selected water quality parameters were continuously monitored. In-stream TMAC concentrations were evaluated weekly. A 3-week acclimation and colonization phase preceded an 8-week chemical exposure. During the colonization phase, stream channels received a 10% final WWTP effluent/90% LEFR water at the flow rate to be used during the 8-week exposure. TMAC was not added during acclimation and colonization. Periphyton, protozoans, and invertebrates were colonized or acclimated using the methods described below (see also Belanger et al., in press; Cuffney et al. 1990; McCormick et al. 1991). Biological evaluations of algal and heterotrophic periphyton, invertebrate and protozoan communities, and a suite of endemic single species assays were developed. Each of these areas, with key findings, are presented below.

Water Chemistry
Belanger et al. (in press) summarized water quality monitoring and specific analytical findings in 1989 and 1990 experiments. Background water quality was not affected by TMAC exposure at 50, 250, and 1250 μg/L. TMAC was successfully added to experimental streams based on measured concentrations in Table 1. Measured concentrations of TMAC in streams were in closest agreement with nominal (theoretical) concentrations in fall 1989. Greater disagreement between actual and nominal exposures in the spring 1989 study probably reflects greater total suspended solids levels in the spring versus fall study (TMAC is a sorptive cationic chemical).

Periphytic Algae
A complete discussion of algal responses to TMAC in ESF stream channels is given in Belanger et al. (in press). Autotrophic (algal) and heterotrophic (bacteria, fungi) periphyton (microbes attached to submerged substrata) colonized on tile substrata were evaluated several times during each experiment. Algal communities were extremely diverse with nearly 300 taxa found during the two studies. Figure 4 gives the hypothetical food web

TABLE 1
Theoretical and actual concentrations of TMAC measured in the final pool of ESF streams during fall, 1989 and spring, 1990 experiments. Means ±1 SD are indicated.

Theoretical Concentration (μg/L)	Fall, 1989	Spring, 1990
0 (control stream)	12.9 ± 12.2	0.1 ± 0.1
50	49.7 ± 11.3	43.2 ± 16.9
250	234.8 ± 56.9	185.1 ± 67.2
1250	1151 ± 124	—
LEFR*	—	0.7 ± 1.1
100% effluent	—	0.4 ± 0.1

*Lower East Fork of the Little Miami River.

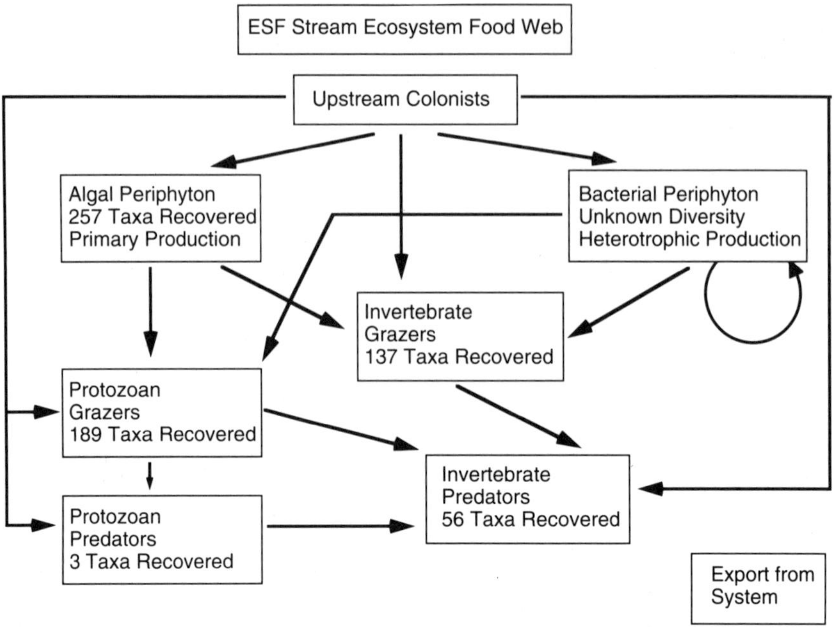

Figure 4. Theoretical trophic linkages in the Experimental Stream Facility (ESF) stream channel mesocosms. Trophic classifications for individual taxa were determined using scientific literature. A total of 642 taxa were identified during the C_{12}TMAC studies.

relationship for aquatic organisms found in ESF stream channels, including the position of algae in the system. Algal diversity was not affected at any concentration. Algal abundance, depending on the parameter (biovolume, cell density, chlorophyll *a*, dry weight), was inconsistently affected. We concluded, based upon chlorophyll *a* biomass, diversity, and community similarity that algal community structure was preserved at 250 μg/L and lower. Algal community function, as indicated by primary productivity on an aerial and biomass specific basis, also indicated a NOEC of 250 μg/L (Fig. 5).

Heterotrophic Periphyton

Heterotrophic periphyton structure (e.g., acridine orange direct bacterial counts, dry weight, protein, carbohydrate) and function (bacterial growth rates, mixed amino acid turnover rate, TMAC turnover rate) were evaluated by standard techniques (Belanger et al., in press; Schermehorn 1990; Schwab et al., in press). Radiotracer techniques were used for function assays. Mixed amino acid and TMAC turnover rates decreased several fold during the course of exposure and were essentially equivalent at 50, 250, and 1250 μg/L (half lives of approximately 5-10 hr). Control turnover rates through the test remained at the initial levels (>1000 hr). Heterotrophic periphyton structure and function (mixed amino acid turnover) were not affected through the highest exposure at 1250 μg/L.

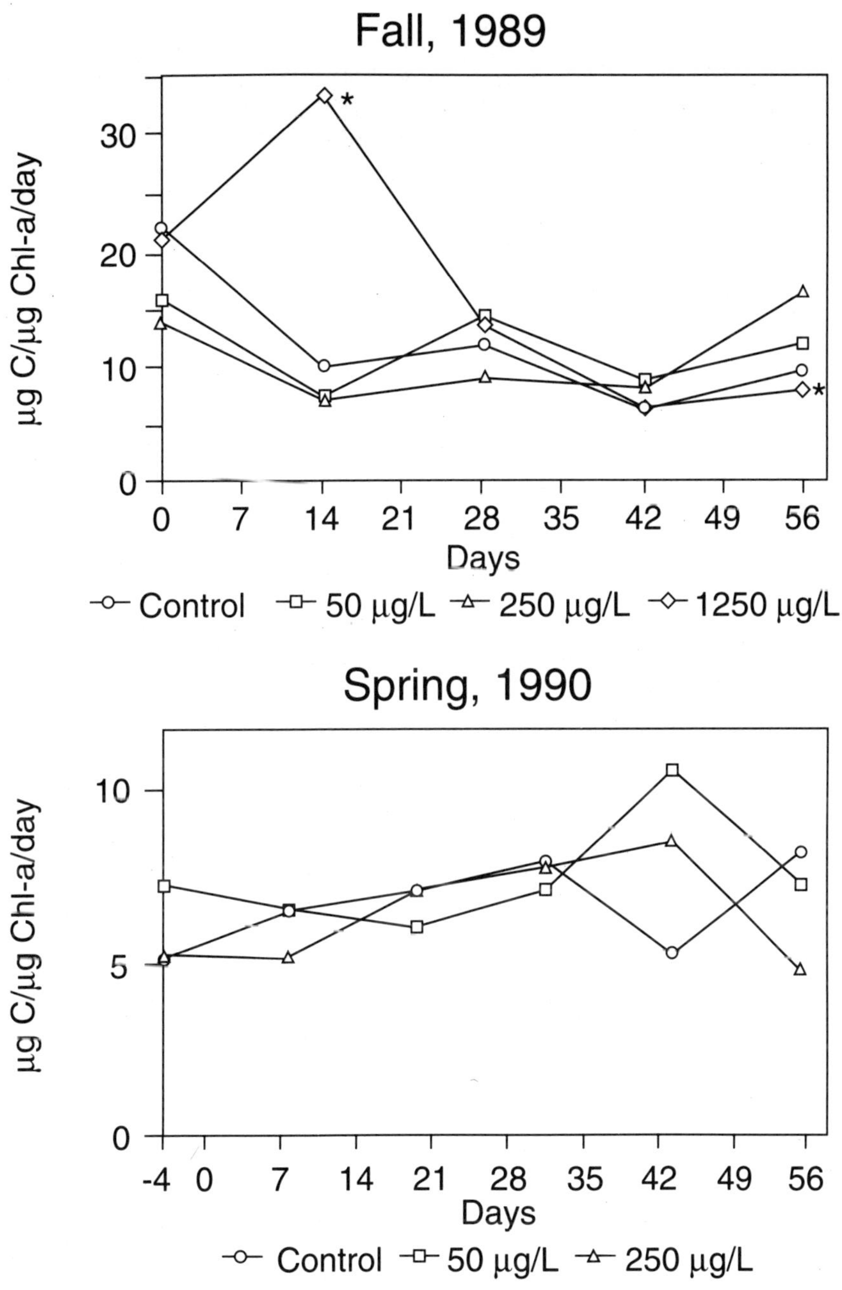

Figure 5. Algal periphyton biomass specific (chlorophyll-a) primary productivity during fall, 1989 and spring, 1990 studies. Treatments that are significantly different from the control are indicated by an asterisk (*, $p<0.05$).

Invertebrate Communities
Invertebrate communities were established in stream channels using either field or mesocosm colonization methods, both of which utilize a tray (approximate dimensions of 26.0 cm long X 15.9 cm wide X 6.4 cm deep) system. Each stream channel accommodated 45 trays. For field colonization, trays were incubated in the source river for 40 days in summer, 1989. They were transferred to the ESF 21 days prior to dosing to allow communities to acclimate to the new environmental conditions. Cuffney et al. (1990) reported this was a suitable means to simulate source river communities. Mesocosm colonization is also suitable (Belanger, S.E., R. G. Bausch, and K. L. Rupe, unpublished P&G data) when trays are allowed to colonize entirely within the stream channels. This method was used in spring, 1990.

The invertebrate communities in the ESF were not affected relative to organism density at 1250 μg/L, but diversity was affected slightly during the fall 1989 experiment when trays were initially colonized in the source river. The numerically dominant invertebrate belonged to the family Naididae, an aquatic oligochaete. Although numerically abundant, this group did not account for much of the invertebrate biomass. Naididae were most abundant at 1250 μg/L, resulting in significantly lower diversity at this concentration (reduced 14% relative to the control, Cuffney et al. 1990). When considering all other taxa, except Naididae, diversity was not affected (Fig. 6). Cuffney et al. (1990) showed drift to be sensitive to TMAC exposure with a NOEC of 250 μg/L. Some populations were affected at 250 μg/L, but trophic feeding group structure was preserved at 250 μg/L. Overall, invertebrate communities tended to be dominated by sediment inhabiting taxa, but mayflies, stoneflies, and trichoptera remained at 10% or greater in abundance through each study. Trophic complexity, including predator taxa, was present at all test concentrations, including the highest test concentration (Table 2).

In the spring study, trays were colonized and exposed solely in the ESF environment. Consistent with the fall study, diversity and density were not affected by TMAC at concentrations up to 250 μg/L (a 1250-μg/L treatment was not included in the spring experiment). Trophic complexity and diversity were higher than in the fall 1989 study. The data clearly demonstrate a broad lack of effects at 250 μg/L and a lack of dose-response.

Protozoan Communities
Protozoan communities were evaluated using methods generally developed by Cairns and associates (Cairns et al. 1979; Pontasch et al. 1989; Pratt and Bowers 1990). Effects upon protozoan communities in ESF stream channels during the 1990 study were summarized by McCormick et al. (1991). The impacts of C_{12}TMAC exposure upon protozoan colonization processes of barren island substrates and protozoan community resilience on mature substrates were tested. Protozoan communities were diverse (Fig. 4) and trophically complex (communities consisted of bactivores, saprotrophs, algivores, and predators; Pratt and Cairns 1985). Trophic

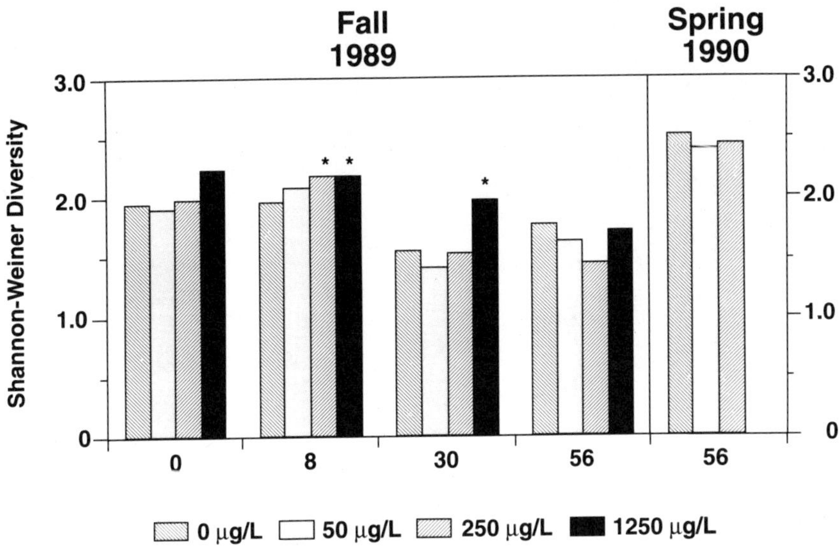

Figure 6. Invertebrate diversity in stream channel mesocosms. The most numerically abundant invertebrate group was Naididae. These organisms were not affected by TMAC exposure. Naidid biomass was small compared to other invertebrate components. For comparison, Naididae were deleted from this analysis. Groups that are significantly different from the control are indicated by an asterisk (*, $p<0.05$).

structure, taxonomic richness, colonization process, and resilience were not significantly affected at 250 µg/L (Fig. 7). In addition to protozoan colonists, the polyurethane foam units (PFUs) provided another substrate to evaluate diatom community structure. As observed with clay tile substrata, diatom taxon richness was not affected (Fig. 7).

TABLE 2
Relative proportions of invertebrate trophic feeding groups (without Naididae) following exposure to C_{12}TMAC during spring, 1990 experiments. Samples are from day 56. Groups that are significantly different from the control are indicated by an asterisk (*, alpha = 0.05, n = 18).

	Exposure Concentration (µg TMAC/L)		
Invertebrate Group	0	50	250
Collector-Filterer	16.8 ± 1.4	16.7 ± 1.5	14.9 ± 1.9
Collector-Gatherer	49.4 ± 2.6	52.3 ± 2.5	48.0 ± 2.8
Predator	9.7 ± 1.2	5.4 ± 0.6*	7.9 ± 1.0
Scraper	3.7 ± 0.5	5.6 ± 0.6	7.3 ± 1.1
Shredder	6.2 ± 0.6	5.3 ± 0.6	5.2 ± 0.6
Unknown	14.1 ± 1.1	14.5 ± 1.0	16.6 ± 1.4

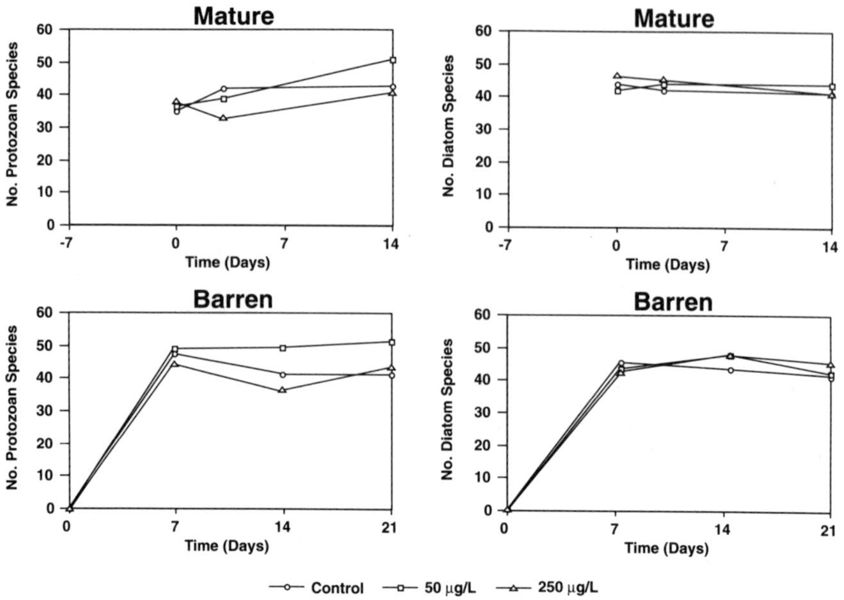

Figure 7. Protozoa and diatom taxa richness on barren foam substrates placed in the channels at the initiation of dosing and on mature substrates colonized for 14 days prior to dosing.

Single Species Bioassays
Ceriodaphnia dubia (a daphnid), fathead minnow (*Pimephales promelas*), and Asiatic clam (*Corbicula fluminea*) bioassays were tested in order to benchmark against community level investigations. *Ceriodaphnia dubia* and Asiatic clam assays employed standard techniques for these species (USEPA 1989; Belanger 1991; Belanger et al., in review). The fathead minnow exposures consisted of extended exposures (2 weeks) to young (14-day old) juveniles by a protocol similar to the protocol of the Organization for Economic Cooperation and Development (OECD 1981). Clams were the most sensitive taxon tested in the ESF (Table 3).

COMPARISON OF ESF ECOSYSTEM AND SINGLE SPECIES DATA

Woltering and Bishop (1989) summarized earlier data on MAQs, of which TMAC is a member. Based on their summary, environmental concentrations in surface waters under low flow would not exceed 80 ug MAQ/L, and, in 90% of US surface waters, concentrations would be <30 µg/L at low flow in the mixing zone.

Figure 8 summarizes single species effect levels from C_{12}TMAC exposure (left side) compared to ESF responses (right side). The range of known

TABLE 3
Chronic effects of C_{12}TMAC on single species in the ESF environment.

Organism	Exposure Duration	Test Type	End Point	NOEC (µg/L)
Ceriodaphnia dubia	7 day	Static renewal[1]	Reproduction	125
	7 day	Static renewal[2]	Reproduction	500
	7 day	Static renewal[3]	Reproduction	250
Fathead minnow	14 day	Flow-through[4]	Growth	>1250
			Survival	>1250
Asiatic clam	28 day	Flow-through[5]	Growth	250
Asiatic clam	49 day	Flow-through[6]	Growth	250
Asiatic clam	56 day	Flow-through[7]	Growth	250

[1]TMAC tested without effluent.
[2]TMAC tested with effluent.
[3]TMAC tested with effluent.
[4]Aquarium exposure using outflow from the final tail pool.
[5]Test conducted in the stream channel, Ohio population.
[6]Test conducted in the stream channel, Virginia population.
[7]Test conducted in the stream channel, California population.

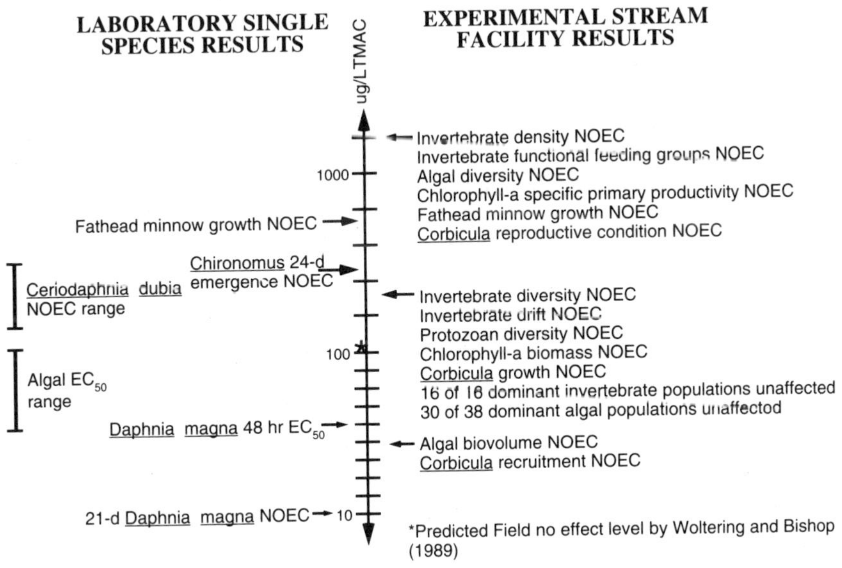

Figure 8. Comparison of acute and chronic single species responses to C_{12}TMAC with stream channel population and community responses in the Experimental Stream Facility. The log-scale vertical line represents C_{12}TMAC concentration in µg/L. Woltering and Bishop (1989) predicted a field no-observed-effect-concentration (NOEC) of 100 µg/L based upon extensive field and laboratory experimental evidence.

single species effects span 10 μg/L (*Daphnia magna* in clear water) to several mg/L (acute effects in river water for insensitive invertebrates such as dragonfly larvae) (Fig. 8). Other previously reported chronic single species responses for algae, invertebrates, and fish span the range of 60-200, 10-350, and 750-1250 μg/L, respectively (Woltering and Bishop 1989). For comparison, ESF responses are given on the right side of the figure for both fall 1989 and spring 1990 studies (the lowest effect level is given for end points that were common to both studies). A large number of end points were not affected at the highest test concentration of 1250 μg/L. These include invertebrate density and changes in feeding group structure and various aspects of algal community structure and function. The true NOEC for these end points is certainly higher than 1250 μg/L given that these responses were essentially identical across concentrations. Other key population and community end points were not affected at 250 μg/L, but were significantly impacted at 1250 μg/L. These include invertebrate and protozoan diversity, invertebrate drift, and most dominant populations (of eight algal populations that were affected, four increased and four decreased with TMAC exposure).

Given the fact that numerous end points are evaluated in such a study, some end points will be significantly affected at the lowest test concentration. As described above, a true ecosystem (mesocosm) level adverse effect requires further interpretation of significant responses in the context of all end points to evaluate ecologically consistent patterns. For example, algal biomass end points, such as dry weight, chlorophyll *a*, and biovolume, provided conflicting data indicating potential effects at 50 μg/L. Further evaluation indicated algal community structure (taxonomy, chlorophyll *a*, and dry weight) and function (primary productivity and biomass specific primary productivity) were not significantly affected at 250 μg/L. Based on the information collected on the diverse array of algal, invertebrate, and protozoan communities in stream channel mesocosms, the NOEC for adverse impacts to the entire system is 250 μg/L.

SUMMARY OF THE EFFECTS OF C_{12}TMAC TO AQUATIC LIFE

These investigations of C_{12}TMAC upon the biota in the ESF were made to allow comparisons with previous studies in laboratory and field systems. Woltering and Bishop (1989) evaluated the effect of TMAC on a model stream system under flow through conditions. Daphnid population structure and density were affected at concentrations above 110 μg/L. Microbial communities adapted to C_{12}TMAC during exposure (Shimp et al. 1989). Both findings are consistent with those presented here. Lewis et al. (1986), Lewis and Hamm (1986), and Lewis (1986) determined effects of C_{12}TMAC to phytoplankton communities enclosed in microcosm vessels *in situ* and to stream algal periphyton in flow through exposure units. Population and community no-effect-levels were 400 to 6100 μg/L for 3-4 week exposures. TMAC exposures to natural stream communities in Howesville Ditch,

Indiana, also resulted in no-effect-levels of 250 μg/L (Lewis et al. 1986). In general, however, single species data are more conservative than effects observed at higher levels of biological organization.

Previous information and ESF studies are consistent, providing assurance that stream channel communities are representative of real world systems. The diversity of the biota and the species distributions, as well as measured functional characteristics of periphyton and invertebrate communities, ensure the stream channel biota are testable, healthy, and representative of a broad range of stream systems.

VALUE AND LIMITATIONS OF STREAM MESOCOSMS

Mesocosms provide a means for experimentally evaluating ecosystem level responses to disturbance and are, therefore, useful in making risk assessments. Mesocosms should not replace field or laboratory based investigations. Laboratory evaluations of toxicity provide a historical, well established data base for comparisons among chemicals. Further, toxicity tests provide a common level of evaluation for regulatory decision making. Field investigations and biomonitoring provide confirmatory evidence for ecosystem changes. However, field evidence is often variable and not easily replicated. Mesocosms provide an experimentally useful bridge between these extremes. Further, the mesocosm provides a system to integrate population, community, and ecosystem level responses that are relevant to real world ecosystems. Because these test systems can be experimentally manipulated, questions pertaining to the predictive value of single species toxicity data can be addressed. For example, recently an article by Okkerman et al. (1991) and a series of related papers have described methods to derive hazardous concentrations (and confidence limits) of chemicals to entire ecosystems based upon limited chronic single species data by statistical extrapolation. The scientific merits of these types of approaches can be best understood by comparing extrapolated predictions with integrated ecosystem responses.

Caution should be exercised when evaluating mesocosm evidence of effects, particularly for scaling or temporal complications. Scaling complications include "container effects" (i.e., spurious findings that are peculiar to the physical test system structure). Generally, mesocosms are designed to model relevant structural and functional aspects of larger and true ecosystems. To provide an environment that can be manipulated, the system must be "scaled down" with the intent that the smaller version retains the required attributes of ecosystems. By investigating the representativeness of the mesocosm relative to the source ecosystem, one can determine the potential impact of scaling on the mesocosm. In the ESF, natural biotic assemblages of the source river are well represented in stream channels. Long-term investigation of the source system will aid the understanding of scaling effects.

In addition to scaling, temporal complexities of the mesocosm should be understood. Ecological or successional trajectories should be known in a test. The mesocosm evaluation should span a period of time long enough to understand effects of chronic chemical exposure. In complex systems, populations may respond by recovering or acclimating to exposure. Certain populations may experience natural seasonal phenomena such as diapause, pupation, and emergence. These types of responses complicate interpretation of effects. Unperturbed systems undergo change in the absence of chemical perturbations as a result of natural perturbations (e.g., rain, variable light levels, invasion of influential or exotic species [such as zebra mussels, grass carp, and *Bythotrephes cederstroemi*, the daphnid recently discovered in the Great Lakes]). One advantage of stream mesocosms is the ability to separate natural (imposed on all streams equally) from chemical (applied to determine exposure-response) perturbation. This is accomplished at the ESF by using experimental control systems with frequent and coordinated sampling.

The use of mesocosms has been recognized by regulatory agencies to negate the presumption of risk (Touart 1988). Mesocosms reduce the uncertainty in evaluating risk of chemical exposure to ecosystems. This is accomplished by evaluating dose-response relationships with diverse aquatic assemblages using appropriate and complementary end points. These end points are relevant to ecosystems that are supposed to be protected. The exposures are chemically characterized well and occur in a once through, flow through environment that physically resembles habitats of the ecosystem. Natural biotic and abiotic interactions proceed in the mesocosm that cannot occur in single species exposures.

The USEPA (1984) applies uncertainty factors to toxicity data by the quotient method. These are then compared to predicted environmental concentrations of a chemical of interest. The uncertainty factor is a function of the complexity of the data set. For field or mesocosm data, an uncertainty factor of 1 is applied to the established ecosystem NOEC. An uncertainty factor of 10 is applied to the lowest chronic NOEC for algae, invertebrates, and fish. The lowest chronic NOEC for C_{12}TMAC (10 μg/L for *Daphnia*) is 25 times lower than the ESF stream channel NOEC of 250 μg/L. The daphnid NOEC would be divided by 10 and compared to the predicted environmental concentration after dilution. This would yield a USEPA-determined safety factor less than 10. However, the ESF and field studies demonstrate a NOEC of 250 μg/L, yielding a safety factor that is 250 times higher (the product of the higher effect level and a lower uncertainty factor).

The ameliorative capacity and the inherent resilience of natural systems present in mesocosms that are not present in traditional laboratory bioassays underscores that, for many chemicals, the latter may be more conservative. When this is not the case (i.e., when ecosystems are more sensitive than the laboratory assay), the utility of mesocosms is even greater. Mesocosm tests are an experimentally useful approach for simultaneously

Mesocosm tests are an experimentally useful approach for simultaneously determining chemical fate, exposure, and effects to assess risks to ecosystems.

ACKNOWLEDGMENTS

This manuscript represents the combined efforts of the Procter & Gamble Stream Ecosystem Research Team including J. B. Barnum, J. B. Guckert, J. W. Bowling, B. G. Hamm, D. H. Davidson, K. L. Rupe, R. G. Bausch, and D. D. Walker. Scientific contributions include many others of the Environmental Safety Department and several academic investigators including D. Hart and F. Acker (Philadelphia Academy of Natural Sciences), J. Cairns, Jr. and P. V. McCormick (Virginia Tech), R. L. Lowe (Bowling Green State University), and R. M. Ventullo (University of Dayton).

REFERENCES

BARNUM, J.B. and WOLTERING, D.M. (1989). Monitoring detergent safety downstream. J. Am. Oil Chem. Soc. 66:383-386.

BELANGER, S.E. (1991). The effect of dissolved oxygen, sediment, and sewage treatment plant discharges upon growth, survival, and density of Asiatic clams. Hydrobiologia 218:113-126.

BELANGER, S.E., BARNUM, J.B., WOLTERING, D.M., BOWLING, J.W., VENTULLO, R.M., SCHERMERHORN, S.D., and LOWE, R.L. (In press). Algal periphyton structure and function in response to consumer chemicals in stream mesocosms. In: *Utilization of Simulated Ecosystems in Ecological Risk Assessment* (R.L. Graney, J H. Rodgers, Jr., and J. Kennedy, eds). Lewis Publishers, Chelsea, MI.

BELANGER, S.E., DAVIDSON, D.H., FARRIS, J.F., REED, D., and CHERRY, D.S. (In review). Effects of cationic surfactant exposure to a bivalve mollusc in stream mesocosms. SETAC Special Symposium on Surfactants.

CAIRNS, J., JR. (1986). The myth of the most sensitive species. BioScience 36:670-672.

CAIRNS, J., JR. (1988). What constitutes field validation of predictions based upon laboratory evidence? In: *Aquatic Toxicology and Hazard Assessment: Tenth Volume*, STP971 (W.J. Adams, G. A. Chapman, and W. G. Landis, eds), pp. 361-368. American Society for Testing and Materials, Philadelphia, PA.

CAIRNS, J., JR. and DICKSON, K.L. (1977). Elasticity, inertia, and resiliency of ecosystems. In: *Proceedings of Symposium Terrestrial and Aquatic Ecological Studies of the Northwest*, pp. 371-381. Eastern Washington State College Press, Cheney, WA.

CAIRNS, J., JR., KUHN, D., and PLAFKIN, J. (1979). Protozoan colonization of artificial substrates. In: *Methods and Measurements of Periphyton Communities, A Review*, STP690 (R.L. Wetzel, ed), pp. 34-57. American Society for Testing and Materials, Philadelphia, PA.

CAIRNS, J., JR., SMITH, E.P., and ORVOS, D.R. (1988). The problem of validating simulation of hazardous exposure in natural systems. In: *Proceedings of the 1988 Summer Computer Simulation Conference* (C.C. Barnett and W. M. Holmes, eds), pp. 448-454. Society for Computer Simulation International, San Diego, CA.

CUFFNEY, T.F., HART, D.D., WOLBACH, K.C., WALLACE, J.B., LUGTHART, G.J., and SMITH-CUFFNEY, F.L. (1990). Assessment of community and ecosystem level effects in lotic environments: The role of mesocosm and field studies. Presented at the North American Benthological Society Technical Information Workshop: Experimental Ecosystems-Applications to Ecotoxicology, May 25, 1990. Blacksburg, Virginia. 40 pp.

HEDTKE, S.F., HERMANUTZ, R.O., ALLEN, K.N., and ROUSH, T.H. (1990). Effects of selenium on aquatic life in outdoor experimental streams. Society of Environmental Toxicology and Chemistry, Tenth Annual Meeting, Toronto, Ontario, Canada.

HURLBURT, S.H. (1984). Pseudoreplication and the design of ecological field experiments. Ecol. Monogr. 54:187-211.

KIMBALL, K.D. and LEVIN, S.A. (1985). Limitations of laboratory bioassays: The need for ecosystem-level testing. BioScience 35:165-171.

LEWIS, M.A. (1986). Comparison of the effects of surfactants on freshwater phytoplankton communities in experimental enclosures and on algal population growth in the laboratory. Environ. Toxicol. Chem. 5:319-332.

LEWIS, M.A. and HAMM, B.G. (1986). Environmental modification of the photosynthetic response of lake plankton to surfactants and significance to a laboratory-field comparison. Water Res. 20:1575-1582.

LEWIS, M.A., TAYLOR, M.J., and LARSON, R.J. (1986). Structural and functional response of natural phytoplankton and periphyton communities to a cationic surfactant with considerations on environmental fate. In *Community Toxicity Testing*, STP920 (J. Cairns, Jr., ed), pp. 241-268. American Society for Testing and Materials, Philadelphia, PA.

MCCORMICK, P.V., CAIRNS, J., JR., BELANGER, S.E., and SMITH, E.P. (1991). Response of protistan assemblages to a model toxicant, the surfactant C12-TMAC (dodecyl trimethyl ammonium chloride) in laboratory streams. Aquat. Tox. 20:41-70.

ODUM, E.P. (1984). The mesocosm. BioScience 34:558-562.

OHIO EPA (ENVIRONMENTAL PROTECTION AGENCY). (1987). Amendments to the Ohio Water Quality Management Plan: Little Miami River Basin, Comprehensive Water Quality Report for the East Fork of the Little Miami River. 197 pp. + appendices.

OKKERMAN, P.C.E., PLASSCHE, J.V., SLOOF, W., VAN LEEUWEN, C.J., and CANTON, J.H. (1991). Ecotoxicological effects assessment: A comparison of several extrapolation procedures. Ecotoxicol. Environ. Safety 21:182-193.

ORGANIZATION FOR ECONOMIC COOPERATION AND DEVELOPMENT (OECD). (1981). OECD Guidelines for Testing of Chemicals. Paris, France.

PONTASCH, K.W., NIEDERLEHNER, B.R., and CAIRNS, J., JR. (1989). Comparisons of single-species, microcosm, and field responses to a complex effluent. Environ. Toxicol. Chem. 8:521-532.

PRATT, J.R. and BOWERS, N.J. (1990). A microcosm procedure for estimating ecological effects of effluents and mixtures. Tox. Assess. 5:189-205.

PRATT, J.R. and CAIRNS, J., JR. (1985). Functional groups in the protozoa: Roles in differing ecosystems. J. Protozool. 32:415-423.

RAPAPORT, R.A. (1988). Prediction of consumer product chemical concentrations as a function of POTW type and riverine dilution. Environ. Toxicol. Chem. 7:107-115.

SANDERS, F.S. (1982). Guidance for the use of artificial streams to assess the effects of toxicants in freshwater systems. Environmental Science Division Publication No. 1800, Oak Ridge National Laboratory, Oak Ridge, TN. 83 pp.

SANDERS, F.S. (1985). Field validation. In: *Fundamentals of Aquatic Toxicology* (G.M. Rand and S.R. Petrocelli, eds), pp. 601-619. Hemisphere Publishing Company, NY.
SCHERMEHORN, S. (1990). The response of freshwater periphyton communities to synthetic organic chemicals. M.S. Thesis, University of Dayton, Dayton, OH. 125 pp.
SCHWAB, B.S., MARUSCIK, D.A., VENTULLO, R.M., and PALMISANO, A.C. (In press). Adaptation of periphytic communities to a quaternary ammonium surfactant. Can. J. Fish. Aquat. Sci.
SHIMP, R.J., SCHWAB, B.S., and LARSON, R.J. (1989). Adaptation to a quaternary ammonium surfactant by suspended microbial communities in a model stream. Environ. Toxicol. Chem. 8:723-730.
SUTER, G.W., ROSEN, A.E., LINDER, E., and PARKHURST, D.F. (1987). Endpoints for responses of fish to chronic toxic exposures. Environ. Toxicol. Chem. 6:793-809.
TOUART, L.W. (1988). Aquatic Mesocosm Tests to Support Pesticide Registrations, EPA 540/09-88-035. Hazard Evaluation Division, Technical Guidance Document, Office of Pesticide Programs, U. S. Environmental Protection Agency, National Technical Information Service, Springfield, VA. 18 pp.
UNITED STATES ENVIRONMENTAL PROTECTION AGENCY (USEPA). (1984). Estimating "concern levels" for concentrations of chemical substances in the environment. Environmental Effects Branch Health and Environmental Review Division, Washington, DC.
UNITED STATES ENVIRONMENTAL PROTECTION AGENCY (USEPA). (1989). Short-term Methods for Estimating the Chronic Toxicity of Effluents and Receiving Waters to Freshwater Organisms, EPA/600/4-89/001. Environmental Monitoring Systems Laboratory, Cincinnati, OH.
VERSTEEG, D.J. and WOLTERING, D.M. (1990). A laboratory scale model for evaluating effluent toxicity in activated sludge wastewater treatment plants. Water Res. 24:717-723.
WOLTERING, D.M. and BISHOP, W.E. (1989). Evaluating the environmental safety of detergent chemicals: A case study of cationic surfactants. In: *The Risk Assessment of Environmental Hazards* (D.J. Paustenbach, ed), pp. 345-389. John Wiley & Sons, NY.

APPENDIX: CURRENT ESF END POINTS

End points used during TMAC studies are indicated with an asterisk (*)

Algal Periphyton End Points: Evaluated on cobble (river incubated), tile (ESF incubated*), and polyurethane foam units (see also protozoans, ESF incubated*)

Structural End Points

- taxonomic identification*
- species diversity*
- community similarity*
- biovolume biomass*
- chlorophyll biomass*
- ash free and dry weight biomass*
- lipid profile

Functional End Points
primary productivity*
lipid biosynthesis
storage/membrane lipids

Heterotrophic Periphyton End Points: Evaluated on cobble (river incubated) and tile (ESF incubated*)
Structural End Points
cell density (acridine orange)*
ash free and dry weight biomass*
lipid profile
Functional End Points
heterotrophic productivity*
lipid biosynthesis
storage/membrane lipids
biodegradation/acclimation to test chemical*

Protozoan Community End Points: Evaluated on PFUs
Structural End Points
taxonomic identifications*
density*
species diversity*
community similarity*
Functional End Points
trophic feeding group structure*

Invertebrate Community End Points: Evaluated in trays (river or mesocosm incubated*)
Structural End Points
taxonomic identifications*
density*
species diversity*
community similarity*
biomass*
drift*
Functional End Point
trophic feeding group structure*

Single Species Assays in ESF Environment
Species include:
Asiatic clams (growth, survival, reproduction)*
Fathead minnows (growth, survival)*
Ceriodaphnia (reproduction, survival)*
Selected invertebrates method development (amphipods, caddisflies, mayflies, snails)

Chemical Monitoring Systems
- *ESF Sensor System*
  - temperature*
  - dissolved oxygen*
  - conductivity*
  - pH*
  - light*
  - flow*
- *Water Quality Monitoring*
  - priority pollutants
  - major and minor nutrients*
  - suspended and settled solids*
  - test chemical*
    - stock stability
    - stock tank concentration
    - in-stream concentration
    - settled solids concentration

CHAPTER 14

ECOLOGICAL RISKS ASSOCIATED WITH RADIOACTIVE MATERIALS

Bernd Kahn
Environmental Resources Center
Georgia Institute of Technology
Atlanta, Georgia

This chapter presents an overview of the impact of radionuclides on the ecosystem. The types and concentrations of natural and man-made radionuclides in the environment are described and their sources are indicated. Research is summarized on the pathways of radionuclides from their sources in nature, at nuclear facilities, and from nuclear explosions to points of exposure to humans, animals, or plants. The adverse effects on humans of ionizing radiations emitted by these radionuclides are reviewed—both immediate tissue damage from the deposition of large amounts of energy and stochastic damage to chromosomes that can cause delayed harm in the form of genetic mutations and cancer—from the point of view of establishing radiation dose limits to protect the most exposed person. Parallel efforts are described that determine effects on animals and plants and set safe radiation dose rates to protect populations.

INTRODUCTION

One major focus of radioecological studies has been delineating the pathways of radioactive materials from their sources to humans. Information is gathered on transport and dispersion of materials in air, surface water, and groundwater and transfer from one environmental medium to the next. Computers are then programmed with calculational models. Given the radioactivity source term, these programs predict radionuclide concentrations at any point in the modelled environment and radiation doses to persons for specified behavior patterns and locations.

A second and distinctly different focus has been the effect of radiation on animals and plants. Test subjects are exposed to selected radiation doses for a range of dose rates and intervals to determine the incidence of physiological, pathological, or reproductive effects, or death. Results are used to predict harm as a function of radiation dose and associated conditions of irradiation.

---

1. Corresponding Author: Bernd Kahn, Environmental Resources Center, Georgia Institute of Technology, Atlanta, GA 30332.
2. Key Words: radionuclides, radioecology, radiation effects, environmental radioactivity.

Extensive research has been devoted throughout the world to both topics—in the United States, mostly under the Atomic Energy Commission (AEC) sponsorship from the 1940s to 1960s. The relatively recent reactor accident at Chernobyl has reactivated such studies, especially in Europe.

The findings from radiation studies are applied for protection by setting radiation exposure limits and controlling releases of radioactive materials to meet these limits. For humans, controls are intended to assure that the limits protect every individual; for other species, the aim is to maintain a balanced healthy ecosystem—to protect the community rather than any individual.

The obvious complexities of radioactive material moving through the environment and the associated radiation affecting the system have led users of these findings to resort to broad simplifying principles. For pathway predictions, experience has indicated a set of critical radionuclides, pathways, and exposed population groups for the more widely encountered nuclear facilities and radioactive contamination events. These three critical groupings are considered first, although priorities must be shifted if site-specific circumstances so dictate. For ecosystem protection, safe levels for humans have been taken to imply safe levels for all biota. Here also, care must be taken to test this assumption for each site.

Approaches for considering the environmental movement and impact of radioactive materials parallel those for other pollutants. Radioactive materials are just one source in the broad spectrum of ecological risks. Very specific complexities are associated with radioactive materials, however: radiation hazards can be external or internal; radionuclides do not change form (although radioactive materials do) except by decay at an immutable decay rate; the difference between a radiation dose that has a barely observable physiological effect and one that is lethal is less than two orders of magnitude; and radiation effect can be both immediate and long delayed.

This discussion of the ecological behavior and impact of radioactive material attempts to summarize pertinent knowledge for a reader who is not a specialist in radiation protection. An initial section presents radiation fundamentals, and the following three sections describe the radiation environment, radioactive material transport in the environment, and radiation impacts on biota in the ecosystem.

RADIATION FUNDAMENTALS

Radionuclides and Radiation

Radionuclides are isotopes that have excess nuclear energy compared to the stable isotopes that constitute the familiar set of chemical elements. They are chemically similar—for most purposes, identical—to their stable isotopes, but differ by emitting their excess nuclear energy as high-velocity radiations to attain a more stable form. A radionuclide is characterized by the types of radiation it emits, the energy of the radiations, the relative intensities of the various types and energies of the radiations, the fraction decaying per unit time (the decay constant), and, in some instances, its origin from, or decay to, another radionuclide.

The common radiations are alpha particles, beta particles, and gamma rays. Alpha particles are rapidly moving helium ions that deposit their energy and come to rest within a very short distance—typically 100-200 μm—in soft tissue. Beta particle is the term given to an electron emitted in a group with a continuous distribution of kinetic energy, for which the maximum range in tissue is 0.1-2 cm. Gamma rays are photons that typically are 10,000-1,000,000 times as energetic as visible light; passage through 20 cm of tissue stops about 95% of the more energetic gamma rays and a higher percentage of the weaker ones. The path length of alpha and beta particles and the fraction of gamma rays stopped in a given distance depend on the energy of the radiation and the electron density of the material. Tables, equations, and graphs of these values are given in nuclear physics and radiation detection texts (Knoll 1989).

Certain radionuclides also emit other radiations—positrons, X rays, conversion and Auger electrons, and neutrinos. Positrons are electrons with a positive instead of negative charge. Upon slowing down, each positron combines with an electron to convert their mass to two gamma rays. Emission of each beta particle or positron is accompanied by a neutrino, which has near-zero mass and deposits essentially no energy in tissue. Conversion electrons are atomic electrons that are emitted in parallel with gamma rays to carry away energy. Residual energy on the atomic level as a result of emitting conversion electrons is released by X rays, which are low-energy photons of energies characteristic to a particular element, and by low-energy Auger electrons.

Radiation energy is reported in kiloelectron volt (keV) and megaelectron volt (MeV). One MeV is 1.6×10^{-13} joule. The decay constant is usually characterized by the half life, in units of time, which is 0.693 divided by the decay constant. A radionuclide may decay directly to a stable isotope or to a second radionuclide; in some cases, there is a whole chain of radionuclides until finally a stable isotope is reached. The decay characteristics of radioisotopes are tabulated in compilations that are periodically brought up to date (Brown and Firestone 1986; Kocher 1981).

The decay rate of a radionuclide is in units of becquerel (Bq), where 1 Bq equals 1 disintegration per second. An earlier unit still in use is the curie (Ci), where 1 Ci equals 37 billion disintegrations per second. The activity (A) in Bq is related to the mass (m) in gram of a radionuclide by

$$A = (6.02 \times 10^{23} \times 0.693 \times m) / (T \times M) \quad [1]$$

where T is the half life in seconds, M is the isotopic mass in gram per gram atom of the radionuclide, and 6.02×10^{23} is Avogadro's number. Equation [1] shows that readily measurable decay rates of 0.1-1000 Bq result from unweighably small amounts of radionuclides in most cases. For example, 1×10^{-12} gram of the radionuclide strontium-90, which has an isotopic mass of 90 and a half life of 28 years (9×10^{8} s), emits 5 Bq. The only radionuclides that have measurable weights at these activities have half lives of the order of a billion years or more.

Hundreds of radioisotopes have been observed. Some occur in nature; others are man-made. Nuclear reactors and weapons produce fission products, which result from splitting uranium or plutonium by neutrons, and activation products, which are formed by neutrons or other nuclear particles bombarding an atom. Other radionuclides are produced for research or medicine by particle accelerators.

Radionuclides of possible concern in the environment are limited to a few dozen because the brief half lives of most radionuclides prevent them from reaching and impacting environmental biota. At the other extreme are radionuclides that have half lives of more than 10 billion years—so long that the radiations emitted by the usual amount of the radionuclide are too few during a lifetime to cause significant damage.

Tissue may be irradiated by radionuclides that are either external or internal. The extent of irradiation by external radionuclides may be reduced by distance, attenuation in material between the radionuclide and the tissue, or removal of either source or target. Internal radiation is decreased by excretion and radioactive decay.

There are external sources of ionizing radiation other than radionuclides. Cosmic rays originate in the stars, including the sun. Man-made radiations can reach the environment from their sources.

Cosmic radiation exists at high levels in interstellar space, but the earth's surface is shielded from most of it by the atmosphere. Cosmic radiation arrives at the stratosphere as enormously energetic protons (87%), alpha particles (11%), heavier ions (1%), and electrons (1%) (National Council on Radiation Protection and Measurements [NCRP] 1987a). The protons interact with air to create muons and neutrons. Muons are short-lived fundamental constituents of matter, charged negatively like electrons but 207 times as heavy. Protons are hydrogen ions. Neutrons are constituents of the atomic nucleus like protons, but are 0.14% heavier, without charge, and therefore more penetrating.

Cosmic radiations occur everywhere on earth, but the flux rate increases at higher latitudes and reaches maxima near the two poles. The flux rate varies by a few percent with time because emissions by stars fluctuate, for example, with sunspots.

Man-made external radiation such as neutrons and gamma rays may be found in the vicinity of reactors. In addition, X rays may be found near medical X-ray machines, particle accelerators, and high-voltage electrical transformers. The intensity depends on the degree of proximity and shielding.

Internal radiation is due to radionuclides within the body. Some radionuclides are present at birth, having reached the fetus from the mother; others are ingested, inhaled, or pass through the skin. These radionuclides are distributed among organs and tissues to the same extent as their stable isotopes and have the same retention and transfer behavior. Special studies have provided these values for radionuclides that have no stable isotopes. The patterns of internal exposure depend on circumstances: intakes can be one time only, periodic, or continuous, while the amounts of a radionuclide in an organ may increase throughout life, reach a constant value, or decrease at a rate that is the sum of the biological turnover rate and the radioactive decay constant.

Effects of Ionizing Radiation

Alpha and beta particles, as well as other charged particles such as positrons, muons, and protons, are ionizing radiation. Gamma rays, X-rays, and neutrons, although not directly ionizing, also are included in this category because their main (although not all) interactions with matter immediately generate ionizing radiation.

The energy absorbed per unit mass from ionizing radiation is measured in units of gray (Gy); 1 Gy equals 1 joule per kilogram. Common fractional units are the milligray (mGy) and microgray (μGy), respectively 10^{-3} and 10^{-6} Gy. Measurements are typically made of the dose rate, in μGy/hr, and summarized as mGy/year.

An earlier unit of radiation dose is the rad, where 100 rad = 1 Gy. Another early unit, still in use, is the roentgen, a measure of the ionization created in air; one roentgen generates ions carrying 1 statcoulomb of charge in 1 cc air at 1 atmosphere and 0°C and is equivalent to 8.77 mGy in air.

The energy deposition rate in lungs by alpha particles emitted by polonium-218 and polonium-214 has another special unit, the working level (WL). One WL is deposited by these decay-chain radionuclides when in equilibrium with 3700 Bq/m^3 of the parent radon-222; however, the decay chain is usually only at 0.2-0.5 of equilibrium in air indoors and 0.6-0.8 outdoors because the particulate daughters of radon tend to deposit on surfaces. The energy emission of 1 WL equals 1.3×10^5 MeV of alpha-particle energy. Based on a complex model of the lungs, the bronchial dose rate is 26 μGy/hr per WL, subject to some uncertainty (United Nations Scientific Committee on the Effects of Atomic Radiation [UNSCEAR] 1988). The WL unit was initially applied to uranium miners, for whom the monthly dose was computed as the working-level month (WLM), which is the WL value multiplied by the number of months, at 170 working hours per month.

In passing through or stopping in matter, ionizing radiation ionizes or excites molecules by, respectively, removing electrons or placing them in elevated energy states. Radiation can break molecular bonds and accelerate electrons and atoms within the material. In tissue, these interactions break bonds in macromolecules and also form ions, hydrated electrons, and free radicals that initiate chemical reactions in the aqueous phase to break such bonds. These effects can alter the DNA in chromosomes and disrupt or destroy cells.

The effect is not dependent solely on the amount of deposited energy, so that the Gy is not necessarily a measure of damage. The measure of the biological effect of energy deposition, defined as the dose equivalent, in sievert (Sv), is the product of the dose in Gy and the relative biological effectiveness (RBE) for tissue.

Alpha particles, protons, and fast neutrons, which deposit much of their energy in very short distances, have a much more disruptive effect on tissue than gamma rays and beta particles. This effect of a high linear energy transfer (LET) is quantified in the quality factor (Q). Recommended Q values are 1.0 for X-rays, gamma rays, electrons, and beta particles, and 20 for alpha particles, fast neutrons, and protons (NCRP 1987b), but other Q values have been used for high-LET radiation.

Various other factors, for which multiplying factors like Q have not been assigned, contribute to the RBE of ionizing radiation in tissue. The total dose, the pattern in which the dose is partitioned, the dose rate, and the conditions under which the dose is received can also affect the impact of energy deposition in tissue. The relative impact varies to some extent with the tissue and the end point under consideration.

Non-linear increases of effect (E) with dose equivalent (D) generally are described by a quadratic term in addition to a linear term:

$$E = aD + bD^2 \qquad [2]$$

where a and b are constants. Lesser damage at lower dose rates can be attributed to the functioning of repair mechanisms until they are overwhelmed at higher doses, or to damage from both single and two-track interactions with ionizing radiation (Committee on the Biological Effects of Ionizing Radiations [BEIR] 1990).

Both immediate and delayed harm have been observed in persons after overexposure to ionizing radiation. Early experimenters with X-rays at excessive doses suffered promptly from skin erythema, which led to infections, tumors, and loss of limb in extreme cases. Workers who ingested large amounts of radium while painting luminous dials developed bone cancer after intervals of years.

In more recent incidents and accidents, persons have been exposed to high radiation levels that resulted in prompt effects ranging from cataract formation and temporary sterility to hematopoietic, gastrointestinal, and central nervous system syndromes that, at the higher levels of radiation exposure, resulted in prompt death. These effects are associated with cell injury and are considered to be nonstochastic, i.e., to occur only above a threshold radiation dose equivalent and to become more serious at higher dose equivalents. A relatively narrow range exists between the detection limit at approximately 0.1 Sv of changes at the cellular level and the midlethal dose (LD_{50}) for acute radiation exposure of approximately 5 Sv in humans.

Delayed stochastic effects result in genetic mutation or carcinogenesis. These effects are attributed to radiation-induced damages in DNA that are not eliminated by cell removal or damage repair. Expression of the radiation-initiated change as cancer after a number of years appears, in most cases, to require an intermediate promoting step.

The major source of information on delayed somatic effects of external radiation is the ongoing study by the Atomic Bomb Casualty Commission (since 1975, Radiation Effects Research Foundation) of survivors of the atom bomb explosions at Hiroshima and Nagasaki in 1945. Mortality from leukemia, multiple myelomas, and cancers of the breast, colon, esophagus, lung, ovary, thyroid, salivary glands, stomach, and bladder has increased with radiation dose. Prenatally exposed persons suffer from increased incidences of cancer and mental retardation. Absence of observed increases in genetic effects permits estimation of an upper limit on radiation-induced mutation rates, which is consistent with the results of animal studies.

Cancer incidence and mortality in humans also have been estimated from epidemiological studies of radiotherapy patients exposed to X radiation or fluoroscopy. Results of studies of occupationally exposed groups have been negative or ambiguous.

Effects of internal exposure have been observed in various epidemiological studies. The populations include the radium dial painters (bone cancer); uranium miners (lung cancer); patients given thorotrast, a thorium compound, as X-ray contrast medium (liver cancer); and radiotherapy patients given iodine-131 (thyroid cancer).

Numerous animal studies contribute information on the effects of external or internal irradiations over various time periods and during various life stages. They are the sources of information on genetic effects and on relative biological effectiveness with respect to somatic end points.

Expert advisory bodies estimate the incidence of stochastic effects of radiation and provide guidance to regulatory agencies for setting dose limits.

TABLE 1
Lifetime excess cancer risk estimate[a]

Cancer	Continuous lifetime exposure 1 mSv/year (deaths per 100,000[b])		Instantaneous exposure to 100 mSv (deaths per 100,000[b])	
	Male	Female	Male	Female
Leukemia	70	60	110	80
Nonleukemia	450	540	660	730

[a]From BEIR (1990).
[b]Exposed persons.

Limits are based on extrapolating the probability of harm from observations at higher values to lower dose equivalents where no effects are detected. To be conservative, the agencies extrapolate by assuming linearity of effect with dose and absence of threshold. Under this linear hypothesis, no assurance can be given of zero probability of harm at any dose, no matter how low. Dose limits are set so that radiation risks are comparable to other commonly accepted risks in life.

The most recent estimate (Table 1) of the risk of excess cancer mortality in humans by an expert group (BEIR 1990) shows about 0.07 deaths per Sv at relatively low doses and dose rates during a lifetime. Risk values for the various solid cancers were summed to calculate the nonleukemia risk added to the leukemia risk and extrapolated to a relatively low dose and dose rate. They were obtained with a relative risk model, which relates excess risk of radiogenic cancer to the existing cancer mortality rate per age intervals. The values differ from those obtained with an absolute risk model, which was considered less appropriate to the data, and have changed several fold during the past 20 years.

The risk of genetic effects estimated by the same group (BEIR 1990) is 0.01-0.02 additional cases per Sv when equilibrium has been reached after a number of generations, as shown in Table 2. This value is for a doubling dose of 1 Sv at chronic radiation exposure and the spontaneous genetic burdens of humans. Tables 1 and 2 indicate that carcinogenesis is a higher risk than mutagenesis and that protecting against the former will also protect against the latter.

It is probably fair to say that the mechanisms of radiation damage are understood in broad outlines and the magnitudes of effects are known, that estimates of damage have improved over the past 25 years with a corresponding change in risk coefficients, and that much still needs to be learned concerning mechanisms and effects. In particular, conclusions concerning cancer from atomic bomb exposure will be tentative while many persons in the study population still live; many of the important epide-

TABLE 2
Estimated genetic effects of 0.01 Sv per generation[a]

Type of disorder	Current incidence per million liveborn offspring	Additional cases/10^6 liveborn offspring/0.01Sv/generation First generation	Equilibrium
Autosomal dominant			
Clinically severe	2500	5-20	25
Clinically mild	7500	1-15	75
X-linked	400	<1	<5
Recessive	2500	<1	Very slow increase
Chromosomal			
Unbalanced translocations	600	<5	Very little increase
Trisomies	3800	<1	<1
Congenital abnormalities	20,000-30,000	10	10-100
Other disorders of complex etiology			
Heart disease	600,000		
Cancer	300,000	Not estimated	Not estimated
Selected others	300,000		

[a]From BEIR (1990).

miological studies suffer from uncertainty concerning the dosimetric data; conclusions from a number of such studies are under debate; and some leap of faith is required in applying animal data to humans.

For radiation protection of humans, expert bodies such as the International Commission on Radiological Protection (ICRP) and NCRP have recommended limits that they consider will provide acceptable safety (ICRP 1991; NCRP 1987c). These groups also maintain the principle that, both for radiation workers and the general public, levels be kept as low as reasonably achievable (ALARA), taking into account social and economic factors.

Limits recommended recently are given in Table 3. Stochastic effects are given as effective dose equivalents, based on a system that permits summing the dose equivalents received by various organs and tissues. The dose equivalent to each organ or tissue is multiplied by a weighting factor that is the ratio of its sensitivity to cancer induction relative to that in the whole body; these products are then added. A dose level has also been proposed in Table 3 that is so low compared to existing levels that the added risk is negligible.

For quantifying environmental radiation doses, the concepts of collective dose and dose commitment are applied. The collective dose equivalent, in

TABLE 3
Recommendations on limits for exposure to ionizing radiation[a]

Type of exposure		Annual dose equivalent, mSv
Occupational	- stochastic[b]	50
	- nonstochastic, lens of eye	150
	- nonstochastic, all others	500
Public	- continuous or frequent[b]	1
	- infrequent[b]	5
	- lens of eye, skin, extremities	50
Embryo/fetus[c]		5
Negligible individual risk level[b,d]		0.01

[a]From NCRP (1987c).
[b]Effective dose equivalent.
[c]Total, not to exceed 0.5 mSv per month.
[d]Per source or practice.

units of person-Sv, is the sum of dose equivalents to individuals in a population. If the linear hypothesis is applicable, then the collective dose equivalent is a measure of incidence of harm. More appropriately, values of collective dose equivalents are used to compare radiation impacts of various facilities or operations.

Dose commitment arises when a present practice can cause potential radiation exposure during the future. Long-lived radionuclides deposited on the ground can be direct radiation sources for years to come or may enter the food cycle in future years. The dose equivalent commitment in Sv from a year of practice is the sum of the projected annual dose equivalents for the life of the potentially exposed person. A parallel concept is the committed dose equivalent, which pertains when a radionuclide taken into the body in the year under consideration will remain in it beyond that year.

ENVIRONMENTAL RADIATION AND RADIOACTIVITY

The Natural Radiation Environment

Naturally occurring ionizing radiation provides the radiation exposure baseline for biota. Presumably, species extremely sensitive to harm from radiation at the background level have not survived. For any hypothesized effect, calculation of the risk coefficient at the dose equivalent rate for natural radiation yields an upper limit attributable to ionizing radiation. For example, the annual collective dose equivalent for a U.S. population of 260 million of 800,000 person-Sv, based on the average dose equivalent presented later in this section, can be applied to the risk estimates in Tables 2 and 3 for calculating radiogenic cancer and mutation incidence for comparison with the available statistics on these effects.

Because the natural radiation background differs from place to place, numerous persons are exposed to levels that are several fold higher or lower than the mean. Hence the possibility exists for determining the magnitude of radiation effects near background levels. The most detailed of these studies compared the effects of lifetime exposures at 3-4 mSv/year and 1 mSv/year for two groups of 70,000 persons in China. Cancer mortality did not differ significantly, although chromosome aberration rates were higher at the elevated dose equivalent rate (BEIR 1990). To be convincing, such comparisons require very large matched populations and thorough examination of possible confounding variables such as exposure to medical radiation or chemical carcinogens.

Natural background radiation is thoroughly described in a monograph (NCRP 1987a) and a series of conference proceedings (Gesell and Lowder 1980). It has two main components: terrestrial radionuclides that remain from the creation of the universe, and cosmic radiation released continuously from stars. Uranium and thorium, among the long lived terrestrial radionuclides, are the parents of shorter lived radionuclides in decay chains. Cosmic radiation also interacts with molecules, particularly nitrogen, in the upper atmosphere to create radionuclides that reach the earth's surface. Spontaneous fission by uranium and the interaction of cosmic radiation in the ground also generate radionuclides, but at levels of negligible dose import.

The four terrestrial radionuclides listed in Table 4 contribute virtually the entire radiation dose from this group. These radionuclides are in rocks and soil at the indicated typical levels. Approximately 20 other long lived radionuclides, including the uranium-235 decay chain, exist at much lower concentrations—some so low that they are difficult to detect.

Of the major terrestrial radionuclides near ground surface (shown in Table 4), three emit gamma rays that result in typical radiation doses 1 m above

TABLE 4
Gamma radiation from terrestrial radionuclides in soil[a]

Radionuclide	Typical concentration elemental, ppm	radionuclide, Bq/kg	Absorbed dose rate in air,[b] mGy/year
potassium-40	15,000	400	0.17
rubidium-87	65	50	0
thorium-232	9	37	0.22[c]
uranium-238	1.8	22	0.085[c]

[a]Based on NCRP (1987a).
[b]1 m above ground.
[c]Dose rate due to gamma rays from progeny in equilibrium with this parent radionuclide.

ground of 0.06 μGy/hr (0.5 mGy/year). Dose rates can be much lower where little uranium, thorium, and potassium occur in surface rocks and soil, and several orders of magnitudes higher above uranium and thorium ore deposits.

The radionuclides in soil that cause these doses in air must be within about 25 cm of ground surface because gamma rays from deeper lying sources are almost completely stopped by the soil. Similarly, beta particles emitted from terrestrial radionuclides will only reach the air from about the upper centimeter of soil, and will then be attenuated by air so that only the most energetic ones will reach 1 m above ground. At 0.5 m above ground, the gamma ray dose rate is not much different from the 1 m level, and the dose due to beta particles equals that from gamma rays (NCRP 1987a). The only alpha particle dose rate in air from alpha particles emitted by the uranium-238 and thorium-232 chains in the ground originates at the very soil surface. The travel distance in air of these alpha particles is at most 10 cm.

Gamma radiation levels in buildings are typically within 20% of those outside. Indoor radiation doses are higher from construction materials such as brick, stone, clay, or concrete that contain terrestrial radionuclides. Low radionuclide materials such as wood partially shield occupants from terrestrial radiation.

Natural radionuclides in soil enter the life cycle through uptake at roots or surfaces of plants and through ingestion or inhalation by animals. Animals can take up terrestrial radionuclides directly with soil particles, or indirectly, after dissolution in water, through plants or passage from other animals in the food chain.

Typical radionuclide concentrations and dose equivalents in human tissue are given in Table 5. Potassium is homeostatically controlled in humans, and potassium-40 is a constant fraction (0.0117% by weight) of the element; hence, the potassium-40 concentration and dose contribution are unaffected by the amount ingested. Potassium in humans is associated with body lean tissue: it decreases with age beyond the teens and is lower in females. Rubidium-87 is chemically homologous with potassium, and its retention may be affected by potassium intake.

Uranium and thorium do not appear to be homeostatically controlled and are taken into the body as a function of the amounts ingested and the physical and chemical forms. The degree of uptake also may be affected by the extent to which elements that follow similar metabolic paths are present. Uranium concentrates in the kidney and thorium in bone.

Radon-222 (half life = 3.8 days) in the uranium-238 decay series and radon-220 (half life = 55 seconds) in the thorium-232 series present special exposure features, being the only gases among the major terrestrial radio-

TABLE 5
Radionuclide levels and radiation doses in human tissue[a]

	Concentrations (mBq/kg wet) and dose equivalent rates (μSv/year) in tissues									
	Lung		Gonad		Bone surface		Bone marrow		Soft tissue	
Radionuclide	Conc.	Dose	Conc.	Dose	Conc.	Dose	Conc.	Dose	Conc.	Dose
Terrestrial										
K-40	63	-	63	-	15	140	130	230	59	180
Rb-87	8	-	18	-	24	14	7	7	7	3
U-238[b]	12	11	5	4.6	7	3	5	0.4	-	4.6
Ra-226[b]	3	3	-	-	170	90	3	15	-	3
Rn-222	-	200	-	-	-	14	-	14	-	7
Po-218[c]	-	24,000	-	-	-	-	0.15	0.4	-	0.1
Pb-210[d]	300	140	300	140	3000	700	300	140	110	140
Th-232	20	8	0.15	0.06	4	2	0.5	0.4	0.15	0.1
Ra-228[e]	4	4.5	4	1.5	50	120	4	22	4	1.5
Rn-220	-	0.1		-	-	-	-	-	-	<0.1
Pb-212[f]	-	400	<0.1	-	-	-	-	-	-	-
Cosmogenic										
H-3	-	-	-	-	-	0.01	-	0.01	0.3	0.01
C-14	-	-	-	-	-	8	-	30	30	10

[a]Based on NCRP (1987a).
[b]Plus immediate daughters.
[c]Dose is to bronchial epithelium due to polonium-218 plus polonium-214 from inhaled radon-222.
[d]Lead-210 plus polonium-210 dose equivalents.
[e]Radium-228 plus actinum-228 and thorium-228 dose equivalents.
[f]Dose is to bronchial epithelium from lead-212 and polonium-212 from inhaled radon-220.

nuclides. The radon isotopes formed by radioactive decay can move with soil gas or groundwater in soil interstices and rock fractures to the surface within the limits set by their short half lives. Once in surface air, radon diffuses upward, but can accumulate at ground level during atmospheric inversions or precipitation. Radon accumulation in buildings, particularly poorly ventilated ones, is a special problem that currently concerns radiation protection officials.

The most important radiation dose impact of radon-222 is deposition of its short lived progeny—polonium-218, lead-214, bismuth-214, and polonium-214—in lungs. The two polonium isotopes emit alpha particles that are believed to be the cause of lung cancer in uranium miners exposed to elevated levels of radon-222 and its progeny. The relatively high average dose to lungs in the population is shown in Table 5.

Radon-222 concentrations in outdoor air are generally in the range of 4-20 Bq/m^3, and radon-220 is at approximately 5% of these levels (NCRP

1987a). The U.S. Environmental Protection Agency currently recommends remedial measures if average radon or progeny concentrations in indoor air exceed 150 Bq/m^3 or 0.02 WL, respectively; a considerable fraction of U.S. homes—possibly one-sixth—may exceed these values.

A lesser effect of the movement of radon gas in near-surface ground and in ground-level air is the fluctuation of the gamma radiation level. Changes in the concentrations of the lead-214 and bismuth-214 progeny of radon-222 and the lead-212, bismuth-212 and thallium-208 progeny of radon-220, all of which emit gamma rays, will affect the effective dose equivalent near the ground by about 0.01 μGy/hr.

The radon-222 decay chain ultimately reaches long lived lead-210 and its daughters bismuth-210 and polonium-210; radon-220 produces lead-212. When formed in air, these particles may remain suspended and be inhaled, or may deposit on vegetation and then be consumed; concentrations of lead-210 in air are approximately 5×10^{-4} Bq/m^3.

The mixture of highly energetic particles that is cosmic radiation changes in composition as it passes through and is attenuated by the atmosphere. The radiation dose at sea level is due mostly to muons with some contribution from electrons and neutrons; at an elevation of 10 km, electrons, protons, and neutrons are the main contributors. Dose equivalent rates are 0.30 mSv/year at sea level, 0.38 mSv/year at 1 km altitude, 1.11 mSv/year at 3 km, and 29 mSv/year (3.3 μSv/hr) at 10 km (O'Brien and McLaughlin 1972). Most persons in the United States are exposed in the range of the first two values, while airline personnel and passengers can be exposed at the latter levels. These average dose equivalent rates are outdoors at 55 N latitude and solar minima. Because the particles reaching the earth's surface are very penetrating, dose equivalent rates within buildings are reduced by only a few percent.

The radionuclides produced by cosmic radiation that are at highest concentrations near ground level are tritium (hydrogen-3), carbon-14, and beryllium-7. Although approximately 20 other radionuclides of this origin have been found (NCRP 1987a), most are difficult to detect near ground level because their generation rate is low or their decay rate is low (long half life) or they have mostly decayed (short half life) while moving downward through the atmosphere. Of these radionuclides, carbon-14 and tritium are widely measured in environmental studies, carbon-14 moving through the carbon cycle and tritium in the water cycle as useful tracers.

Because tritium and carbon-14 emit only low energy beta particles and no gamma rays, their radiation impact on humans is through intake, not external exposure. The concentrations and associated radiation doses shown in Table 5 for soft tissues (and applicable to the total body) indicate a low dose due to carbon-14 and a lesser dose from tritium.

The natural carbon-14/carbon ratio is 0.25 Bq/g. Carbon dating has shown fluctuations of about 10% attributed to variations in the cosmic radiation flux. Releases of non-radioactive carbon dioxide from burning coal and oil have decreased this ratio, although carbon-14 in fallout from fusion devices increased the ratio temporarily to a peak value of 0.4 Bq/g in 1964. The current level is about 0.27 Bq/g (NCRP 1985).

The natural production rate of tritium is estimated to result in a tritium/water ratio in persons of 0.46 Bq/L; the concentration is slightly higher in air and lower in water (NCRP 1979). Tritium concentrations are also measured in tritium units (tritium atoms per 10^{18} hydrogen atoms); in water, 1 Bq/L = 8.4 tritium units. Fallout tritium concentrations reached a peak of about 100 Bq/L in 1964 and have since decreased to about 2 Bq/L.

Beryllium-7 in ground level air has a concentration of about 0.001 Bq/m^3. Levels increase in spring due to mixing at the interface of the troposphere and stratosphere. Deposition on vegetation results in levels of about 100 Bq/kg.

The summary of effective dose equivalents from the natural radiation environment, in Table 6, indicates that radon-222 inhaled in the home is the major source. The value of 0.4 mSv/year from terrestrial radionuclides in the body is mostly due to potassium-40, with minor contributions from lead-210 plus polonium-210. The value of 0.01 mSv/year for cosmogenic radionuclides in the body is due to carbon-14. External radiation dose rates from cosmic radiation and terrestrial radionuclides are almost equal, with a combined value of 0.55 mSv/year. These low values occur because most persons live at low elevations and over soil for which terrestrial radionuclide concentrations are much lower than in Table 4.

Individuals can change their dose rates by moving to a different elevation, to an environment with different levels of terrestrial radionuclides, or,

TABLE 6
Estimated average annual effective dose equivalent from background radiation to persons in the United States and Canada[a]

Source	Effective dose equivalent rate, mSv/year
Cosmic	0.27
Cosmogenic in body	0.01
Terrestrial	0.28
Inhaled	2.0
In body	0.40
Total	3.0

[a]From NCRP (1987a).

most importantly, they can modify their home to change radon-222 concentrations in air. Changes in food intake usually have minor impact on doses.

Man-made Radionuclides

Radioactive materials have been released to the environment in the United States from three major sources: first, government facilities that developed, produced, and tested nuclear weapons and reactors; later, nuclear devices tested in the atmosphere; and more recently, commercial nuclear power plants. Radionuclides are also widely produced, applied, and released in nuclear medicine. On a lesser scale, such material is used in industry and at educational and research institutions.

The government facilities were built during and soon after World War II. The nine larger installations, such as the Hanford (WA), Idaho (ID), and Oak Ridge (TN) laboratories and the Savannah River Site (SC), are large and relatively isolated areas within which reactors, radioactive material processing facilities, and waste storage sites are located distant from the boundary. Numerous smaller facilities contributed to nuclear weapons production and to nuclear research and development during and after World War II.

These facilities discharged radioactive materials to air, surface water, and the ground; operated structures that may be radioactively contaminated; and stored radioactive wastes above and below ground. Operations were and are required to meet radiation protection regulations that limit radiation doses to the most exposed person on- and off-site. Currently recommended annual limits are given in Table 3; years ago, some limits were higher.

At these facilities, man-made radionuclides and radiation generally are readily detectable on site but are barely detectable, if at all, beyond the fenceline. This is true even when large amounts of radionuclides are discharged; the Savannah River Site, for example, in 1990, released to air and water—mostly to air—about 1×10^{16} Bq/year tritium, 1×10^{15} Bq/year krypton-85, and 5×10^{10} Bq/year of all other longer lived radionuclides, but the dose to the maximum exposed person off site was less than 0.002 mSv/year (Cummins et al. 1991). Releases at this and other facilities generally were higher in earlier years when the facilities were more active.

Exceptions to the above cited controls occurred early in the operation of some facilities. At Hanford, approximately 2×10^{16} Bq of airborne iodine-131 were discharged from December 1944 to May 1947 (Pacific Northwest Laboratory 1991). At the time, the critical pathway for iodine-131 through milk consumption (see next section) had not been recognized. A preliminary estimate for the current study of resulting radiation doses in the vicinity is that the thyroids of a few persons were exposed to several Gy.

and many more to about 0.05 Gy. Iodine-131 releases were reduced below 1×10^{14} Bq/year later in 1947 and are zero now that fissionable material production and recovery have been stopped at the site.

The radionuclides (listed in Table 7) commonly discharged at these facilities to air, surface water, and the ground are the major longer lived fission and activation products. The spectrum differs among facilities; it reflects the activity at a facility in producing or processing material and the extent to which waste treatment removes radionuclides from effluent. Information concerning radionuclide discharges and environmental monitoring results is published annually for the Department of Energy by each facility.

Tests of nuclear weapons in the atmosphere, most by the United States and USSR from 1945 to 1962, distributed fission products throughout the world. In 360 tests, the yield was 513 megatons (UNSCEAR 1988). A megaton of fission produces approximately 1×10^{19} Bq of longer lived fission products.

Part of this fallout is deposited on the ground near the test locations (for U.S. tests, in Nevada and the South Pacific). A second fraction enters the

TABLE 7
Major radionuclides released to the environment

Nuclear weapon tests in atmosphere ("fallout")			
Long lived		*Intermediate lived*	
hydrogen-3	(12.4 year)	strontium-89	(50 day)
carbon-14	(5760 year)	zirconium-95	(65 day)
strontium-90	(28 year)	niobium-95	(35 day)
ruthenium-106	(1.0 year)	ruthenium-103	(39 day)
antimony-125	(2.5 year)	iodine-131	(8.0 day)
cesium-137	(30 year)	barium-140	(12.8 day)
cerium-144	(244 day)	cerium-141	(32 day)
Nuclear materials production facility			
Airborne		*Waterborne*	
hydrogen-3		hydrogen-3	
carbon-14		phosphorus-32	
krypton-85		zinc-65	
iodine-129		strontium-90	
iodine-131		cesium-137	
Nuclear power station			
Airborne		*Waterborne*	
hydrogen-3		hydrogen-3	
iodine-131		chromium-51	
xenon-133		manganese-54	
		cobalt-58	
		cobalt-60	
		cesium-134	
		cesium-137	

Note: Lists do not include short lived radionuclides (half life less than 5 days).

troposphere and is deposited on the ground in a matter of weeks. The rest is carried into the stratosphere where it circles the earth and gradually deposits on its surface over the years. These fission products were deposited everywhere on earth, but not uniformly. Extensive information has been published on fallout radionuclides and their levels over the years (UNSCEAR 1982).

Radionuclide composition in fallout can be categorized in terms of three periods after a nuclear test—within days, months, and years—on the basis of radionuclide half lives. The major radionuclides with intermediate and long half lives are listed in Table 7. In addition to the major fission products, tests since 1952 of fusion ("hydrogen bomb") devices associated with fission devices also introduced considerable tritium into the environment, as well as some carbon-14, transuranium elements, and activation products such as cobalt-60.

These radionuclides cause human radiation exposure externally while in air, on the ground, in water, and in biota, and internally when inhaled or ingested. The cumulative radiation dose estimated from fallout, due to the major test series (to 1970) and considering dose commitment to many generations, is estimated to be 1.4 mSv in the United States (NCRP 1987a).

The low dose is due to worldwide efforts to end nuclear tests in the atmosphere that reached fruition in the test-ban treaty of 1963. The 63 more recent nuclear tests with 32.5 megaton yield in the atmosphere by China and France, who did not sign the treaty, added minor amounts of fallout between 1964 and 1980. No further tests in the atmosphere occurred since then, although underground testing continues.

The pattern of fallout at ground level consists of peaks after each test, with the highest amounts from relatively frequent and intensive tests in 1961-1962. At present, only the long lived fission products are readily detected at levels of a few Bq/kg in surface soil and vegetation. The concentrations in water of cesium-137 is about 3×10^{-4} Bq/L and that of strontium-90 is 3×10^{-3} Bq/L (Cummins et al. 1991). Without new tests, levels of these radionuclides will continue to decrease.

Another source of worldwide fallout is an isotopic power device named SNAP 9A for a U.S. satellite that was destroyed upon re-entering the atmosphere in April, 1964. It distributed 6×10^{14} Bq plutonium-238 throughout the air. Its levels in air have decreased with a 14-month half life. Concentrations are of the same magnitude in ground level air—approximately 1×10^{-7} Bq/m^3—as plutonium-239 from nuclear testing, which also cycles through the environment.

Routine discharges from nuclear power facilities contribute mostly local environmental radioactivity in nearby air and surface water, and, in some cases, in groundwater. About 70 of these facilities in the United States are nuclear power plants. Other links in the nuclear fuel cycle are uranium ore mines, ore mills and refineries, chemical conversion and isotopic enrichment plants, fuel fabrication plants, spent fuel storage sites and reprocessing plants, and waste storage facilities.

Radiation dose equivalent rates to persons from releases for the fuel cycle are limited by regulations in the United States to 0.25 mSv/year. Facilities at the front end of this cycle discharges uranium-238, uranium-235, and their progeny. These effluents increase the terrestrial radiation background in some sparsely populated locations. Mill tailings have been a public health problem when used as fill beneath homes due to radon-222 emanation and gamma radiation and in tailings piles that are sources of radon-222 and airborne radioactive particles.

Nuclear power stations release airborne and liquid effluents that contain fission and activation products such as those listed in Table 7. Median airborne radionuclide releases in 1986 at boiling and pressurized water reactors in the United States were about 1×10^{14} Bq noble gases, 2×10^{11} Bq tritium, and 1×10^{8} Bq iodine-131 and particles; releases in liquids were approximately the same for tritium and 1×10^{10} Bq for all other radionuclides. Tritium releases were two orders of magnitude higher at pressurized water reactors. Maximally exposed persons living near these stations receive doses of the order of 0.001-0.02 mSv/year (NCRP 1987c).

In the United States at present, all spent fuel from commercial nuclear power stations is stored in temporary facilities, and no reprocessing is planned. Long lived gaseous radionuclides—tritium, carbon-14, krypton-85, and iodine-129 (NCRP 1987c)—are released during fuel reprocessing at the Savannah River Site, were released at Hanford and West Valley in the United States, and are discharged at fuel reprocessing plants in other countries. These are ultimately distributed worldwide.

Iodine-129 concentrations were approximately 1 Bq/g iodine (2×10^{-7}g iodine-129/g iodine) in air, rain, surface water, and grass in the early 1970s due to fallout. The specific activity in air corresponded to an iodine-129 concentration of 4×10^{-9} Bq/m^3 air. Specific activity levels were 50- to 100-fold higher from fuel reprocessing effluents at Hanford, and the resulting dose equivalent rate was estimated to be about 0.001 mSv/year. Levels also were elevated near the other two U.S. reprocessing sites. Iodine-129 also is produced by cosmic rays and natural fission, but prebomb levels in samples of thyroids and other iodine-containing samples were below 1×10^{-10} g/g iodine (Brauer et al. 1974).

Krypton-85 levels in air increased from below 0.02 Bq/m^3 in 1955 (NCRP 1975) to 0.8 Bq/m^3 in 1979 (Office of Radiation Programs 1982) mostly due to fuel reprocessing. Predictions of concentrations in air of about 20 Bq/m^3 by 1990 (NCRP 1975) have not come true because only about 5% of the fuel from power reactors is being reprocessed worldwide, and none in the United States. Krypton-85 releases at the three European reprocessing plants had increased from 0.7×10^{17} Bq in 1979 to 1.1×10^{17} Bq in 1985 (UNSCEAR 1988), hence airborne krypton-85 levels would have increased by 1985 to somewhat more than 1 Bq/m^3. The dose equivalent corresponding to 1 Bq/m^3 is 0.0004 mSv/year (NCRP 1975). Doses to persons from these gaseous radionuclides are much below the natural radiation background at this time but will increase if nuclear power production and reprocessing expand.

Three accidents at nuclear reactors have been sources of radioactivity in the environment. A fire at Windscale (now called Sellafield) in England at a plutonium production reactor started on October 12, 1957, during carbon-annealing heating of the reactor. The damaged reactor released 6×10^{14} Bq iodine-131, about 7% as much cesium-137, and lesser amounts of other radionuclides. These radionuclides contaminated vegetation, milk, and water in nearby regions of England, particularly in the southeastern direction toward which the wind was blowing; traces were found for considerable distances. Although consumption of contaminated milk was restricted, doses to the thyroid were as high as 0.1 Sv (Eisenbud 1987).

A reactor accident in the United States at the Three Mile Island II nuclear power plant, a pressurized water reactor, on March 28, 1979, destroyed the reactor core but not the vessel. The airborne radionuclides consisted almost entirely of 1×10^{17} Bq radioactive noble gases, with about 1×10^{12} Bq iodine-131. External doses in the immediate vicinity were estimated to be somewhat below 0.7 mSv, and internal doses were lower.

The serious accident in the USSR at Chernobyl nuclear power station destroyed a graphite moderated and water cooled reactor on April 26, 1986, by explosions and fires. Approximately 3×10^{18} Bq, 3.5% of its contents of fission products, were released into the air. More than one-half the radioactivity was deposited in the immediate vicinity (within 30 km), and most of the rest was deposited in nearby Russia, Scandinavia, and other countries in Europe (UNSCEAR 1988). Persons within the immediate vicinity were estimated to be exposed to effective dose equivalents as high as 3 Sv before being evacuated. Long lived radionuclides such as Cs-137 from Chernobyl have been observed in Russian and European surface soil and vegetation during subsequent years.

Airborne radionuclides from Chernobyl arrived in the United States on May 5, 1986, and persisted at readily detectable levels for about 2 months. The radionuclides were mostly low levels of cesium-137, -136, and -134, iodine-131, ruthenium-103, and tellurium-132. Some of these radionuclides were also detected in vegetation and foods, notably iodine-131 in milk. The radionuclides resulted in a maximum committed dose equivalent of 0.5 mSv to a child's thyroid from iodine-131, and much less to other organs and from the other radionuclides (Broadway et al. 1988).

Many smaller users of radionuclides in medicine, research, and education discharge radionuclides, typically into the air and sewer. Iodine-131 and technetium-99m used for therapeutic purposes at hospitals have been detected at low levels in municipal sewage treatment plants and in downstream bodies of water.

Human intervention in the natural environment by mining, construction, and manufacturing can also be a source of radiation exposure when terrestrial radionuclides are taken from the ground and exposed, separated, or concentrated. This pertains to minerals used for a variety of purposes, such as phosphate fertilizers, refractories, and building and paving materials. Water and natural gas pumped from the ground may carry elevated levels of radionuclides. In coal and wood, natural radionuclides at low levels may be concentrated by ashing. Manufactured products with elevated radionuclide contents may be returned to the environment as waste.

Solid and solidified liquid radioactive waste in drums or boxes from both small users and major facilities are placed in near-surface, low-level waste repositories that are controlled, maintained, and monitored. Materials are stored in large trenches that are backfilled and capped when full. In the recent past, some U.S. repositories were not operated satisfactorily. Several received damaged packages in poorly constructed trenches that collected rain and groundwater. Waste from damaged waste containers leaked into these trenches and then into the environs. As a result, three of the six repositories were closed, although doses to nearby persons were low—in the 0.01 mSv/year range. Recent regulations, issued in response to these poor waste storage practices, specify repository design and operation to control environmental contamination.

PATHWAYS OF RADIONUCLIDES THROUGH THE ENVIRONMENT

Radiation doses to persons ideally are determined by directly measuring the external radiation to which they are exposed and their body burdens of internally deposited radionuclides. In many circumstances, however, such measurements are difficult to perform, give uncertain results, or cannot be distinguished from the natural background.

Since the 1950s, pathway calculations are recognized as a useful alternative to such measurements of dose to persons, especially in large groups. By this technique, internal and external exposure is predicted from information on the release of radionuclides at the source or from measurements along the pathway at intermediate points. Calculations provide the extent of dispersion, of transfer from one medium to the next, of uptake by plants and animals in the food chain to man, and, finally, of the dose to humans. Several good texts discuss the behavior of radionuclides in the environment (Whicker and Schultz 1982; Eisenbud 1987) and present pathway models with compilations of transfer and dose coefficients (see Table 8).

TABLE 8
Information for radionuclide pathway and radiation dose calculations

Airborne dispersion and deposition:	Eisenbud (1987); NRC (1977a); Turner (1970); Till and Meyer (1983)
Direct radiation exposure from air and ground surface:	NRC (1977c); Till and Meyer (1983)
Radionuclide uptake by agricultural products:	Ng (1982); NRC (1977c); Till and Meyer (1983)
Subsurface movement in water:	Till and Meyer (1983)
Aquatic dispersion:	NRC (1977b); Till and Meyer (1983)
Radionuclide uptake by edible aquatic organisms:	NRC (1977c); Thompson et al. (1972); Till and Meyer (1983)
Radiation dose related to intake by humans:	NRC (1977c); Till and Meyer (1983)

The Nuclear Regulatory Commission (NRC) has described calculations for airborne dispersion and deposition (NRC 1977a) and for dispersion and dilution in water (NRC 1977b) that the agency uses with annual data of radionuclide releases to estimate doses due to effluents from nuclear power stations. These models were developed by NRC for radionuclides but are applicable to other pollutants and are based on generally used dispersion and transfer concepts. They include calculations for external radiation doses to persons on the ground or in water surrounded by airborne or waterborne radionuclides, respectively, or standing on radioactively contaminated soil or sediment (NRC 1977c).

The possible pathways taken by radionuclides in water or on the ground are many, as shown in Figure 1 (Platt et al. 1971). An initial estimate of transfer is obtained with a factor for predicting uptake by a second medium from the first. The bioaccumulation factor for fish from water, in units of Bq/kg fish per Bq/kg water, for example, is multiplied by the known concentration of a radionuclide in water to give its concentration in fish.

The next step in a simple path may then be consumption of food, such as fish, by humans. The calculation requires a value of food consumption (e.g., of fish) in kg/year by a person in order to compute annual radionuclide intake with this food in Bq per year and a dose factor in terms of Sv/Bq intake to compute the annual dose. The dose factor is unique for each radionuclide, and both intake and dose factors differ for generic population groups such as infants, children, teenagers, and adults. These calculations are also demonstrated in the NRC Regulatory Guide (NRC 1977c).

Many paths are less direct and can involve several alternative or sequential transfers. For each transfer, information is required on the radionuclide

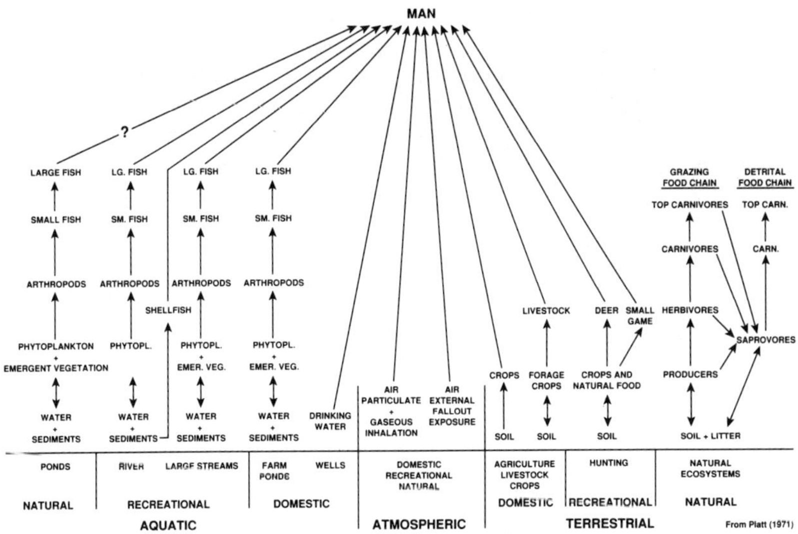

Figure 1. Ecosystem pathways (from Platt et al. 1971).

concentration in the initial medium and the equilibrium transfer factor or the transfer rate and the fraction transferred per intake.

Generic values of transfer factors for each element that has a radioisotope can be obtained from a variety of sources. One is the ratio of elemental concentrations in the two media between which the transfer occurs. Another is the ratio of radionuclide concentrations occurring in the environment. A third is the ratio of the radionuclide concentrations determined by tracer studies. If none of these are available, ratios are inferred from chemically similar elements.

Such values are useful for focusing calculations of radiation exposure on a few critical radionuclides, pathways, and most exposed groups of persons in what would otherwise be an almost impossibly complicated situation. Radiological monitoring guidance commonly advises such initial narrowing of effort, and radioecological studies have concentrated on these critical items.

Reality is too complex, however, for definition by generic transfer factors. Generic values are selected from experimental data that often range over several orders of magnitude. An element may have several chemical or physical forms that behave differently; pathways are affected by environmental conditions; and generic transfer or uptake factors are crude approximations for a specific animal or plant species or soil type. Biota age and physical condition, temperature, season, and nutrient availability all influence transfers.

For example, the widespread studies of cesium-137 uptake in food animals encountered unexpectedly high levels in feed under several circumstances. Accumulation of fallout deposited over years by northern lichens and mosses was responsible for elevated levels in reindeer and caribou, while high uptake by certain vegetation from nutrient-deficient soils caused elevated levels in deer and dairy cows in the southeastern U.S. coastal regions (Whicker and Schultz 1982).

An important pathway for airborne radionuclides to humans is by deposition on pasture grass, consumption of grass by cows, transfer to milk, and consumption of milk by humans. This pathway is critical for iodine-131 (half life = 8 days), but is also important for radioactive cesium, strontium, and barium. Cows consume grass from a large area when on pasture and transfer a large fraction of ingested iodine to milk, which reaches the consumer in only 2-3 days.

This pathway presents an example of the extent of radioecological studies. These studies have examined deposition of radionuclides in precipitation *versus* dry deposition; retention on vegetation surfaces as a function of shape and condition; distribution of deposition between vegetation and soil; uptake by direct deposition on surfaces *versus* uptake by roots from soil for longer lived radionuclides; movement between vegetation and ground as a function of time; intake with feed by the animal in fresh *versus* stored feed; and transfer of radionuclides into the cow's or goat's lactating system and radionuclide concentration in the milk as function of milk production.

In the freshwater ecosystem, drinking water is the direct pathway to humans, but fish consumption may be the critical pathway because fish accumulate certain radioelements. The bioaccumulation factor from water to fish characterizes the pathway when the fish takes much of the radioelement from water, as in the case of calcium. For radio-phosphorus, among others, the pathway is through the food chains shown in Figure 1, and bioaccumulation depends largely on the extent of food availability, feeding practices, and radioactive half life. Any cited bioaccumulation factor only documents an observed overall value for the element in fish relative to water, not necessarily the pathway.

Consideration of the varied paths of radionuclides in aquatic systems requires the information identified in Figure 2 (Strauch 1971). Sorption on suspended solids and deposition in sediment reduces the concentration of radionuclides dissolved in water, but the radionuclides may still move in insoluble form, return to soluble status, or be taken up by biota.

Movement of radionuclides in groundwater is controlled by the water flow rate and the degree of sorptive retention of radioelements on soil and rock. Only if the medium is sand-like with well defined ion-exchange properties are the usual groundwater models fully applicable. More commonly, the

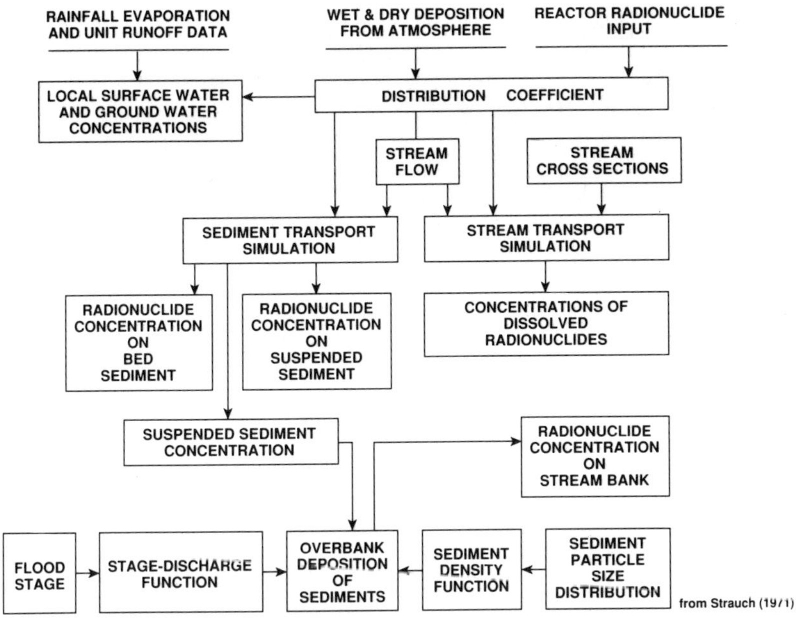

Figure 2. Radionuclide transport model in water (from Strauch 1971).

ground is a highly nonuniform medium for water flow that is defined only in a limited way by observations at wells.

Dispersion in air and surface water can be observed in much greater detail, but is also highly subject to local features. Conditions of release, nearby structures, the weather, and topography all affect dispersion and dilution.

A detailed numerical approach to calculating radionuclide transfer through the environment considers each medium to be a compartment, as illustrated in Figure 3 (Booth et al. 1971). Each compartment is defined in terms of its mass, amount of stable elements, connection with other compartments, and rates of inflow and outflow along each connecting path. In Figure 3, S is the fraction of fallout depositing on the compartment, D adjusts the dietary intake of a generic food component to the compartment, and T is the transfer rate. These characteristics may be constant or may change in a specified pattern.

Differential equations are written for each compartment. Given information on radioelement input and boundary conditions, groups of equations can be solved to obtain concentrations and specific activities (Bq of radioisotope/kg of element) as a function of time in a compartment.

Although compartmental models enhance understanding of complex systems, numerically they work best for just a few compartments and relatively

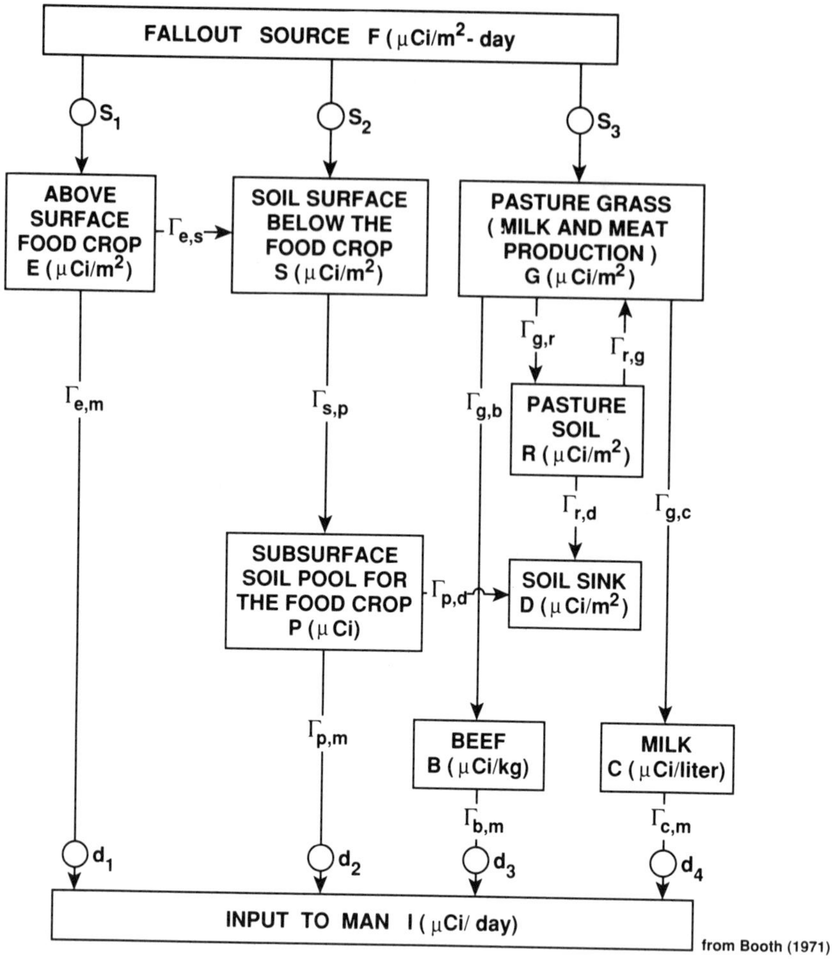

Figure 3. Coupled compartment model for radioactive fallout in terrestrial food paths (from Booth et al. 1971).

simple conditions. It is too much to expect that responses of biota to their environment fit into neat mechanical patterns. Such models do give useful predictions of changing radionuclide levels in media in response to brief contaminating events and other nonequilibrium circumstances.

Compartment models are also applied successfully to predict internal transfers within a system, for example, a plant or animal. Radionuclide flows between roots and leaves, blood and various tissues, and surface and subsurface ground are clarified by such models.

The described transfer calculations usually are formulated as computer programs, from simple to highly intricate, for application to brief or continuous releases, single or multiple sources, and local or regional exposures. Use of computers allows consideration of many radionuclides, pathways, and exposed persons. A wide selection of generic and site-specific factors can be stored for use, and the impacts of numerous source and condition parameters can be compared. The output of such computer analyses can be impressive, but the calculated radionuclide concentrations and radiation doses are appropriate to a situation only to the extent that the algorithms, factors, and source input are appropriate.

The usual pathway calculation at a nuclear facility, such as a nuclear power station or research laboratory, starts with the radionuclide discharge. Gaseous and particulate radionuclides are commonly discharged through stacks and vents into air, and liquid effluent with dissolved and particulate radionuclides are discharged into bodies of water. Regulations require reporting effluent radionuclide amounts and concentrations at least annually and limiting releases to concentrations that will not result in radiation exposures above regulatory limits to any person off site.

The model begins by predicting radionuclide concentrations in air beyond the site boundary and radionuclide concentrations in water at the point of discharge. Atmospheric dispersion of routinely discharged radionuclides conventionally is calculated (NRC 1977a) with a Gaussian distribution model to determine radionuclide concentrations in ground-level air for sectors in 16 wind directions, incrementally from the boundary to 80 km distant. Dispersion factors averaged over a year are determined on the basis of the release height, wind directions and speeds measured continuously at a meteorological tower on site, and atmospheric stability categories inferred from temperature gradients measured at the tower. Complications are introduced into these calculations by topographical features such as hills and valleys and by the necessity of matching nonuniform radionuclide releases to varying meteorological conditions.

These radionuclide concentrations are combined with breathing rates and dose factors to yield the inhalation dose, with immersion dose factors to yield the external radiation dose from beta particles and gamma rays, and with the deposition velocity for particles and gases to yield the amount of radionuclide per area of the ground surface. This information for deposited radionuclides is then used to calculate the external radiation and to initiate following the movement of these radionuclides through the food chain to humans.

Calculation of aquatic dispersion from discharge canals or submerged diffusion pipes utilizes the models cited earlier (NRC 1977b) in the receiving stream, reservoir, or bay. Significant deviations from predictions can occur when the radionuclides are in relatively warm effluent coolant water that may rise to the surface, the discharge is under tidal influence, or flow in the receiving stream is affected by islands or other barriers.

The predicted radionuclide concentrations are used for calculating external radiation doses to swimmers, boaters, and persons on the banks or shore, and internal radiation doses from drinking water—as modified by water treatment—if a water supply intake receives the diluted effluent. These concentrations are also the beginning of food chain calculations for consumers of fish and other aquatic biota.

Airborne effluents can contribute to aquatic pathways by deposition on bodies of water and by runoff, both surface and subsurface. Conversely, application of irrigation water or deposition of dredge material transfers aquatic radionuclides to the terrestrial ecosystem.

Another category of radionuclide sources is subsurface radioactive waste storage sites. There are existing and planned near-surface burial facilities for low-level radioactive waste and deep repositories for high-level radioactive and transuranium wastes. Models estimate leaching of radionuclides from solids in containers by groundwater and their subsequent movement through soil.

Uncontrolled access to radioactively contaminated sites represents a possible source of radiation dose for the future. Scenarios are considered of persons inadvertently intruding on above-ground or subsurface repositories of radioactive material. Major changes in climate and topography are potential factors in exposing such repositories to human access in the distant future.

Radioecological studies to delineate pathways and predict radionuclide concentrations in air, water, and food for estimating doses to humans were performed on a large scale in the United States to trace effluents from AEC facilities in the 1940s and 1950s. Extensive study programs existed at the Hanford, Idaho, Oak Ridge, and Savannah River sites, among others. Site operators monitored discharges by sampling airborne and liquid effluents, and the environment by sampling airborne particles, deposition from air, surface and groundwater, soil and sediment, vegetation, food crops and milk, livestock and wildlife, and by measuring radiation exposure. Special studies were performed to relate radionuclide discharge rates to monitored concentrations and to identify major exposure pathways and concentrating or accumulating media.

Aquatic pathway studies, notably in the Columbia, Clinch, and Savannah Rivers, observed the distribution of radionuclides between water and sediment and the movement through lower aquatic life forms to fish. Radionuclide uptake patterns by fish were related to feeding patterns, water quality, and the chemical form of the various radionuclides. The three rivers provided widely different aquatic systems for study, and an opportunity for a long-term study under relatively controlled conditions existed in several impoundments, particularly White Oak Lake at Oak Ridge.

Studies at AEC facilities also examined the deposition of airborne radionuclides on vegetation and soil and the subsequent movement of the radionuclides into the root system or deeper into the soil. The movement of radionuclides from liquid waste ponds and seepage pits through the ground was examined with numerous test wells.

Nuclear weapon tests were followed by extensive radioecological studies by AEC contractors near test sites in the Nevada desert and on South Pacific coral reef atolls. Because of the worldwide distribution of fallout, groups in many countries studied the movement of the major longer lived radionuclides through the environment.

Consolidation of data from widespread environmental monitoring of fallout, both airborne and at ground level, identified patterns of transfer that can be used to predict radionuclide burdens, compartment amounts, turnover rates, and transfer factors. Upper atmosphere air flows were identified that pick up radioactive particles from the nuclear explosion plume and transport them across the United States and around the world. Transfer patterns to the troposphere with maxima in spring were observed, as was a residence time of about 1 year in the atmosphere. Precipitation intersecting with that air path was found to bring much more radioactivity to ground level than dry deposition.

Studies of food intake and human body burdens yielded calculational models that predict radionuclide concentrations in human tissue and organs from concentrations in foods and identify the critical pathways for the major fallout radionuclides to humans. Fallout radionuclide concentrations, decreasing from the maxima in the mid-1960s in foods and human tissues, are still being followed. The models consider the combined effect of direct surface deposition from air and the indirect soil-root pathway.

The recent accident at the Chernobyl nuclear plant renewed interest in environmental transport studies in the countries where radionuclide levels are elevated. The data expand the information accumulated in fallout studies by observing radionuclides in different physical forms, at relatively high levels over a wide range of ambient conditions throughout Europe (UNSCEAR 1988).

Operation of nuclear power plants and associated nuclear fuel cycle facilities in the United States and other countries is accompanied by extensive environmental monitoring and some special environmental studies in response to regulations for determining and controlling environmental impacts. Studies at the first commercial U.S. plants in the late 1960s and early 1970s traced plumes of radioactive noble gases in air and followed iodine-131 airborne deposition from pasture to dairy cows to milk for human consumption. Radionuclides in the aquatic pathway were followed to sediment, food fish, and water supplies (Blanchard and Kahn 1973, 1979). Reduction of radioactive effluents in new power plants has made detection of radionuclides in the environment for study purposes increasingly difficult over the years, while, of course, reducing radiation exposures to the population.

To comprehend environmental radionuclide transfers in complex systems, even via a clearly specified route, monitoring and *in-situ* studies have been supported by laboratory and controlled field experiments to isolate the component factors. Although both controlled experiments and *in-situ* measurements have obvious drawbacks, combining them has contributed considerably to understanding how substances move through the environment and what factors most influence their transfer.

It is important to recognize that, even after many years of radiological monitoring and environmental study, surprises still occur when radionuclides are found in unexpected media at unexpected concentrations. Although transfer and uptake factors are presented for every element and most common pathway interfaces, some values have had little evaluation and others are based on observations that may not be applicable to the circumstances under consideration.

RADIATION EFFECTS IN THE ECOSYSTEM

The effects of radiation on the various components of the ecosystem have been studied during the past 40 years to avoid irreversible harm by limiting the dose from man-made radiation and by remedying the results of accidents. Studies of radiation effects and radionuclide pathways, for obvious reasons, have been focused on humans, but considerable efforts have also been devoted to animals and plants. A number of surveys have presented the results of ecosystem effects studies (Blaylock and Trabalka 1978; Turner 1975; Whicker and Fraley 1974), and two recent reports by expert committees present excellent overviews (International Atomic Energy Agency [IAEA] 1991; NCRP 1991).

The most appropriate studies of ecosystem effects, but also the fewest, have been performed at the community level. Biota were observed for years in White Oak Lake at the Oak Ridge National Laboratory, which was continually contaminated with mixed fission and activation products. A similar study was undertaken in an unnamed lake in Russia. Terrestrial studies have been performed at locations of elevated natural terrestrial background radiation and also of man-made radionuclide contamination. Irradiations by encapsulated gamma-ray sources have been used to study forests and grasslands within controlled areas and their indigenous small mammals and insects (IAEA 1991; NCRP 1991).

The greatest source of information concerning the effects of radiation on animals and plants is laboratory studies of external radiation or radionuclide uptake. Most were undertaken to support radiation protection of humans by comparing radionuclide distributions and radiation effects among tissues or organs and the impacts of various radiations and dose rates. As a result, more is probably known of radiation responses in small laboratory animals, such as mice, than for humans. Most information on mutagenic effects comes from major studies with *Drosophila* and mice.

Larger animals, such as miniature pigs, beagles, and primates, were used in laboratory studies for better extrapolation of radionuclide uptakes and radiation effects to humans. Animals were selected for similarity to humans in the function or response of specific organs or systems.

Concern for protecting the human food supply was another stimulus for animal studies. The effect of radiation on fisheries, particularly in oceans, was considered soon after the development of nuclear weapons (NRC 1957; Templeton et al. 1971), as was the effect on domestic animals, with focus on livestock (Bell et al. 1971; Garner and Barber 1966).

Biota were exposed to acute irradiation to determine the immediate impact of an accident and to chronic radiation exposure to evaluate the effects of lifetime exposure to elevated levels. In each case, doses and dose rates were presented at several levels to observe effects ranging from the lethal to the barely detectable. In practice, acute exposures were not always as brief compared to the biological response, and chronic exposures, not always for the lifetime of the subject, as their definitions would imply. In some cases, chronic dose rates were not as uniform as would be desired. Partitioned exposure—repeated acute exposures at lower doses—bridged the two extremes to a certain extent.

The overview of acute radiation exposure effects in Table 9 shows that small mammals have a range of lethal response in terms of LD_{50} that

TABLE 9
Range of LD_{50} for organisms under acute radiation exposure, in Gy

Plants[a]	
trees	3-150
lettuce	50
barley and wheat	20-30
Vertebrates[a]	
small mammals	5-11
domestic livestock	4-7
birds	5-30
Aquatic organisms[b]	
fish	10-100
amphibians	1-10
crustaceans	10-200
molluscs	200-2000
microorganisms	10-5000

[a]From IAEA (1991).
[b]Estimated from graph in NCRP (1991).

includes the value of 5 Sv for humans. The LD_{50} value appears to be up to an order of magnitude higher for fish and up to two orders of magnitude higher for microorganisms. This value is also about two orders of magnitude higher for insects (IAEA 1991) and for mosses and lichens. The information on plants is from field studies, while all other data are from the laboratory. Doses rather than dose equivalents (units of Gy, not Sv) are reported for these animal and plant studies, but, because the ionizing radiation is generally gamma rays or beta particles, these values may be compared numerically to the dose equivalent for humans.

One concern in these studies has been to identify species within taxa that are especially radiation sensitive, and a low LD_{50} value has been considered one indicator of such sensitivity. Thus, pine trees appear to be most sensitive among trees, and the estimated LD_{50} in Table 9 at the low end of the range for trees applies to pines.

Various conditions of the irradiation, the subject, and its environment affect LD_{50} values. Extended irradiations, partitioned or chronic, generally decrease the impact of the radiation (i.e., elevate the LD_{50} value). Biota in their developmental stages are most sensitive to radiation. The influence of the environment is especially important for poikilothermic animals and requires, for example, that the definition of death within 30 days—$LD_{50/30}$—for many aquatic biota be changed to $LD_{50/60}$.

The common pathological effects found in humans—systemic malfunctions, tissue injury and destruction, and tumor induction—were observed in animals and plants over a range of sublethal doses. At still lower doses and dose rates, infertility or reduction in fecundity occurs. The effects detected at the lowest doses—in the range of 0.2 to 2 Gy—are genetic mutations and cytogentic observations of chromosome changes. These are the most sensitive indicators of radiation effects.

Communities irradiated at sublethal levels show changes in species abundance, composition, and diversity. These changes appear at doses of 3-30 Gy in certain trees and grasses (IAEA 1991).

Chronic exposure studies examining the onset of radiation effects significant to a population were performed at radiation dose rates ranging from the natural radiation background to about 100 Gy/year. The average radiation background from cosmic and terrestrial radiation to humans in the United States and Canada (see Table 6) is about 0.6 mSv/year (2 μGy/day). This value depends on the elevation of the locale with respect to cosmic radiation and on the concentration of radionuclides in soil for terrestrial radiation. Moreover, the major component of the dose rate, as in humans, may be radioactive material within the organism. The immediate environment of the organism can also contribute additional radiation (i.e., the soil) or shield from radiation (i.e., a body of water). A range of 0.7-1.5 μGy/day has been estimated for aquatic organisms (Whicker and Schultz 1982).

Field studies of small mammals exposed to chronic irradiation showed no changes in the reproduction rate at dose rates of 10-20 mGy/day, where fecundity is considered to be the most sensitive indicator of response to radiation at the population level. Effects at lower dose rates, such as a decline in the sperm count at 4 mGy/day but not 1 mGy/day, did not appear to affect the population (IAEA 1991). Studies of *Gambusia affinis* in White Oak Lake, exposed to radionuclides that yielded doses decreasing from above 10 to 0.6 mGy/day in about 17 years, showed a higher fecundity attributed, however, to high productivity in the water (NCRP 1991).

The two expert committees that reviewed radiation effects data both made recommendations concerning appropriate levels for protecting the ecosystem. The NCRP Scientific Committee on the Effects of Ionizing Radiation on Aquatic Organisms proposed 10 mGy/day (0.4 mGy/hr) (NCRP 1991) as a reasonable criterion for protecting populations. The IAEA Committee on the Effects of Ionizing Radiations on Plants and Animals concluded that limiting the dose rate to 1 mGy/day for terrestrial populations provides adequate protection because it is unlikely to cause measurable detrimental effects (IAEA 1991).

The committees paid closest attention to the studies of populations in the field because of the need to consider radiation impacts in natural settings with contributions of multiple factors. Results of laboratory studies, with the benefits of control over radiation levels, subject lineage and health, and conditions of exposure, can indicate the expected magnitude of the effect; wide differences from field results call attention to the need to examine study conditions for explanations.

For example, effects may be lower than observed in the laboratory because of recruitment from nearby to replace affected individuals, increased activity by unaffected members of the population, or adverse radiation effects on predators. Conversely, effects may be greater than predicted if habitat or food supply is adversely affected.

Most study results examined by the two committees showed no pathological responses in individuals at or below the recommended dose rates and no adverse somatic or genetic effects on populations. Although cytogenetic changes could be observed at levels below 10 mGy/day in some organisms, it was concluded that these would not harm the population.

Several reports of more severe radiation effects observed at lower dose rates were also considered. Some, in both aquatic and terrestrial systems, are at dose rates so near the natural radiation background that the reported adverse effects were unconvincing. For some reports, efforts at replications did not confirm the reported effects; for others, it was speculated that the dose was higher than reported due to radiations other than the measured gamma rays. In the aquatic study in Russia, insufficient information concerning water quality is available to assure that any observed effects were induced by radiation.

These concerns call attention to some of the problems inherent in field studies. Effects from other causes, either separately or in synergism with radiation, are always a possibility. In many cases, the dose in a complex or inhomogeneous radiation field is poorly defined.

In considering radiation protection for animals and plants at the population level, effects at low radiation dose rates may lose the significance that they have for individuals. Some adverse effects can be counterbalanced, as indicated by observations of lower incidence of fertility balanced by larger litters. For species where the egg or newborn survival fraction is very small, the probability of transmitting mutations is also small.

The two committees tested the assumption made by radiation protection agencies that protecting the human population against undue radiation exposure from effluents at nuclear facilities also protects the environment. They applied the pathway models discussed in the preceding section to calculate doses to terrestrial and aquatic biota from discharged radionuclides when the discharges were limited to a dose of 1 mGy/year to the most exposed person. The calculational results showed that some biota in the paths from radionuclide sources to human would have higher dose rates, as expected, but generally supported the assumption that the highest dose rates were well below the acceptable levels suggested above, of 3.6 Gy/year for aquatic biota and 0.36 Gy/year for terrestrial biota.

The committees cautioned that site-specific calculations and possibly supporting studies would be needed if such calculated levels approached or exceeded the recommended limits at a particular location, as occurred for a few sample calculations (NCRP 1991). The generic nature of the transfer coefficients and pathways considered in the evaluation may not be appropriate to the site. The location may have media that accumulate or concentrate radionuclides in unexpected ways or biota that are unusally sensitive to radiation or stresses on the ecosystem that act synergistically with radiation.

The principle of limiting the radiation dose rate to protect an animal or plant population is not applicable to animals and plants that are rare, threatened, or otherwise valued. Under this circumstance, the potential dose to the most exposed individual must be calculated for comparison with information on radiation effects in individuals and limited accordingly.

CONCLUSION

Consideration of risks from radioactive materials in the environment has focused on protecting humans. Sources of radionuclides are monitored, pathways and the extent of transfer from the source to the point of exposure are determined, and the radiation dose to the most exposed individual is considered relative to doses known to cause harm. Limits on radionuclide releases, concentrations, or intakes by and dose to the individual are then established to reduce the predicted incidence of harm to an acceptably low level. Information for setting the limits has been accumulated from many radioecological studies of radionuclide movement and radiobiological and epidemiological studies of radiation effects. In this way, risks from ionizing radiation have been evaluated at least as thoroughly as from any other pollutant, and possibly more so.

Animals and plants have also been the subjects of extensive radiological studies. Many of these were experiments of radionuclide uptake or radiation effects for application to humans or tests to assure that the human food supply is safe at given levels of radiation exposure. Others, however, were designed specifically to determine the impact of radionuclides and radiation on the terrestrial or aquatic community under consideration. Used for these purposes were sites that had elevated natural radiation or were contaminated with man-made radionuclides in the course of routine facility operation, accidentally, or intentionally for the experiment. Encapsuled sources of gamma radiation were also used for radiation effects studies.

The pathway studies provide a wide range of transfer and bioaccumulation factors for terrestrial and aquatic ecosystems. These permit development of lists of generic factors for predicting the concentration of any element—hence, any radionuclide—in particular animal or plant taxa, given the source and the setting. The radiation bioeffects studies specify the magnitudes of radiation doses for such taxa that result in harm ranging from the lethal to systemic injury, infertility, reduced fecundity, tumors, or mutations. In combination, this information permits calculating radiation effects throughout the ecosystem due to specified levels of radioactive materials and setting protective levels at which these effects occur at an acceptably low incidence.

The conventional approach to establishing such protective levels for animals and plants is different than for humans in that the aim is to maintain a healthy community rather than protect the individual. Hence, the annual protective levels recommended by expert committees are higher than the levels for humans of 1 mSv by a factor of 400 for terrestrial plants and animals and of 4000 for aquatic biota.

The expert committees also tested the commonly applied hypothesis that controls to protect humans serve to protect the rest of the ecosystem by performing pathway and dose calculations for a variety of sources of radionuclides entering the environment. In most instances, the hypothesis was confirmed, but the occurrence of a few instances of doses to animals that exceeded the recommended limits cautions that doses to all parts of the ecosystem must be considered in addition to the dose to humans.

It would be excessively optimistic to assume that applying calculational models with a set of generic transfer and bioaccumulation factors and dose response values will assure protection of any natural community. Comparisons of radiation effects in field and laboratory studies have often shown differences due to site-specific conditions that could not have been predicted. Wherever the potential for elevated radiation exposure exists, exposure of the ecosystem must be considered, supported by appropriate studies. Among special needs are identifying the population most sensitive to radiation, factors that could act in combination with radiation, and individuals that must be protected because they are rare.

REFERENCES

BROWN, E. and FIRESTONE, R. (1986). Radioactivity Handbook. Interscience Press, NY.

BELL, M.C., SASSER, L.B., and WEST, J.L. (1971). Simulated fallout radiation effects on livestock in the event of nuclear war. In: *Survival of Food Crops and Livestock in the Event of War*, Report No. CONF-700909 (D.W. Benson and A.H. Sparrow, eds), pp. 193-207. U.S. Atomic Energy Commission (now Department of Energy), Washington, DC.

BLANCHARD, R.L. and KAHN, B. (1973). The fate of radionuclides discharged from a PWR nuclear power station into a river. In: *Environmental Behavior of Radionuclides Released in the Nuclear Industry*, pp. 195-204. International Atomic Energy Agency, Vienna, Austria.

BLANCHARD, R.L. and KAHN, B. (1979). Abundance and distribution of radionuclides discharged from a BWR nuclear power station into a marine bay. Nucl. Safety 20:190-205.

BLAYLOCK, B.G. and TRABALKA, J.R. (1978). Evaluating the effects of radiation on aquatic organisms. In: *Advances in Radiation Biology 7* (J.T. Lett and H. Alder, eds), pp. 103-152. Academic Press, NY.

BOOTH, R.S., KAYE, S.V., and ROHWER, P.S. (1971). A systems analysis methodology for predicting dose to man from a radioactively contaminated terrestrial environment. In: *Proceedings of the Third National Symposium on Radioecology*, Report No. CONF-710501 (D.J. Nelson, ed), pp. 9-17. Oak Ridge National Laboratory, Oak Ridge, TN.

BRAUER, F.P., SOLDAT, J.K., TENNY, H., and STREBIN, R.S., JR. (1974). Natural iodine and iodine-129 in mammalian thyroids and environmental samples taken from sites in the USA. In: *Environmental Surveillance around Nuclear Installations*, pp. 43-66. International Atomic Energy Agency, Vienna, Austria.

BROADWAY, J.A., SMITH, J.M., NORWOOD, D.L., and PORTER, C.R. (1988). Estimates of radiation dose and health risks in the United States population following the Chernobyl Nuclear Plant accident. Health Phys. 55:533-539.

COMMITTEE ON THE BIOLOGICAL EFFECTS OF IONIZING RADIATIONS (BEIR). (1990). Health Effects of Exposure to Low Levels of Ionizing Radiation. National Research Council, Washington, DC.

CUMMINS, C.L., MARTIN, D.K., and TODD, J.L. (1991). 1990 Savannah River Site Environmental Report, Report WSRC-IM-91-28. Westinghouse Savannah River Co., Aiken, SC.

EISENBUD, M. (1987). Environmental Radioactivity, 3rd ed. Academic Press, NY.

GARNER, R.J. and BARBER, D.A. (1966). Effects of radioactive substances on animals and plants. In: *Radioactivity and Human Diet* (R.S. Russell, ed), pp. 159-170. Pergamon Press, NY.

GESELL, T.F. and LOWDER, W.M. (1980). Natural Radiation Environment III, Report No. CONF-780422. U.S. Department of Energy, Washington, DC.
IAEA. (1991). Effects of Ionizing Radiation on Plants and Animals at Levels Implied by Current Radiation Protection Standards, Tech. Rep. Ser. International Atomic Energy Agency, Vienna, Austria.
ICRP. (1991). 1990 Recommendations of the International Commission on Radiological Protection, Publ. No. 60. Pergamon Press, NY.
KNOLL, G.F. (1989). Radiation Detection and Measurement, 2nd ed. John Wiley and Sons, NY.
KOCHER, D.C. (1981). Radioactive Decay Tables, DOE/TIC-11026. National Technical Information Service, Springfield, VA.
NCRP. (1975). Krypton-85 in the Atmosphere, Report No. 44. National Council on Radiation Protection and Measurements, Bethesda, MD.
NCRP. (1979). Tritium in the Environment, Report No. 62. National Council on Radiation Protection and Measurements, Bethesda, MD.
NCRP. (1985). Carbon-14 in the Environment, Report No. 81. National Council on Radiation Protection and Measurements, Bethesda, MD.
NCRP. (1987a). Exposure of the Population in the United States and Canada from Natural Background Radiation, Report No. 94. National Council on Radiation Protection and Measurements, Bethesda, MD.
NCRP. (1987b). Recommendations on Limits for Exposure to Ionizing Radiation, Report No. 91. National Council on Radiation Protection and Measurements, Bethesda, MD.
NCRP. (1987c). Public Radiation Exposure from Nuclear Power Production in the United States, Report No. 92. National Council on Radiation Protection and Measurements, Bethesda, MD.
NCRP. (1991). Effects of Ionizing Radiation on Aquatic Organisms, Report No. 109. National Council on Radiation Protection and Measurements, Bethesda, MD.
NG, Y.C. (1982). A review of transfer factors for assessing the dose from radionuclides in agricultural products. Nucl. Safety 23:57-71.
NRC. (1957). The Effects of Atomic Radiation on Oceanography and Fisheries. National Research Council, Washington, DC.
NRC. (1977a). Methods for Estimating Atmospheric Transport and Dispersion of Gaseous Effluents in Routine Releases from Light-water-cooled Reactors, Regulatory Guide 1.111. U.S. Nuclear Regulatory Commission, Washington, DC.
NRC. (1977b). Estimating Aquatic Dispersion of Effluents from Accidental and Routine Reactor Releases for the Purpose of Implementing Appendix I, Regulatory Guide 1.113. U.S. Nuclear Regulatory Commission, Washington, DC.
NRC. (1977c). Calculation of Annual Doses to Man from Routine Releases of Reactor Effluents for the Purpose of Evaluating Compliance with 10CFR50, Appendix I, Regulatory Guide 1.109. U.S. Nuclear Regulatory Division, Washington, DC.
O'BRIEN, K. and MCLAUGHLIN, J.E. (1972). The radiation dose to man from galactic cosmic rays. Health Phys. 22:225-232.
OFFICE OF RADIATION PROGRAMS (ORP). (1982). Environmental Radiation Data, Report 30, USEPA Report No. EPA 520/5-6-83-006. Office of Radiation Programs, Montgomery, AL.
PLATT, R.B., RAGSDALE, H.L., and SHIVE, D.J. (1971). An ecological approach to the evaluation of radioactivity within the man-environment ecosystem. In: *Proceedings of the Third National Symposium on Radioecology*, Report No. CONF-710501 (D.J. Nelson, ed), pp. 9-17. Oak Ridge National Laboratory, Oak Ridge, TN.

PACIFIC NORTHWEST LABORATORY. (1991). Air Pathway Report, Phase 1 of the Hanford Environmental Dose Reconstruction Project, Report No. PNL-7412 HEDR, Rev. 1. Pacific Northwest Laboratory, Richland, WA.

STRAUCH, S. (1971). The year 2000 - nuclear power and man. In: *Proceedings of the Third National Symposium on Radioecology*, Report No. CONF-710501 (D.J. Nelson, ed), pp. 9-17. Oak Ridge National Laboratory, Oak Ridge, TN.

TEMPLETON, W.L., NAKATANI, R.E., and HELD, E.E. (1971). Radiation effects. In: *Radioactivity in the Marine Environment*, pp. 223-239. National Research Council, Washington, DC.

THOMPSON, S.E., BURTON, C.A., QUINN, D.J., and NG, Y.C. (1972). Concentration Factors of Chemical Elements in Edible Aquatic Organisms, Report No. UCRC-50564, Rev. 1. Lawrence Livermore Laboratory, Livermore, CA.

TILL, J.E., III and MEYER, H.R. (1983). Radiological Assessment, Report No. NUREG/CR-3332. U.S. Nuclear Regulatory Commission, Washington, DC.

TURNER, D.B. (1970). Workbook of Atmospheric Dispersion Estimates. U.S. Environmental Protection Agency, Research Triangle Park, NC.

TURNER, F.B. (1975). Effects of continuous irradiation on animal populations. In: *Advances in Radiation Biology 5* (J.T. Lett and H. Adler, eds), pp. 83-144. Academic Press, NY.

UNSCEAR. (1982). Sources, Effects and Risks of Radiation. United Nations Scientific Committee on the Effects of Atomic Radiation, United Nations, NY.

UNSCEAR. (1988). Sources, Effects and Risks of Radiation. United Nations Scientific Committee on the Effects of Atomic Radiation, United Nations, NY.

WHICKER, F.W. and FRALEY, L. (1974). Effects of ionizing radiation on terrestrial plant communities. In: *Advances in Radiation Biology 4* (J.T. Lett, H. Adler, and M. Zelle, eds), pp. 317-366. Academic Press, NY.

WHICKER, F.W. and SCHULTZ, V. (1982). Radioecology: Nuclear Energy and the Environment. CRC Press, Boca Raton, FL.

CHAPTER 15

PREDICTING ECOSYSTEM RISK: GENESIS AND FUTURE NEEDS

John Cairns, Jr. and B. R. Niederlehner
University Center for Environmental and Hazardous Materials Studies
Virginia Polytechnic Institute and State University
Blacksburg, Virginia

INTRODUCTION

In the past, ecosystem risk has not been of much concern. Society believed that the supply of some resources was infinite, that ecosystem destruction was a necessary component of progress, that ecosystems not being harvested were of no benefit to society, that any use was harmless until evidence of harm became direct and dramatic, and that technological advances would continue to correct any problems resulting from destruction of natural resources. These beliefs have been challenged in the last part of this century, and public attitudes in developed countries have undergone a remarkable transformation. Public priorities have shifted to embrace environmental values. In the United States, over 70% of people questioned identified themselves as environmentalists and acknowledged personal concern for pollution (Gallup 1990). Attitudes in developing countries have not changed as dramatically, and for good reason: these societies must remain focused on day-to-day personal survival. Even under such circumstances, however, resources and services found in natural systems are increasingly appreciated worldwide.

Even with a general increase in awareness of the benefits of ecosystem services, the potential difficulties in re-creating what is destroyed, society's dependence on the life support system of its planet, and the frontier attitude (if an ecosystem is damaged one can move elsewhere and find relatively unspoiled territory) persist. These attitudes are evidenced by the comments of a boat tour narrator in Fort Lauderdale, Florida, who stated that not many years ago this area that is now filled with expensive homes, boats, docks, and green lawns was a "useless" swamp infested with alligators, frogs, and mosquitoes. In this mind set, wetlands are useless and rivers are primarily waste conduits. Pollution is not a problem as long as the next downstream user is unaffected. For example, water laws in Colorado as-

---

1. Corresponding Author: John Cairns, Jr., 1020 Derring Hall, University Center for Environmental and Hazardous Materials Studies, Virginia Polytechnic Institute and State University, Blacksburg, VA 24061-0415.

2. Key Words: aquatic toxicology, ecosystem services, biological hierarchy, landscape ecology, assessment end points, measurement end points.

sumed, until recently, that a beneficial use of water required removal from the stream and a demonstrable economic benefit. The thriving business of whitewater rafting clouded the issue somewhat, but, in many parts of the West, Spanish water law, of appropriative use having top priority, still prevails. However, shifts in attitude have occurred in almost all developed countries, and, in some cases (e.g., Sweden), they are more dramatic for the country as a whole than for the United States. This chapter focuses primarily on changes in attitude and the accompanying development of methods for predicting ecosystem risk in the United States and areas in which these approaches can be improved.

PREDICTING RISK IN THE AQUATIC ENVIRONMENT: AN EXAMPLE

Sloof (1983) states "Around the turn of the century, aquatic toxicology was born as an illegitimate child of classical (mammalian) toxicology." In classical toxicology, the object of protection is specific and unambiguous, i.e., human health. In aquatic and wildlife toxicology, the focus has progressively broadened from one test species (parallel to classical toxicology) to thousands of interacting species that make up a natural community. Some organisms upon which mankind depends (domestic stock) can be protected by using risk estimates based on toxicity tests similar to those utilized for humans. This has been extended to commercially or recreationally valuable organisms in the environment, initially for the aquatic ecosystems—fish. In the early 1950s, Cairns had the idea that picking "representative" organisms from different levels of the food chain (e.g., algae and invertebrates as well as fish) would more faithfully display the range of response to toxicants than fish alone (e.g., Cairns 1956a,b). The extension of environmental management goals from protection of fish, a harvestable resource with a market value, to the more inclusive goal of protecting biotic integrity for recreation, aesthetics, and the intrinsic worth of a diverse and functioning community is a much more recent development, perhaps starting with the Water Quality Act Amendments of 1972 (PL 92-500), and the extension is not yet complete.

The process of predicting ecosystem risk can be broken down into three components (e.g., Cairns et al. 1978; National Research Council [NRC] 1983; Suter 1990). First, the problem must be defined. This is a somewhat subjective process, codifying societal values. Next, a scientific risk assessment compares a prediction of the level of stress that will occur in the environment after some human activity to a prediction of likely biological effects at that level of stress; this comparison is used to estimate the probability of an adverse effect. After the scientific assessment, risk is considered, along with benefits, costs, options, and political will, to arrive at a risk management plan. A subsequent, but important step, is the comparison of predictions of risk to observed effects in natural systems for the purposes of quality control and validation of methods for risk prediction.

Over the years, several factors have been observed as affecting the accuracy of a risk assessment in an aquatic environment: environmental factors modifying toxicity, interactions between stresses, differences in sensitivity between ecosystems, differences in sensitivity between species, interaction between species, and lack of correspondence between the object of protection and the response monitored.

Environmental Factors Modifying Toxicity

Simple differences in receiving water flow and chemistry can make large differences in risk. Many toxicity tests of the 1950s and 1960s focused on factors that affected toxicity, such as water quality, hardness, temperature, pH, food quality, etc. The differences in toxicity that result from natural variations in chemical and physical properties of the receiving system can be biologically important. For example, in the Shayler Run studies (Geckler et al. 1976), fluctuations in water chemistry, stream flow, and seasonal temperature brought the predictions of a safe concentration of copper perilously close to the 96-hr LC50 at certain times of the year.

To address this source of error, the influence of some environmental parameters, most notably temperature, hardness, and pH, have been explored, modeled, and incorporated into management instruments (e.g., United States Environmental Protection Agency [USEPA] 1980). However, these models are often based on data for only one species or only acute effects. There are few chemicals for which several mitigating factors have been simultaneously considered (e.g., Paulauskis and Winner 1988).

In absence of appropriate multivariate models, these changes in toxicity with environmental conditions have been dealt with by estimating risk for the worst case condition. Risk is predicted for the highest concentration in the lowest flow at the most disadvantageous pH and temperature. Although this approach is stringent, the margin of error involved remains uncharacterized.

Interactions Between Stresses

Although toxicity testing has focused on single chemicals, most natural communities are subject to a number of stresses from many sources. Not only are point source discharges typically complex and widely varying mixtures, but the receiving community is also subject to upstream discharges, storm runoff, habitat changes, flow extremes, etc. The cumulative impact of all these stresses through space and time may be different from a simple summation of effects. Individually minor stresses can be collectively significant.

The interaction of various chemical stresses has been the subject of much research (e.g., Calabrese 1991; Sprague 1970; Vouk et al. 1987). However, many techniques focus on a single proportional mixture of chemicals rather than independently varying levels of each stress. In addition, because the sensitivity to chemical stress varies over species, knowing the joint action of toxicants for one species does not necessarily predict the response of others to the same combination. Testing whole mixtures as they vary over time with a sample of species is a technique used to circumvent these problems. Even this technique is limited to existing mixtures. For predictive purposes, response surface methodology can be used to develop multivariate models to predict the response of organisms to complex and varying mixtures (Allus et al. 1988; Cairns et al. 1990). Such techniques are limited to well defined mixtures, and there are practical limitations to the number of stress factors that can be considered simultaneously.

Interactions between chemical stresses and natural stresses have been examined. Factors such as competition, harvesting rate, system heterotrophy, food quality and quantity, and parasitism have all been shown to affect toxicity (e.g., Burnett and Liss 1990; Guth et al. 1977; Mosser et al. 1972).

Differences in Sensitivities Between Ecosystems

Ecosystems respond differently to stress. Some ecosystems are perturbation dependent (e.g., Vogl 1980), although the perturbations upon which they depend are natural—fires, floods, and the like. Other ecosystems are classified as perturbation independent (e.g., tropical rain forests) and, once placed in disequilibrium, are not likely to recover in times of interests to humans or even societies. In addition to a prediction of a deleterious effect from a proposed course of action, an ecosystem risk assessment should have estimates of the elasticity or resiliency of an ecosystem, that is, the ability to recover from disequilibrium of either functional or structural norms.

In the summary of the conference on Recovery and Restoration of Damaged Ecosystems, Cairns et al. (1977) state that tropic terrestrial ecosystems had a slow rate of recovery measured in geologic time and that the temperate terrestrial ecosystems had a comparatively fast rate of recovery, again measured in geologic time. However, at that time, restoration to predisturbance condition seemed to be much more probable, especially with management intervention, than it does today. Presently, this seems highly improbable (e.g., Cairns 1989).

Differences in Sensitivity Between Species

The sensitivity of organisms to stress varies considerably. Differences of two orders of magnitude are common, and differences of seven orders of magnitude have been reported (Kenega 1987). Given this variability, predictions of risk based on the response of a single organism might not protect other organisms. Although it is possible to extrapolate from the response of one species to the response of another closely related species (e.g., Maki 1979; Mayer et al. 1987; Peakall and Tucker 1985), the ability to do so with sufficient accuracy for management decisions is far from robust.

In order to limit any error inherent in extrapolating from a single organism to all potentially affected organisms, an early suggestion was to test the most sensitive species. The USEPA (1973) suggested testing "significant species, the assumption being that by so doing, the entire system is protected." Further, USEPA states "The most sensitive species provide a good estimate of the range of sensitivity of all species" (USEPA 1973). This is an appealing idea, but, unfortunately, impossible in practice (Cairns 1986; Tebo 1986). First, identifying the most sensitive species for any given situation is impossible. Not only does the sensitivity of organisms vary considerably, but the same species is not consistently most sensitive to all toxicants, and a change in receiving water chemistry can reorder sensitivities from one day to the next. As Cairns (1986a) notes, an enormous search for a lamprycide for the Great Lakes resulted in finding one chemical (3-trifluoromethyl-4-nitrophenol) to which the sea lamprey is slightly more sensitive than is a more valuable species such as lake trout. It seems intuitively reasonable that, given a concerted effort, a chemical could be found for which any given species was most sensitive. Second, in addition to difficulties in identifying the most sensitive species, the most sensitive or important response is also impossible to identify. In the Shayler Run studies (Geckler et al. 1976), toxicity tests successfully predicted mortality and reductions in reproductive success, but a more important determinant of the ecological outcome in the natural system was avoidance, yet avoidance responses are not commonly measured in predictive protocols. Third, even if it were possible to identify the most sensitive species and response, it is most likely that no acceptable method for testing them would exist. Despite the relatively sizable number of species that have been proposed as test organisms, only a relatively few (i.e., less than 20) have been tested often.

Although the most sensitive species approach has not been completely abandoned, it has been largely replaced by the representative array approach. As previously mentioned, the basic assumption in testing representatives from major taxonomic or functional groups (microbial, photosynthetic, invertebrate, vertebrate) is that the range of response of aquatic organisms in general would be better displayed. It is possible to test an array of species, observe a range of sensitivities, develop a model for the distribution of tolerances of aquatic organisms in general, and use this model to predict likelihood of effects on untested organisms (e.g., USEPA 1980). However, this approach is compromised by several problems. The organisms tested are not a random sample but, instead, are highly selected for tolerance to laboratory conditions (Seegert et al. 1985). As such, their tolerances to other stresses may not be representative of organisms in general. There can be some difficulty combining information from the array of organisms in order to make management decisions. For example, is a 50% reduction in production of algal cells over a 96-hr period equivalent in ecological importance to a 50% mortality of fish? Additionally, the whole is more than the sum of its parts. The tolerance of organisms tested in isolation from competitors and predators is not the same as the tolerance of an interacting community.

Interactions Between Species
Communities are more than just a collection of populations, independently growing, reproducing, and dying. Organisms in natural systems also are subject to competition for food and habitat; predation; disease; and naturally occurring environmental stress from changes in light, temperature, flows, etc. The response of a natural system to stress depends on both population specific effects and effects on the interactions between populations and between populations and abiotic components of their environment. Interactions between species can be affected by stress at concentrations lower than that affecting survival and reproduction of the individual species (e.g., Mosser et al. 1972). Errors in predictions of risk attributable to changes in interactions between species have been described. For example, although diflubenzuron had no effect on the alga *Selenastrum* in a single species test, algal biomass and diversity were significantly affected in more complex test systems (Hansen and Garton 1982). This discrepancy was attributed to effects on interactions, i.e., release from grazer pressure and subsequent algal competition for nutrients and substrate. Because changes in interactions can significantly affect ecological outcome, toxicity tests examining the interaction between species have been developed (e.g., Lawrence and Holoka 1979; Taub et al. 1986). However, for predictive purposes, there is the same problem of selection as with test species. Which of thousands of affected interactions are the important ones that will affect ecological outcome and merit evaluation? Investigators began to use what are now called microcosms or mesocosms, that is, models with some but not all the parts and interactions of natural systems. Generally, behavior at any level of biological organization (i.e., cell, organism, population, community, ecosystem, landscape) is explained in terms of the level below; however, its significance is found in the level above. As such, responses at the community level should integrate effects of stress on interactions between species without selecting particular interactions for mechanistic description. The significance of changes in interactions between populations can be evaluated by monitoring end points at the community or ecosystem level. Important interactions at a specific site would be selected for more intense scrutiny just as important single species are.

Correspondence Between Assessment and Measurement End Points
The important question of how test results predict environmental outcome has not been pursued as vigorously as might be expected. In order to improve risk assessment methods, the relative success of available predictive techniques in reference to management goals must be evaluated. Unfortunately, there is sometimes confusion about what management goals are, and results of the most common test methods are confused with management goals. Suter (1990) clearly describes the relationship between goals and measurements. He suggests that there are ultimate values relating to quality of life. From this amorphous goal, assessment end points are defined as formal expressions of actual environmental values that are to be protected. These operationally defined goals for environmental management may include adequate standing crop and production of important species,

dominance of preferred species in the community, retention of soil and nutrients in the ecosystem, maintenance of waste processing, and preservation of landscape aesthetics or biodiversity. In contrast, a measurement end point is an expression of an observed or measured response to the hazard. Suter points out a number of ways that measurement end points can be related to assessment end points.

(1) The measurement end point and the assessment end point may be the same. Identity of measurement and assessment end points is most desirable since it circumvents the error involved in extrapolating from one response to another.
(2) The measurement end point may be a surrogate for the assessment end point. This is far more common. A simple, cost-effective, and reliable test is used to characterize the relative risk presented by various human activities. However, in order to make a prediction of the probable effects on the environment from these tests, an explicit model for extrapolation is needed and validation studies are necessary to characterize the margin of error inherent in the extrapolation.
(3) Measurement end points may be part of the causal link between a disturbance and the assessment end point. An example can be found in the use of biomarkers, a cellular or physiological process that precedes clinical expression of harm, as early warnings of ecosystem damage. By measuring damage before it becomes grave enough to stand alone as an assessment end point, an early warning capacity is gained, but the same problems with extrapolation that occur with surrogate measurement end points also apply to measurement end point early in the causal chain.

Although conventional measurement end points have focused almost exclusively on responses of populations to stress (parallel to the practice in classical toxicology), management goals in predicting risk to entire ecosystems clearly include protection of attributes at various levels of biological hierarchy. Not only is it important to maintain populations of preferred fish species, but also to maintain biodiversity for resilience and a genetic pool, and to maintain ecosystem services such as nutrient and waste transformation, flood control, and aquifer recharge.

Because close correspondence between assessment and measurement end points is desirable, there is a persuasive case for obtaining toxicological evidence at different levels of biological organization, especially when risk is high and tolerance for error is low (Cairns 1983). When it is impossible to measure the assessment end point, the model used to extrapolate from a surrogate or early causal chain end point to an assessment end point must be explicit and its assumptions, reasoning, calculations, and supporting data must be subject to critical review (NRC 1983). In addition, the success of the model at predicting ecosystem risk must be determined by comparing predictions to observed effects on natural systems. From this comparison, the magnitude and importance of errors in the predictions can be characterized and evaluated in respect to costs resulting from acting on faulty information. If predictions indicate probable harm where there is, in fact, none, treatment facilities may be constructed at substantial cost with no resultant biological benefit. If predictions suggest no risk where there is, in fact, damage, the loss of natural resources, cleanup, and rehabilitation costs result. A model for extrapolation must be judged adequate for regulatory or management purposes based on the size and importance of its predictive error as well as by the apparent soundness of its reasoning. Validation of a predictive model requires experimental evidence deliberately gathered for this purpose. In the past, validity has occasionally been surmised from an absence of gross harm when there has been only a limited attempt to find deleterious effects and no attempt to characterize the magnitude of errors. Predictions were often further reduced by a safety, uncertainty, or application factor. This approach provided some protection as evidenced by a lack of dramatic damage, but provided no assurance that more subtle or long-term damage was not occurring and no indication of the margin of error of the predictions. Clearly, without a validated model for extrapolation, evidence should be gathered at all the levels of biological organization for which protection is being inferred, especially in situations where the risk is high either in terms of potential loss of a valuable, unresilient system or in terms of treatment costs.

ESTIMATES OF RISK AT DIFFERENT LEVELS OF BIOLOGICAL ORGANIZATION

Predicting Risk at the Population Level

Assessment end points at the population level focus on the abundance, production, and condition of valued species. Although population level toxicity tests are used to address this goal, the vast majority of single species toxicity test information is used in a surrogate or early warning capacity. Often, the model used for extrapolation from the surrogate or early warning end point to an assessment end point does not exist or is not explicit. In every case, the margin of error of extrapolation has been inadequately characterized. Although good individual studies have been conducted (see review by LaPoint et al. 1989), almost no consistent information characterizing the magnitude of errors of prediction across various systems is available (e.g., Parkhurst et al. 1990). Part of the problem is that optimal design of validation studies is still being debated (e.g., Cairns and Smith 1989). Another factor is that the scope of protective efforts is not always clearly defined.

Because we have expressed concern about the state of the art of models for extrapolation, it has been suggested that we are hostile to single species toxicity tests. This is not the case. Single species toxicity tests provide a convincing demonstration of the potential for harm. We believe conventional single species tests to be the most effective way of determining survival, growth, reproductive success, respiratory function, and a wide variety of other attributes characteristic of individual organisms and laboratory populations. We believe an array of tests, including single species tests, is an efficient way to rank chemicals, conditions, and processes for relative toxicity. We believe single species toxicity tests with valued species are essential, but not sufficient, to protect aquatic ecosystems. We believe that single species toxicity test data will be useful components of validated models for predicting ecosystem risk.

Predicting Risk at the Community and Ecosystem Levels

While single species toxicity tests provide convincing evidence for potential for harm, the most direct and convincing evidence of actual harm to aquatic ecosystems is field surveys of the natural systems. The focus of most field surveys to determine ecological effects of human use is at the community level. Many explicit management goals are described in terms of healthy communities. Field surveys of aquatic systems typically describe the biodiversity and condition of fish, macroinvertebrates, or periphyton and the dominance of preferred species. Increasingly, field surveys also describe functional characteristics of these systems. This information is put into perspective by comparison with normal operating ranges determined in regionally appropriate unimpacted reference sites. These are the features that make currently developing operational definitions of biotic integrity a valuable tool in monitoring ecological health (e.g., Karr 1981; Plafkin et al. 1989). The development of composite measurement end points that operationally define ecological health has also been influenced by the development of general theories of ecosystem response to stress (e.g., Odum 1985; Pratt 1990; Rapport et al. 1985). Together, these approaches are defining ecosystem health in operational terms that can be applied to management problems.

In attempts to improve the correspondence between end points in laboratory tests used for prediction and field monitoring, the end points chosen for field surveys have been applied to portions or models of natural systems isolated for experimentation (e.g., Cairns 1985, 1986b; Giesy 1980). Microbial (e.g., Seidler 1991), periphytic (e.g., Cairns et al. 1985), macroinvertebrate (e.g., Pontasch and Cairns 1989), and occasionally fish communities (e.g., Touart 1989) have been transplanted from natural systems to surrogate systems and used in designed experiments. End points used in indices of biotic integrity and those thought to be general indicators of ecosystem response to stress have been directly applied to these micro- and mesocosm toxicity tests. Measurement end points used in these tests include species richness and abundance, dominance of preferred species, functional group distributions, condition of organisms, loss of functional redundancy, loss of homeostatic properties, loss of productive capacity, and loss of nutrients. Species composition and richness of small, rapidly reproducing and easily dispersed members of the community have been shown to be particularly sensitive and consistent indicators of stress in both microcosms (Cairns et al. 1988a) and whole lake experiments (Schindler 1987). When the same end points are used in both laboratory and field assessment of risks, the problems of extrapolating from one end point to another are minimized, although extrapolations over size, time span, and complexity remain. Validation studies characterizing the margin of errors for these predictions are equally essential (e.g., Tolle et al. 1985).

The acceptance of more complex systems for determining risk in regulatory applications varies greatly. Mesocosm studies have been accepted as an appropriate tool in studies for pesticide registration (e.g., Touart 1989), premanufacture notification (USEPA 1987), and evaluation of products of biotechnology (Seidler 1991). These applications have in common that they are predictive. Community and ecosystem level tests are not as accepted in regulatory applications where a problem already exists in a natural system, e.g., effluent permitting (Tebo 1986) or hazardous waste sites (Parkhurst et al. 1989). In that situation, a field survey of effects in the natural system, including those at the community and ecosystem level, provides the assessment of harm. Chemical and physical analyses suggest a causative agent, and toxicity tests are used to link the agent to the adverse effect.

Predicting Risk in the Landscape Ecology Context

Although the term *landscape ecology* appears new to many people, it was first used by Troll (1939) and arose from European traditions of regional geography and vegetative science. Reviews of the historical development may be found in Naveh (1982) and Naveh and Lieberman (1984). Risser et al. (1984) state that landscape ecology is not a distinct discipline or simply a branch of ecology, but rather is the synthetic intersection of many related disciplines that focus on the spatial-temporal pattern of the landscape. They further believe that, conceptually, landscape ecology considers the development and maintenance of spatial heterogeneity, the spatial and temporal interactions and exchanges across heterogeneous landscapes, the influences of heterogeneity on biotic and abiotic processes, and the management of that heterogeneity. Finally, they suggest that because of the spatial patterning of landscapes, flows and transfers between spatial components assume special importance, and the process of redistribution of organisms, materials, and/or energy among landscape components is an essential feature of landscape ecology (Risser et al. 1984). Most important to this discussion, landscape ecology is already well integrated into land use planning and decision making in Europe (e.g., Buchwald and Engelhart 1968; Ruzicka 1987; Ruzicka et al. 1988; Schreiber 1977; van der Maarel 1978; Vink 1983). Ruzicka et al. (1988) espouse the use of landscape studies as a basis for determining the optimal uses of land across whole regions.

The theoretical basis of landscape ecology emphasizes ecological processes (e.g., Forman and Godron 1986; Risser et al. 1984; Turner 1987, 1989; Urban et al. 1987). In addition, a variety of practical applications are developing concurrently with the development of underlying theory (e.g., Baker 1989; Forman 1986; Hayes et al. 1987; Joyce et al. 1987; Noss 1983). Turner and Gardner (1991a) have indicated that quantitative methods that link spatial patterns and ecological processes at broad spatial and temporal scales are needed both in basic ecological research and in solving applied environmental problems.

Incorporating the landscape context into risk estimation is important for another reason. Generally in science, the typical procedure is to first propose a general theory that is not intrinsically restricted to a narrow range of species. The applicability of this theory is not disclosed until a substantial data base is generated. When ecosystems are involved, this requires substantial field work. In many scientific fields, including ecology, simple cornerstone theories and models have undergone a marked loss of credibility because they are lacking in their ability to describe observable nature (e.g., Hall 1988; Hastings 1988; Ulanowicz 1988). In ecology, this may have happened because the models do not include the most critical factors influencing population dynamics (DeAngelis 1988). Worse yet, the inclusion of such factors often renders the models analytically intractable (Hall 1988). It seems almost inevitable that, as ecotoxicology comes of age, calls will be made for statements of general hypotheses and the validation thereof. As a consequence, environmental toxicology will be regarded as having a defensible scientific basis only if rigorous scientific procedures are followed. Even the use of simple laboratory toxicity tests should be validated in the spatial and temporal dynamics of ecological landscapes if that is the level at which predictions are made.

It is important to distinguish between ongoing risk assessments in support of risk management and the scientific field of ecotoxicology because the demands and driving factors are sometimes different. Risk management must occur despite uncertainty in risk assessment. Decisions cannot wait, and so areas with insufficient information are bridged by inference and best professional judgment. Making management decisions on the basis of these inferences is likely to be less costly than failing to act. So, decision making tools that lack complete scientific rigor are still useful (e.g., see figure 4 in Cairns 1980). For the sake of efficiency and fairness, these methods must not be subject to trivial tinkering; however, risk assessment methods should not be static. Areas of inadequate information can be addressed by appropriate research. New statistical methods can facilitate decisiveness and model development. New test methods can improve the correspondence between assessment and measurement end points. Field validations can suggest areas in which predictions are inadequate. Nevertheless, some problems and uncertainties will not be quickly resolved (Cairns et al. 1988b). Research on the nature of ecological health at various spatial scales does clarify operational definitions of management goals. The obvious goal for both risk assessment and environmental toxicology is a system for predicting ecosystem risk that is completely grounded in current scientific theory. It is not necessary to wait until then to make any decisions, but, neither is it necessary to defend current methods by the assertion that they need not be evaluated or improved.

REGULATION OF RISK—PUBLIC OR PRIVATE SECTOR MODEL

Most people assume that the government (whether local, state, or national) will assume control of the regulation of ecosystem risk, that is, the determination of how much is acceptable and the enforcement that predetermined quality conditions are being met. Society also carries considerable mistrust of government, especially big government, as a bumbling giant with endless rules to protect the bureaucrats. Often the rules are incomprehensible and, sometimes, when they are comprehensible, contradictory. Fear that unique local attributes will be overlooked in the search for a universally applicable policy or regulation is also considerably justified. In England and Wales, and possibly other parts of the United Kingdom, a very definite move has begun toward privatization of certain activities that, until relatively recently, would have been regarded by most as being strictly in the public sector. The public sector activities being privatized include public water supplies and sewerage treatment plants. This is an intriguing experiment to determine whether the private sector can provide a persuasive demonstration that the needs of the citizens can be better met and, at the same time, costs fixed for a multiple year stretch. In view of this trend toward privatization of public sector activities, the question of management and enforcement of ecosystem risk can be examined in this light.

In the private sector model, the process has two types of control. In the first, treasury control, the primary drive is to reduce taxes and also interest rates on indebtedness for substantial periods of time. The main interest is money. In reducing ecosystem risk (or any risk for that matter), money is, of course, always a factor; perhaps it should not be as dominant a factor as it is under treasury control.

In addition to treasury control, there is political control. Political control in an electoral system with frequent elections is characterized by a short-term interest in long-term problems. Unfortunately, except for catastrophic events, coping with ecosystem risk is definitely a long-term problem. The time of service by elected officials is usually not great enough to deal with this situation.

In the private sector model, standards must still be set by government. It is difficult to visualize how ecosystem risk criteria and standards for acceptable risk could be turned over entirely to the private sector. Risk is a probablistic determination requiring scientific evidence. However, the determination of safety is a value judgment (including, among other considerations, cost benefit ratios) having many facets, not all amenable to the scientific method. On the other hand, the cost of determining criteria for risk and setting standards based on these criteria is a very different process more amenable to peer review than the actual implementation.

Independent regulators could possibly check compliance. If the criteria and standards are set in an approved scientific and democratic process, regulation could be assigned to private companies.

To facilitate private sector regulation of risk, the question arises: How does one facilitate competition? Of course, the essence of privatization is improved performance because presumably performance can be better rewarded in the private than in the public sector. This has definitely been seen in the privatization of garbage and solid waste collection and disposal in the United States. The matter is worth exploring in other sectors as well. On the other hand, competition in ecosystem risk regulation is different and may be accomplished only by comparing the activities of a company in one area with the activities of a company in another. If we can compare performance of other activities in different regions, why not this activity?

SUMMARY

Attitudes about environmental risk have changed considerably in the past two decades. More ambitious management goals require better techniques for predicting ecosystem risk. Among the challenges we see for improving these techniques are: (1) improving correspondence between assessment and measurement end points; (2) incorporating differences in ecosystem resilience into risk assessments; (3) incorporating landscape scale assessments into the picture; and (4) assessing the performance and error of models currently used for predicting ecosystem risk.

ACKNOWLEDGMENTS

The authors wish to thank Darla Donald for her editorial assistance and Teresa Moody for typing drafts of this chapter.

REFERENCES

ALLUS, M.A., BRERETON, R.G., and NICKLESS, G. (1988). Chemometric studies of the effect of toxic metals on plants: The use of response surface methodology to investigate the influence of Tl, Cd, and Ag on the growth of cabbage seedlings. Environ. Pollut. 52:169-181.

BAKER, W.L. (1989). Landscape ecology and nature reserve design in the Boundary Waters Canoe Area, Minnesota. Ecology 70:23-35.

BUCHWALD, K. and ENGELHART, W. (eds). (1968). *Handbuch fur Landschaftpflege und Naturschutz*. Bd. 1. Grundlagen. BLV Verlagsgesellschaft, Munich.

BURNETT, K.M. and LISS, W.J. (1990). Multi-steady-state toxicant fate and effect in laboratory aquatic ecosystems. Environ. Toxicol. Chem. 9:637-647.

CAIRNS, J., JR. (1956a). Effects of heat on fish. Ind. Wastes. 1:180-183.

CAIRNS, J., JR. (1956b). Effects of increased temperatures on aquatic organisms. Ind. Wastes. 1:150-152.

CAIRNS, J., JR. (1980). Estimating hazard. BioScience 30(2):101-107.

CAIRNS, J., JR. (1983). The case for simultaneous toxicity testing at different levels of biological organization. In: *Aquatic Toxicology and Hazard Assessment: Sixth Symposium*, STP802 (W.E. Bishop, R.D. Cardwell, and B.B. Heidolph, eds), pp. 111-127. American Society for Testing and Materials, Philadelphia, PA.

CAIRNS, J., JR. (ed). (1985). Multispecies Toxicity Testing. Pergamon Press, NY.

CAIRNS, J., JR. (1986a). The myth of the most sensitive species. BioScience 36:670-672.

CAIRNS, J., JR. (ed). (1986b) Community Toxicity Testing, STP920. American Society for Testing and Materials, Philadelphia, PA.

CAIRNS, J., JR. (1989). Restoring damaged ecosystems: Is predisturbance condition a viable option? Environ. Prof. 11:152-159.

CAIRNS, J., Jr. and SMITH, E.P. (1989). Developing a statistical support system for environmental hazard evaluation. Hydrobiologia 184:143-151.

CAIRNS, J., JR., DICKSON, K.L., and HERRICKS, E.E. (1977). The recovery and restoration of damaged ecosystems, a challenge for action: Symposium analysis. In: *Recovery and Restoration of Damaged Ecosystems* (J. Cairns, Jr., K.L. Dickson, and E.E. Herricks, eds), pp. 522-525. University Press of Virginia, Charlottesville, VA.

CAIRNS, J., JR., DICKSON, K.L., and MAKI, A.W. (1978). *Estimating the Hazard of Chemical Substances to Aquatic Life*, STP657. American Society for Testing and Materials, Philadelphia, PA.

CAIRNS, J., JR., PRATT, J.R., and NIEDERLEHNER, B.R. (1985). A provisional multispecies toxicity test using indigenous organisms. J. Test. Eval. 13:316-319.

CAIRNS, J., JR., PRATT, J.R., NIEDERLEHNER, B.R., and BOWERS, N.J. (1988a). Structural and Functional Responses to Perturbation in Aquatic Ecosystems, NTIS AD-192 071/9/GAR. National Technical Information Service, Springfield, VA.

CAIRNS, J., JR., SMITH, E.P., and ORVOS, D.R. (1988b). The problem of validating simulation of hazardous exposure in natural systems. In: *Proceedings of the 1988 Summer Computer Simulation Conference* (C.C. Barnett and W.M. Holmes, eds), pp. 448-454. The Society for Computer Simulation Intenational, San Diego, CA.

CAIRNS, J., JR., NIEDERLEHNER, B.R., and J.R. PRATT. (1990). Evaluation of joint toxicity of chlorine and ammonia to aquatic communities. Aquat. Toxicol. 16:87-100.

CALABRESE, E.J. (1991). Multiple Chemical Interactions. Lewis Publishers, Chelsea, MI.

DeANGELIS, D.L. (1988). Strategies and difficulties of applying models to aquatic populations and food webs. Ecol. Model. 43:57-73.

FORMAN, R.T.T. (1986). Emerging directions in landscape ecology and applications in natural resource management. In: *Proceedings of the Conference on Science in the National Parks* (R. Herrmann and T. Bostedt-Craig, eds), pp. 59-88. George Wright Society, Washington, DC.

FORMAN, R.T.T. and GODRON, M. (1986). Landscape Ecology. John Wiley & Sons, NY.

GALLUP, G., JR. (1990). The Gallup Poll 1989. Scholarly Resources Inc., Wilmington, DE.

GECKLER, J.R., HORNING, W.B., NEIHEISEL, T.M., PICKERING, Q.H., ROBINSON, E.L., and STEPHAN, C.E. (1976). Validity of Laboratory Tests for Predicting Copper Toxicity in Streams, EPA-600/3-76-116. National Technical Information Service, Springfield, VA.

GIESY, J.P., JR. (ed). (1980). Microcosms in Ecological Research, CONF-781101. National Technical Information Service, Springfield, VA.

GUTH, D., BLANKENSPOOR, H., and CAIRNS, J., JR. (1977). Potentiation of toxic strtess caused by parasitic infection of snails. Hydrobiologia 55:225-229.

HALL, C.A.S. (1988). An assessment of several of the historically most influential theoretical models used in ecology and of the data provided in their support. Ecol. Model. 43:5-31.

HANSEN, S.R. and GARTON, R.R. (1982). Ability of standard toxicity tests to predict the effects of the insecticide diflubenzuron on laboratory stream communities. Can. J. Fish. Aquat. Sci. 39:1273-1288.

HASTINGS, A. (1988). Food web theory and stability. Ecology 69:1665-68.

HAYES, T.D., RISKIND, D.H., and PACE, W. L., III. (1987). Patch-within-patch restoration of man-modified landscapes within Texas state parks. In: *Landscape Heterogeneity and Disturbance* (M. G. Turner, ed), pp. 173-98. Springer-Verlag, NY.

JOYCE, L. A., HOEKSTRA, T.W., and ALIG, R.J. (1987). Regional multiresource models in a national framework. Environ. Manage. 10:761-71.

KARR, J.R. (1981). Assessment of biotic integrity using fish communities. Fisheries 6:185-197.

KENEGA, E.E. (1987). Methods for assessing the effects on non-human biota of mixtures of chemicals as applied to specific taxonomic representatives of individuals or groups of species. In: *Methods for Assessing the Effects of Mixtures of Chemicals* (V. Vouk, G. Butler, A. Upton, D. Parke, and S. Asher, eds), pp. 395-408. John Wiley & Sons, NY.

LAPOINT, T., FAIRCHILD, J.F., LITTLE, E.E., and FINGER, S.E. (1989). Laboratory and field techniques in ecotoxicological research: Strengths and limitations. In: *Aquatic Ecotoxicology: Fundamental Concepts and Methodologies*, Vol. II (A. Boudou and F. Ribeyre, eds), pp. 239-255, CRC Press Inc., Boca Raton, FL.

LAWRENCE, S.G. and HOLOKA, M.H. (1979). A microbial predator-prey test system. In: *Toxicity Tests for Freshwater Organisms* (E. Scherer, ed), pp. 91-103. Can. Spec. Publ. Fish. Aquat. Sci. 44, Department of Fisheries and Oceans, Winnipeg, Manitoba.

MAKI, A.W. (1979). Correlations bewtween *Daphnia magna* and fathead minnow (*Pimephales promelas*) chronic toxicity values for several classes of test substances. J. Fish. Res. Board Can. 36:411-421.

MAYER, F.L., JR., DEANS, C.H., and SMITH, A.G. (1987). Inter-taxa correlations for toxicity to aquatic organisms. Office of Pesticides and Toxic Substances, United States Environmental Protection Agency, Washington, DC.

MOSSER, J.L., FISHER, N.S., and WURSTER, C.F. (1972). Polychlorinated biphenyls and DDT alter species composition in mixed cultures of algae. Science 176:533-535.

NATIONAL RESEARCH COUNCIL. (1983). Risk Assessment in the Federal Government: Managing the Process. National Academy Press, Washington, DC.

NAVEH, Z. (1982). Landscape ecology as an emerging branch of human ecosystem science. Adv. Ecol. Res. 12:189-237.

NAVEH, Z. and LIEBERMAN, A.S. (1984). Landscape Ecology, Theory and Applications. Springer-Verlag, NY.

NOSS, R.F. (1983). A regional landscape approach to maintain diversity. BioScience 33:700-6.

ODUM, E.P. (1985). Trends expected in stressed ecosystems. BioScience 35:419-422.

PARKHURST, B.R., LINDER, G., McBEE, K., BITTON, G., DUTKA, B.J., and HENDRICKS, C.W. (1989). Toxicity tests. In: *Ecological Assessment of Hazardous Waste Sites: A Field and Laboratory Reference*, EPA/600/3-89/013 (W. Warren-Hicks, B.R. Parkhurst, and S.S. Baker, Jr., eds) pp. 6-1-6-26. National Technical Information Service, Springfield, VA.

PARKHURST, B.R., MARCUS, M.D., and NOEL, L.E. (1990). Review of the Results of EPA's Complex Effluent Toxicity Testing Program. Utility Water Act Group, Washington, DC.

PAULAUSKIS, J.D. and WINNER, R.W. (1988). Effects of water hardness and humic acid on zinc toxicity to *Daphinia magna* Straus. Aquat. Toxicol. 12:273-290.

PEAKALL, D.B. and TUCKER, R.K. (1985). Extrapolation from single species studies to populations, communities and ecosystems. In: *Methods for Estimating Risk of Chemical Injury: Human and Non-human Biota and Ecosystems*, SCOPE 30 (V.B. Vouk, G.C. Butler, D.G. Hoel, and D.B. Peakall, eds), pp. 611-636. John Wiley & Sons, NY.

PLAFKIN, J.L., BARBOUR, M.T., PORTER, K.D., GROSS, S.K., and HUGHES, R.M. (1989). Rapid Bioassessment Protocols for Use in Streams and Rivers: Benthic Macroinvertebrates and Fish, EPA 444/4-89-001. National Technical Information Service, Springfield, VA.

PONTASCH, K.W. and CAIRNS, J., JR. (1989). Establishing and maintaining laboratory-based microcosms of riffle insect communities: Their potential for multispcies toxicity tests. Hydrobiologia 175:49-60.

PRATT, J.R. (1990). Aquatic community response to stress: Prediction and detection of adverse effects. In: *Aquatic Toxicology and Risk Assessment: Thirteenth Volume*, STP1096 (W.G. Landis and W.H. van der Schalie, eds), pp. 16-26. American Society for Testing and Materials, Philadelphia, PA.

RAPPORT, D.J., REGIER, H.A., and HUTCHINSON, T.C. (1985). Ecosystem behavior under stress. Am. Nat. 125:617-640.

RISSER, P.G., KARR, J.R., and FORMAN, R.T. (1984). Landscape Ecology: Directions and Approaches, Special Publ. No. 2. Illinois Natural History Survey, Champaign, IL. 16 pp.

RUZICKA, M. (1987). Topical problems of landscape ecological research and planning. Ekologia-Cssr 5:233-238.

RUZICKA, M., HRNCIAROVA, T., and MIKLOS, L. (eds). (1988). Proceedings of the VIIIth International Symposium on Problems in Landscape Ecological Research, Vol. 1. Institute of Experimental Biology and Ecology, CBES SAS, Bratislava, Czechoslovakia.

SCHINDLER, D.J. (1987). Detecting ecosystem responses to anthropogenic stress. Can J. Fish. Aquat. Sci. 44 (Suppl.1):6-25.

SCHREIBER, K.F. (1977). Landscape planning and protection of the environment: The contribution of landscape ecology. Appl. Sci. Dev. 9:128-39.

SEEGERT, G., FAVA, J.A., and CUMBIE, P.M. (1985). How representative are the data sets used to derive natural water quality criteria? In: *Aquatic Toxicology and Hazard Assessment: Seventh Symposium*, STP854 (R.D. Cardwell, R. Purdy, and R.C. Bahner, eds), pp. 527-537, American Society for Testing and Materials, Philadelphia, PA.

SEIDLER, R.J. (ed). (1991). The Use and Development of Environmentally Controlled Chambers (Mesocosms) for Evaluating Biotechnology Products. United States Environmental Protection Agency, Environmental Research Laboratory, Corvallis, OR.

SLOOF, W. (1983). Biological Effects of Chemical Pollutants in the Aquatic Environment and Their Indicative Value. University of Utrecht, The Netherlands. 191 pp.

SPRAGUE, J.B. (1970). Measurement of pollutant toxicity to fish. II. Utilizing and applying bioassay results. Water Res. 4:3-32.

SUTER, G.W. (1990). Endpoints for regional ecological risk assessments. Environ. Manage. 14:9-23.

TAUB, F.B., KINDIG, A.C., and CONQUEST, L.L. (1986). Preliminary results of interlaboratory testing of a standardized aquatic microcosm. In: *Community Toxicity Testing*, STP920 (J. Cairns, Jr., ed), pp. 93-115. American Society for Testing and Materials, Philadelphia, PA.

TEBO, L.B., JR. (1986). Effluent monitoring: Historical perspective. In: *Environmental Hazard Assessment of Effluents* (H.L. Bergman, R.A. Kimerle, and A.W. Maki, eds), pp. 13-31. Pergamon Press, NY.

TOLLE, D.A., ARTHUR, M.F., CHESSON, J., and VAN VORRIS, P. (1985). Comparisons of pots versus microcosm for predicting acroecosystem effects due to waste amendment. Environ. Toxicol. Chem. 4:501-509.

TOUART, L.W. (1989). Aquatic Mesocosm Tests to Support Pesticide Registrations, EPA 540/09-88-035. National Technical Information Service, Springfield, VA.

TROLL, C. (1939). Luftbildplan and okologische Bodenforschung. Z. Ges. Erdkunde: 241-98, Berlin.

TURNER, M.G. (1987). Landscape Heterogeneity and Disturbance. Springer-Verlag, NY.

TURNER, M.G. (1989). Landscape ecology: The effect of pattern on process. Ann. Rev. Ecol. Syst. 20:171-97.

TURNER, M.G., and GARDNER, R.H. (1991). Quantitative methods in landscape ecology: An introduction. In: *Quantitative Methods in Landscape Ecology*, Ecological Studies 82 (M. G. Turner and R. H. Gardner, eds), pp. 3-14. Springer-Verlag, NY.

ULANOWICZ, R.E. (1988). On the importance of higher-level models in ecology. Ecol. Model. 43:45-56.

UNITED STATES ENVIRONMENTAL PROTECTION AGENCY (USEPA). (1973). Water Quality Criteria 1972, EPA-R3-73-003. U.S. Government Printing Office, Washington, DC.

UNITED STATES ENVIRONMENTAL PROTECTION AGENCY (USEPA). (1980). Guidelines for deriving water qualtiy criteria for the protection of aquatic life and its uses. Fed. Reg. 45:79341-79347.

UNITED STATES ENVIRONMENTAL PROTECTION AGENCY (USEPA). (1987). Toxic Substances Control Act guidelines. Fed. Reg. 52:36334-36371.

URBAN, D.L., R.V. O'NEILL, and H.H. SHUGART. (1987). Landscape ecology. BioScience 37:119-27.

VAN DER MAAREL, E. (1978). Ecological principles for physical planning. In: *The Breakdown and Restoration of Ecosystems*, NATO Conference Series 1 (Ecology), Vol. 3. (M.W. Holgate and M.J. Woodman, eds), pp. 413-50. Plenum Press, NY.

VINK, A.P.A. (1983). Landscape Ecology and Land Use. Longman, London.

VOGL, R.J. (1980). The ecological factors that produce perturbation-dependent ecosystems. In: *The Recovery Process in Damaged Ecosystems* (J. Cairns, Jr., ed), pp. 63-94. Ann Arbor Science Publishers Inc., MI.

VOUK, V.B., BUTLER, G.C., UPTON, A.C., PARKE, D.V., and ASHER, S.C. (1987). Methods for Assessing the Effects of Mixtures of Chemicals. John Wiley & Sons, NY.